5.4.91

Dad,

Keep this copy for your pleasure. Read it, or just look at the pictures — I'm very happy to give it to you. Who knows, it might convey something that I haven't been able to explain in person about what I'm doing.

Much love,

Nick.

Knowledge-Based Systems in Agriculture

BIOLOGICAL RESOURCE MANAGEMENT

A Series of Primers on the Conservation and Exploitation of Natural and Cultivated Ecosystems

Wayne M. Getz, Series Editor
University of California, Berkeley

Carl Walters • *Adaptive Management of Renewable Resources*

Anthony Starfield and A. L. Bleloch • *Building Models for Wildlife Management*

Peter B. R. Hazell and Roger D. Norton • *Mathematical Programming for Economic Analysis in Agriculture*

John P. Workman • *Range Economics*

Joseph Buongiorno and J. Keith Gilless • *Forest Management and Economics: A Primer in Quantitative Methods*

Gordon L. Swartzman and Stephen P. Kaluzny • *Ecological Simulation Primer*

Raoul A. Robinson • *Host Management in Crop Pathosystems*

C. Ronald Carroll, John H. Vandermeer, and Peter M. Rosset (eds.) • *Agroecology*

Chadwick D. Oliver and Bruce C. Larson • *Forest Stand Dynamics*

Knowledge-Based Systems in Agriculture

Richard E. Plant
Department of Agronomy and Range Science
University of California, Davis

Nicholas D. Stone
Department of Entomology
Virginia Polytechnic Institute and State University

McGraw-Hill, Inc.
New York St. Louis San Francisco Auckland Bogotá
Caracas Hamburg Lisbon London Madrid
Mexico Milan Montreal New Delhi Paris
San Juan São Paulo Singapore
Sydney Tokyo Toronto

Library of Congress Cataloging-in-Publication Data

Plant, Richard E.
Knowledge-based systems in agriculture / Richard E. Plant, Nicholas D. Stone.
p. cm.
Includes bibliographical references and index.
ISBN 0-07-050316-8
1. Agriculture—Data processing. 2. Expert systems (Computer science) I. Stone, Nicholas D. II. Title.
S494.5.D3P53 1991
630'.285—dc20 90-25284

1 2 3 4 5 6 7 8 9 0 DOC/DOC 9 7 6 5 4 3 2 1

ISBN 0-07-050316-8

The sponsoring editor for this book was Trev Léger, the editing supervisor was Stephen M. Smith, and the production supervisor was Pamela A. Pelton. It was set in Century Schoolbook by McGraw-Hill's Professional Publishing composition unit.

Printed and bound by R. R. Donnelley & Sons Company.

To our families:
Kathie, Carolyn, and Suzanne
Carol, Eric, Kendall, and Ethan

Contents

Preface

This book is intended for people who develop knowledge-based systems for use in agriculture or who want to learn about them. It is not intended for people who just want to use these systems. Indeed, one of the virtues claimed for knowledge-based systems is that virtually no training is required to use them. Another virtue claimed for knowledge-based systems is that much of the theory behind them is independent of the domain of their application. Our interest is in the management of agricultural crops, and that is the problem domain to which this book is devoted. However, to the extent that domain independence is true, our book should be useful to those working in related fields such as forestry, natural resource management, agricultural economics, animal husbandry, and so forth.

The role of knowledge-based systems is information dissemination. They are designed to substitute for a human expert when such expertise is in short supply. Of course, a computer program cannot replace a human expert, but if such an expert is unavailable and a problem needs to be solved, then a computer program may offer the best alternative. In such cases it is incumbent on the developer of the program to make it the best possible substitute for a human expert, and the knowledge-based system may in many cases be this best possible substitute. The theory of these systems is less than two decades old; their application to agriculture began in the early 1980s and did not reach a high level until the latter part of that decade. It is not surprising, therefore, that the field has suffered from growing pains and setbacks, nor is it surprising that much still needs to be learned. We feel, however, that the use of these systems has matured to the point where a volume on the subject is both feasible and appropriate, and that is why we wrote this book.

This book can be divided into three parts of roughly equal length. The first part, which consists of Chaps. 1 through 4, provides a basic introduction to rule-based expert systems. There is some discussion of applications to agricultural management, but for the most part this section of the book is independent of the problem domain and focuses on explaining the systems themselves. The second part, consisting of Chaps. 5 through 7, describes how one links a rule-based expert system to other sorts of computer programs to address problems of agricultural management. These programs include simulation models, optimization routines, and regression routines. An informal survey of developers of agricultural expert systems indicated that virtually all of them employed such linkage for some aspect of their program. The third part of this book, consisting of Chaps. 8 through 12, is a discussion of advanced topics in knowledge-based systems. These topics include forms of knowledge representation beyond the basic rule base and applications of artificial intelligence beyond the basic expert system. Only a few knowledge-based systems presently employ any of the concepts described in this part of the book, but we feel that this is where the future lies.

Unlike the traditional computer simulation models that have become so widespread in agricultural science, expert systems have played only a small role in original research. They are intended to be widely disseminated to individuals who lack the expertise experts provide. There may be some specialized uses of knowledge-based systems in a research context, for example, in the use of certain laboratory equipment. The primary role of simulation models in research is as a mathematical statement of a hypothesis, and knowledge-based systems will never replace them in this role. The expert system is more of an extension tool than a research tool.

This extension role presents several fundamental obstacles to the successful adoption of expert systems in agriculture. A simulation model can be considered a success if it simply works, i.e., if it adequately performs its simulation function. Expert systems, however, must be judged by a harsher standard. An expert system cannot be called successful just because it works correctly; it must also be employed by at least some of its intended users, or its development will have been an empty gesture. Moreover, the developers of the expert system, if they are university researchers, will often derive their professional satisfaction and benefit primarily from the act of developing the system and the publication of the results; they may be unwilling to persevere in the often tedious and unrewarding endeavor of pushing the system through to general adoption. Finally, the intended adopters, often farmers or pest control advisors, are often intimidated by computers and skeptical of their applicability to problems in the real world. It is no wonder that the adoption of expert systems in agriculture has been painfully slow.

Anyone remotely connected with agriculture knows that it is besieged from all sides. In the developed countries, the public is demanding greater restrictions on the use of inputs such as chemical pesticides and fertilizers while expecting the continued availability of high-quality agricultural products at low cost. In the developing countries, the progress made in past decades by the "green revolution" is being eroded by population growth and yield decline. It would be preposterous to claim that the use of knowledge-based systems can solve these problems, but maybe it can help. We feel that the future of such systems is bright; the problems faced by agriculturists will require information and knowledge at such a high level that there will be virtually no alternative to the use of computers.

Having said what this book is, we should also say what it is not. It is not a catalog of expert system shells or development software. Such a book would be obsolete almost before it became available. There is virtually no discussion of commercially available software. This book is also not a review of all the work ever conducted on agricultural expert systems. We have been quite eclectic in our choice of whose work to discuss, and we have restricted our references to those directly relevant to the topic under discussion.

Expert system development is a team sport, and we have each been the beneficiary of considerable assistance from many people, far too many to acknowledge them all here. We would be remiss, however, if we did not thank Ray Frisbie, Pete Goodell, Tom Kerby, Merry Makela, Hannu Saarenmaa, Ted Wilson, and Julie Young. A portion of R.E.P.'s work on this book was carried out while he was on sabbatical leave at the University of Massachusetts. He would like to thank Dan Cooley and the Department of Plant Pathology for their gracious hospitality during this period. As is always the case, any mistakes in this book are solely the responsibility of the authors. When mistakes are discovered, each author will cheerfully blame the other.

Richard E. Plant
Nicholas D. Stone

Knowledge-Based Systems in Agriculture

Chapter 1

Introduction

The emergence of artificial intelligence (AI) technologies in agriculture has come at a time of precipitous change in agricultural crop management. After years of yield increases fueled by the green revolution and increasing use of chemical inputs, farmers face yield losses, soil erosion, and diminishing market prices from international competition. Chemical pesticide costs and pest resistance are increasing, and the public is becoming increasingly intolerant of food and groundwater contamination from petrochemical use on farmland. Meanwhile, government price supports and federal farm programs have built economic barriers that are hindering the adoption of farming strategies that might solve farmers' biological and environmental problems.

Farmers in the next decade must become expert managers of all aspects of their farming operation (Fig. 1.1). Success will depend more and more on the farmer's ability to balance the need for low-input,

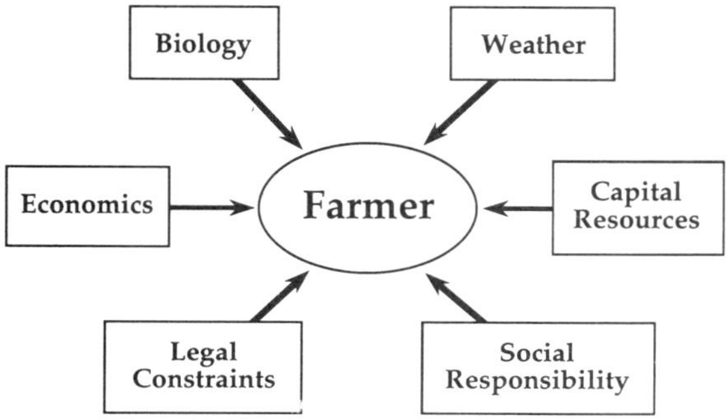

Figure 1.1 The problem of integrated farm management from the point of view of the decision maker, the farmer. Management decisions are made by balancing the influences of all these factors.

sustainable agricultural practices with the financial, legal, and practical constraints imposed by federal economic programs, legislation, and the capital resources of the farm. In short, farmers must manage from a systems perspective. Good management will be more intensive and more demanding in terms of time and expertise, and more critical to farm survival.

Agricultural researchers face the same problems but from a different perspective. Research must address the problems of farm management—developing new cultivars and pest control approaches, but, above all, discovering new management strategies that will promote farm success in the next decade. Traditionally, systems level problems have been solved using systems analysis. In agriculture, this systems-analysis approach was at the heart of the integrated pest management (IPM) projects of the 1970s and early 1980s (Huffaker, 1980; Frisbie and Adkisson, 1986). Recently, however, the use of simulation models and quantitative analyses to solve problems in agricultural crop management and IPM has been questioned (Getz, 1986), especially as more and more emphasis has been placed on practical application of modeling to agricultural decision making (Stone, 1989). Numerical methods have failed in crop systems because so much of our understanding about those systems is qualitative, based on experience; it does not lend itself well to mathematical representation.

Knowledge-based systems offer a new alternative to mathematics as a way to represent our knowledge about cropping systems and to apply that knowledge to solve problems. In a sense, knowledge-based systems are a compromise; their use is an admission that our knowledge is incomplete and uncertain. More to the point, building knowledge-based systems for cropping systems is also a move toward practicality, recognizing that making decisions based on qualitative and sometimes incomplete understanding is still better than making decisions without any understanding. Fortunately, one need not make a choice. Knowledge-based systems can include conventional numerical methods. However, the focus of knowledge-based systems is on reasoning and how those numerical tools can provide the information needed to make a decision.

1.1 Knowledge-Based Systems

1.1.1 Definition of knowledge-based systems

Knowledge-based systems are computer programs that solve complex problems within some defined domain. For example, SOYBUG is a knowledge-based system that gives farmers advice about insect pest management for soybeans in Florida (Beck et al., 1989). Usually, the problem domain is very clearly defined and limited. SOYBUG, for instance, gives advice on only four insect pests in one crop in one area of the country. Knowledge-based systems differ from conventional computer programs because they solve problems by mimicking human reasoning processes, relying on logic, belief, rules of thumb, opinion, and experience. Conventional computer programs solve problems by performing numerical calculations, solving equations, and using optimization algorithms to process large data sets. They have proven very capable in problem domains that are easily and well described by mathematical equations but have not been nearly as successful solving problems in an area like the management of agricultural crops, which is dominated by uncertainty and incomplete understanding.

The realization that human beings are quite capable and sometimes expert at solving complex problems in the face of uncertainty and incomplete knowledge led artificial intelligence researchers to try to mimic this facility of humans in the computer. This work has been underway since the 1940s when Post (1943) proposed the *production system* as a general computational mechanism for problem solving, representing problem-solving knowledge as IF-THEN rules and using a rule-interpreter to sequence those rules to solve problems. However, it was not until the 1970s and 1980s with the development of the first *expert systems* and *expert system shells* that the concept of computer reasoning and problem solving became accessible beyond the field of artificial intelligence.

By far the most commonly known type of knowledge-based system is the rule-based expert system, in which the experience and knowledge of a human expert is captured in the form of IF-THEN rules and facts and used to solve problems. Figure 1.2 shows an elementary expert system, with its main components. At the center of the figure is the inference engine, the part of the program that contains the reasoning capability. The inference engine interacts with a knowledge base. The knowledge base contains information about how to solve problems within the problem domain in general. In the case of expert systems, for example, the knowledge base may contain several hundred rules describing the thought process that a particular human expert uses in solving problems. Juxtaposed to the knowledge base is the global memory. This is where the knowledge-based system records information relating to a specific problem that it is trying to solve. Much of this information comes from the user, but the memory is also used by the inference engine to record its own conclusions and to remember its chain of reasoning. By comparing what it knows about the problem domain in general with what it knows about the specific problem, the inference engine tries to proceed logically toward a solution.

The third component shown in Fig. 1.2 is labeled *demons* or *external*

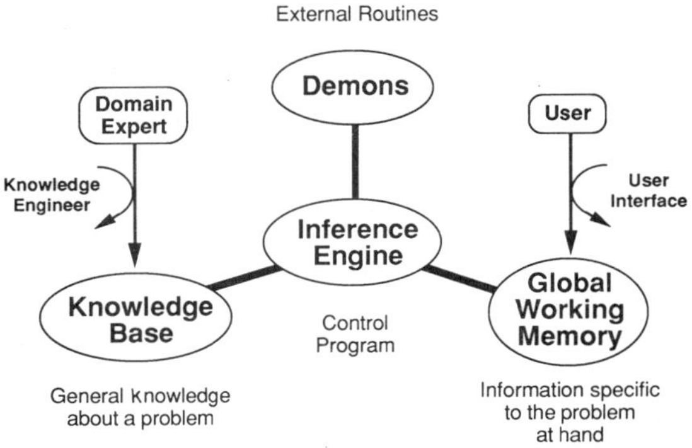

Figure 1.2 Main components of an expert system. The knowledge engineer transcribes the expert's knowledge into computer structures, like IF-THEN rules, stored in the knowledge base. The user provides information about a specific problem, and the inference engine uses its knowledge and (optionally) some external routines to come to a conclusion, give some advice, or achieve a goal.

routines. These are computer programs or algorithms not directly related to the reasoning process but which perform a useful function. The computer science term demon is best interpreted as meaning an invisible helper, rather than some sort of evil spirit. For example, if a pest management expert normally runs a predictive simulation model to determine the first likely insecticide spray date for several fields, an expert system mimicking this decision process would need to call that simulation model as a demon.

One key aspect of knowledge-based systems illustrated by the expert system is that the knowledge is removed from the inference process that acts on that knowledge. Typically, when we build an expert system, we buy the inference engine with an empty knowledge base (an expert system *shell*) and a manual describing the syntax for adding knowledge to the knowledge base. Building an expert system in this way does not require constructing an inference engine, only adding knowledge to the purchased shell. Quite often, the knowledge base accepts knowledge written in English-like *rules,* e.g., if a bee colony is nearby, then do not spray carbaryl, making it relatively easy for even a novice to begin to write an expert system.

Many different types of expert system shells exist, each one differing slightly in the types of knowledge it can include in its knowledge base and in the internal representation used to store knowledge. As a result, knowledge-based systems can be characterized in two ways—by the problem domain and by the type of shell used. For example, suppose two people develop an expert system to recommend weed management options for a specific crop. An agronomist would likely describe the systems based on the problem domain, calling both systems weed-control expert systems. However, a computer scientist would classify the systems by the inference engine's characteristics, so that one might be a case-based reasoning system and the second a truth maintenance system.

This problem is best eliminated by focusing attention on the problem type being addressed. For instance, the problem of diagnosis or identification can be approached with a similar type of knowledge-based system, regardless of whether one is determining what plant diseases are affecting the crop or what insect is under the microscope. In this book, therefore, we refer to problem types (diagnosis, prediction, and planning) and to knowledge-based system approaches (knowledge representations and inference strategies) to describe the use of knowledge-based systems in agriculture.

We will also distinguish among different types of agricultural management decisions. Following Conway (1984) we distinguish between tactical, strategic, and policy decisions and the types of knowledge-based approaches used to address each. Tactical decisions involve the implementation of specific practices to achieve a goal. For example, if a farmer determines that there are nematodes in a field normally planted in soybean, the farmer must decide whether to plant a resistant cultivar of soybeans or to plant another crop in that field. These are alternative tactics, and tactical decisions are made in response to specific problems.

Strategies are best thought of as general rules that help decision makers make tactical decisions. For example, one strategy for managing soybean-damaging nematodes is to grow susceptible varieties of soybeans every third year, in rotation with a resistant cultivar. The strategy is an attempt to prevent the nematodes in the field from evolving a tolerance for the resistant cultivar. Once this strategy has been adopted, the farmer can make the tactical decision of what cultivar to plant in a given year more easily. The tactical decision of which specific variety to plant must still be made, but the general characteristics of the desired cultivar are determined by following a strategy.

Questions of policy can be considered as the next level of abstraction. Policies refer to the types of strategies employed by a decision maker. For example, it may be a farmer's policy to use organic farming methods. This policy decision affects that farmer's choice of strategies, which then affects the tactics used in managing. Few knowledge-based systems have been developed to address policy questions. One, the economic adviser in COTFLEX, gives farmers advice about whether or not to participate in the federal farm program (Helms et al., 1987).

1.1.2 Strengths of knowledge-based systems

As mentioned, one key reason to build knowledge-based systems for crop production is that these systems provide a framework with which to capture and apply nonalgorithmic knowledge. Crop systems are poorly understood compared to, for instance, many physical systems. We have only a general understanding of key soil processes like mineralization of nutrients and the relationship of soil organic matter to different plants in rotation, soil structure, leeching, and runoff. We do not understand insect population dynamics well enough to predict accurately population outbreaks or their effects on yield. We know only approximately what effect the pesticides we apply to crop fields will have. We have no means of predicting the market price of a crop or the quality of a crop's yield.

In all these cases, we have some idea of what these relationships are. We do not operate in a vacuum by any means. However, our sometimes vague and general knowledge about cropping systems has been difficult to corral and integrate because the interactions are boggling. The English-like knowledge structures available in knowledge-based systems give us a way to encode this knowledge and to apply and test it.

Knowledge-based systems are also computer programs—software that can be readily copied and distributed. Human experts in obscure or complex domains, on the other hand, are scarce and access to their time is limited. Because of the recent proliferation of powerful microcomputers, knowledge-based systems will be able to provide dissemination of expertise not possible before.

Other often cited advantages of knowledge-based systems are more equivocal. Knowledge-based systems are said to be user friendly and to provide an explanation of their reasoning on demand. Farmers, who as a group are just joining the computer revolution, are very sensitive to ease-of-use problems. The ability of a knowledge-based system not only to make recommendations but also to tell the farmer why those recommendations were made may be critical to the success of these systems. In theory, providing an explanation is possible since knowledge-based systems solve problems by generating a chain of logic. However, in practice explanation facilities have not yet fulfilled their promise. The ability to explain reasoning comes from the separation of knowledge and inference. For example, in a *rule-based expert system,* all knowledge is represented by IF-THEN rules in the knowledge base. The inference engine chains these rules together into a logical sequence to solve a problem. When asked to explain its reasoning, the inference engine ought to be able to go to the chain of rules and present that chain to the user as an explanation. For example, if asked, "Why did you ask me whether bees are nearby?"

the system could find the appropriate rule and respond, "If bees are nearby, then one should not spray carbaryl." However, very often rules don't provide any insight into the reasoning process. They might be rules that control the flow of the system logic, or they might describe information needed by the computer but taken for granted by the user. For instance, the question, "Why did you ask me if it rained yesterday?" might elicit, "If it rained yesterday, then the ground is wet," a statement of very little interest to the user. We will explore this in more detail in Chap. 2.

Another advantage of knowledge-based systems is their ability to handle uncertain information. This is very important in crop systems because much of the information we can gather about a field is uncertain, i.e., incorporates sampling error, and what we know of the future is dependent on random elements like the weather. Knowledge-based systems have many ways of dealing with decision making under uncertainty; however, there is little consensus about which methods are best (see Chap. 3).

From the researcher's point of view, knowledge-based systems are attractive because of their potential to help organize and synthesize knowledge and information of different types. Using knowledge-based systems as the framework, it is possible to focus and apply diverse avenues of research to solve difficult problems. One can link together quantitative data, simulation models, basic research results, and expert opinion into a knowledge-based model of how difficult decisions ought to be made. Chapter 12 discusses the development of integrated knowledge-based systems for decision support, but this integration is also useful in research planning. Knowledge-based systems allow us to structure our research based on the management problem, i.e., what information is needed to make management decisions and how is that information related. A model of the management problem provides, first, an organizational structure for research and, second, a blueprint for a knowledge-based system developed to solve that problem. We will discuss the issue of organizing and building a system more in Chap. 4.

1.1.3 Failings and criticism

Knowledge-based systems have also been criticized. They are perceived to have two fundamental failings as they have been applied to agriculture. First, knowledge-based systems are sometimes perceived as information delivery tools, not as research tools. It has been argued that institutions and individuals charged with undertaking research in agriculture should therefore not get sidetracked into knowledge-based systems work. Because it does not generate new knowledge but only repackages existing information, the technology is not an appropriate area of study. Solutions to complex problems will come only from the discovery of new knowledge (Berryman, 1987).

This criticism, in only slightly different form, could be used to argue that research institutions should not use mathematical modeling or any sort of systems analysis. After all, these are also techniques that mainly try to represent what is known in new forms. The point is that basic, reductionist research must be complemented by holistic synthesis research into systems and their behavior. Knowledge-based systems are tools for merging systems science with decision analysis.

A second criticism is made at a more fundamental level (Dreyfus and Dreyfus, 1986). This is that knowledge-based systems, and indeed all of artificial intelligence, are really something of a fraud. Specifically, Dreyfus and Dreyfus assert that there is no evidence that any man-made computer system has ever demonstrated anything that remotely resembles human intelligence. They argue that it is not true that expert systems have in any way captured human intuition or reasoning ability. We would answer that these arguments may well be true, but from the perspective of agricultural management they miss the point. The shortage of experts in agriculture and the seriousness of the problems makes the development of some form of computer-assisted management tool imperative. Whether or not these tools really are intelligent is irrelevant provided they do the job. If the best available structure for some applications is the knowlege-based system, then this structure should be used. It is perhaps unfortunate that computer scientists use overblown terms to describe their products (perhaps "imitation intelligence" would be a better term than "artificial intelligence"), but this is no reason to abandon these products.

A third criticism is that agriculture is so full of natural variation and unexpected and misunderstood phenomena that finding bounded problems for which sound expertise is available will be impossible. Agricultural experts disagree, and they are not consistently correct. Why, if we cannot trust our experts, should we trust expert systems that are based on the expert's knowledge? In other words, should we not be putting our efforts and money into trying to improve our understanding, rather than into trying to synthesize and deliver what we know to be imperfect?

1.2 Management and Control

In response to this third criticism, we make the following distinction between agricultural practices aimed at manipulating the resource (control) and agricultural decision making (management). The former is the domain of basic research, the latter is the domain of knowledge-based systems and decision theory. We use the problem of pest management and control as an example, but the same distinction applies to problems of machine maintenance or crop variety selection.

Pest control is the manipulation of pest numbers to further a human goal; pest management is the process of choosing combinations of pest control options to maximize profit and farm survival. Ideally, pest problems can be eliminated directly by improving pest control technology through biological or improved chemical control, developing new crop cultivars resistant to pest attack, designing new cultural practices that ameliorate the problem, etc. When people argue that we should spend our time and money trying to improve basic understanding, they are hoping for this kind of cure.

However, many key pests of agricultural commodities are persistent and difficult to control. Research on such persistent pests has had to focus more and more on the problem of decision making, i.e., how to choose from among strategies and tactics, none of which is totally reliable, affordable, or compatible with the whole crop production system. This problem of decision making in pest control is the essence of pest *management* research.

Pest management is complicated further by the unpredictability of the abiotic and economic systems. The weather governs the growth and development of crops and strongly influences pest population dynamics and plant-pest interactions. Yet the weather is not predictable, even on a relatively short scale. Furthermore, farming is a business; it may be constrained by weather, but it is driven by the economy. Pest control practices must be chosen in accordance with economic principles of resource management. Even if the yield of agricultural production systems could be predicted with certainty, the value of that yield depends on changing and unpredictable prices and market forces.

At the beginning of the chapter we said that successful farmers will manage from a systems perspective, but in the face of all the complexity and uncertainty implied in farming (Fig. 1.1), it is clearly unreasonable, if not impossible, for a farmer to acquire and apply all this knowledge each time a pest management decision is made. Instead, farmers develop heuristics, rules of thumb, that help them make difficult decisions. These rules of thumb take the place of a systems-level understanding of all the complex interactions affecting crops and livestock. Over time, heuristics are tested using a kind of natural selection. The good heuristics deliver more consistent and higher profits to their users. Unfortunately, as agriculture changes and new technologies emerge, heuristics must change as well. It is not practical to require farmers to learn purely from experience. Research must provide not only new technology but also new heuristics.

Researchers and extension agents develop their heuristics based on experience and on controlled experiments. In theory, researchers and extension agents are aware of new technologies and are more objective than farmers. Still, because farmers' heuristics are based on experience with their particular farm, they will sometimes serve better than the researcher's rules. The overwhelming variation of the natural and economic systems virtually ensure that no individual will experience enough of the real world to develop stable and robust heuristics. It should not surprise us that "experts" in agricultural pest control disagree or develop very different rules of thumb. Experts, in the classical sense, do not exist. Instead, there is a spectrum of expertise in agricultural systems representable by the spectrum of decision rules used by individual decision makers.

When we build knowledge-based systems in agriculture, we are developing hypotheses about how systems ought to behave and how they ought to be managed. These are issues of agricultural management science, a discipline that has been largely ignored but that is now emerging as critical to the success of farmers today and in the future. Knowledge-based systems are one of the primary tools useful in agricultural management science. They provide the theoretical core for addressing questions of planning, pest control, and scheduling. We do not see knowledge-based systems as a distraction from agricultural research; we see them as the basis for a revitalization of agricultural management science.

1.2.1 Knowledge-based systems compared to other computer decision aids

Knowledge-based systems are not the only kind of computer decision aid that will play an important role in the next decade. Other computer applications have and will continue to provide information to decision makers to help them make better decisions in cropping systems. These include database systems, management information systems (MIS), information retrieval systems, geographic information systems (GIS), decision support systems (DSS), simulation models, and operations research (OR) techniques. The ultimate goal of all these decision aids is to improve decision making, but they differ in their approach and in what they offer to the user. Their boundaries are also fuzzy. Any particular computer decision aid might contain elements of several of these systems. As with knowledge-based systems, for example, decision support systems often contain simulation models.

The best way to compare these different types of systems is to look at their components and ask, "What do they provide to the user?" The components of all these systems are ultimately a user interface, databases, simulation models, analyzers, and advisers. The goal of database systems, MIS, and GIS is information delivery. They are designed to make it easy for a user to access relevant information. They do not solve a problem, they merely allow access to a variety of information that the user might require to solve the problem. These systems can be quite sophisticated. GIS applications will certainly play an important role as we begin to address landscape-level problems in agriculture. However, it is important to keep in mind that a GIS is still just a database system specialized for map-oriented data. Only by adding a reasoning system, analyzer, or adviser to a GIS can the system solve problems.

Simulation models likewise are not problem solvers. They are simulations of reality that may be very informative and useful in making a decision. Simulations, like database systems, GIS, and MIS, are often components of larger systems that solve problems. For example, mathematical models of crop systems have often been used in optimization analyses to determine such things as optimal cutting dates for alfalfa (Shoemaker and Onstad, 1983), thresholds for insect pests (Gutierrez et al., 1979, 1981), and optimal pest control strategies (Stone et al., 1986b).

Providing information doesn't solve problems, although it can make a problem's solution trivial. Computer programs for analysis and advice are problem solvers. Operations research, for example, includes a wide variety of techniques for representing problems mathematically and determining the optimal solution. The best known OR method is linear programming (LP), used heavily in economic analysis. Economists use LP to maximize the profit function ($\Pi = \textit{revenue} - \textit{yield} \cdot \textit{price}$), subject to other functions describing the constraints on production. LP analyses can be used to come up with an optimal crop mix, for example, if the costs and revenue functions for producing the crops can be described by linear mathematical functions. We discuss optimization techniques several times throughout the remaining chapters. In particular, Chaps. 7 and 9 provide comparisons of numerical and knowledge-based approaches.

On-line identification keys (Pankhurst, 1978, 1984) are another kind of analyzer. They ask the user a set of questions and display descriptive information that lead the user to identify something, e.g., a plant, an insect, or a disease, from a list of potential choices. Recently, with the availability of low-cost, high-resolution graphics and large computer memories, on-line identification devices have become quite powerful. At each turn in an identification problem they can show pictures of the user's choices, greatly facilitating an identification. These systems appear very similar to knowledge-based systems because they ask the user a set of questions designed to narrow down the possible answers to an identification question. The difference is that on-line programs use a quantitative algorithm to select questions. Typically they try at each step to eliminate roughly half of the remaining possibilities, so that the identification proceeds as efficiently as possible. A knowledge-based approach, on the other hand, might go through an identification by quickly determining a likely candidate based on general characteristics and then trying to verify the hypothesis with more specific questions, thereby being most efficient at identifying common specimens but taking much more time on the unusual ones (see Chap. 2).

Perhaps the distinctions among these different computer decision aids are trivial. Ultimately, the value of a computer program is whether it helps make better decisions in the field. Many people lump all computer decision aids together, calling them all decision support systems. However, there are at least two reasons for maintaining a distinction and for focusing today on knowledge-based systems. First, of all these decision aids, only knowledge-based systems are new to agriculture. Simulation models, for instance, have been used to analyze complex problems in agriculture for almost 30 years. OR methods were developed in the 1940s. Databases are as old as computers. Modern versions of these tools are more sophisticated, surely. GIS systems have brought map-like visualization to database systems, and supercomputers have made it possible to use bigger and bigger simulations in numerical optimization. However, the limitations of these tools are now well documented. Databases can overwhelm the farmer, who must be expert in so many disciplines to achieve success. Farmers cannot examine all the information relevant to their decisions; there isn't time and they do not have the expertise. Simulation models make systems artificially rigid. Our models are not very accurate, certainly not accurate enough to predict what will happen in a particular field. Knowledge-based systems are the new frontier in computer decision aids. They must be judged critically and fairly, but their promise is profound.

The second reason to concentrate on knowledge-based systems is that they can capture knowledge—heuristic knowledge, rules of thumb, experience, and qualitative information—that no other type of system can. Crop systems have defied purely quantitative explanations. There are fundamental laws of nature at work, but the complexity of even simple agricultural ecosystems is daunting. Just understanding the interactions in the soil is beyond our capabilities today. Adding the perverse laws of economics on top of the natural system makes the whole package hopelessly beyond our ability to write equations to describe it. Human beings, however, have shown themselves to be capable of dealing with problem solving in this and similar domains. By adopting problem-solving methods similar to those used by humans dealing with other complex problems, we have the capability to improve agricultural management. Furthermore, what understanding we have is often qualitative. If we are to use what we know, and disseminate that knowledge to farmers, we need computers and we need the power of knowledge-based systems.

This does not mean that databases and simulation models are unimportant. They describe aspects of the agricultural production system very clearly. Some computer decision aids built from these components exist today in agriculture, e.g., AUSSIM in Alabama for soybean management decision support (Backman et al., 1989), and TEXCIM (Sterling and Hartstack, 1987) in Texas for cotton pest management support, but despite their power, their acceptance and practical use has been limited. Such systems tend to overwhelm their users with complexity and information. Perhaps by embedding their power within knowledge-based systems, such programs may become easier to use and interpret.

1.3 Brief History of Knowledge-Based Systems in Agriculture

The first agricultural expert system, PLANT/ds, a system that diagnosed soybean diseases in Illinois, was built by a computer scientist (Michalski et al., 1983) as an application of a machine-learning algorithm. Other systems based on PLANT/ds have followed (Fermanian et al., 1988; Liu and Fermanian, 1988), but because of its emphasis on machine learning, it is really atypical of most subsequent applications in crops, except for its domain: diagnosis. Diagnostic systems have dominated the early development of knowledge-based systems in crops and agriculture in general. Diagnosis includes problems that can be represented as selecting one item from a well-defined list of potential candidates based on observable symptoms. Disease diagnosis means deciding which disease fits the observed symptoms. Pest identification in general is a diagnosis: which known pest fits the observed characteristics? Even the problem of recommending a pesticide can be broken down into a diagnosis and prescription problem. Examples of such systems include identification of key beneficial and noxious organisms (Michalski et al., 1983; Blancard et al., 1986; Woolley and Stone, 1987; Cervo et al., 1988; Hoshi and Kozai, 1988; Le Renard, 1988), pesticide selection (Stone et al., 1987; Beck et al., 1989), tactical recommendations in grape systems (Saunders et al., 1987), and tactical advice in well-defined areas like weed control (Blonningen, 1988; Maran and Pike, 1988; Smith et al., 1989) and variety selection

(Shroger et al., 1987). The general problem of diagnosis and prescription is basically one of analysis of a problem and synthesis of a solution, and we shall often use these terms synonymously.

The emphasis on diagnosis began with the first true expert system, MYCIN, in the late 1970s (Shortliffe, 1976). It was a rule-based expert system that diagnosed bacterial infections in human blood. The reason it became a model system is that its inference engine was the first to be separated fully from the knowledge base. It was this part of MYCIN that was developed into the first expert system shell, EMYCIN, which was the basis for most of the expert systems shells appearing on the commercial market in the 1980s. Since the shell had been developed for a diagnostic problem, applying it to other diagnostic problems was straightforward.

The first knowledge-based system to attract widespread attention in crop systems and agriculture in general was COMAX, the cotton management expert system (Lemmon, 1986). COMAX represented another type of expert system application; it was an integrated system in which the expert system triggered the running of a simulation model, GOSSYM (Baker et al., 1983), using the output from iterative runs of the model to recommend fertilizer and irrigation schedules for a cotton crop. COMAX has since been renamed GOSSYM/COMAX and has been implemented in several states in the United States across the cotton belt as a decision aid. New modules for insect pest control are also being developed (McKinion et al., 1989).

The idea that simulation models could be used by expert systems sparked considerable activity in agriculture. The development of simulation models for major crops in the United States had been one of the key goals of the National IPM Project and the Consortium for IPM Project, but the models developed had become very complex and not easily used for practical farm management and crop production. Even if the models could be run, their output was often difficult to understand and apply to a specific problem. Knowledge-based systems seemed to be a way to deliver a simulation model along with the knowledge and expertise needed to run and interpret it (Stone et al., 1986a). Interface systems (Fig. 1.3*b*) are the simplest example of this integration. For example, when the USDA Soil Erosion Laboratory releases the WEPP simulation model of soil erosion, it will also release an expert system interface to help users construct the input files for the model (Meyer, 1990).

Integrated systems refer to knowledge-based systems that use database systems and/or simulation models to supplement their reasoning component. One example of this level of integration (Fig. 1.3*c*) is the COTFLEX system, which uses an expert system to recommend whether the user should purchase federal crop insurance. This recom-

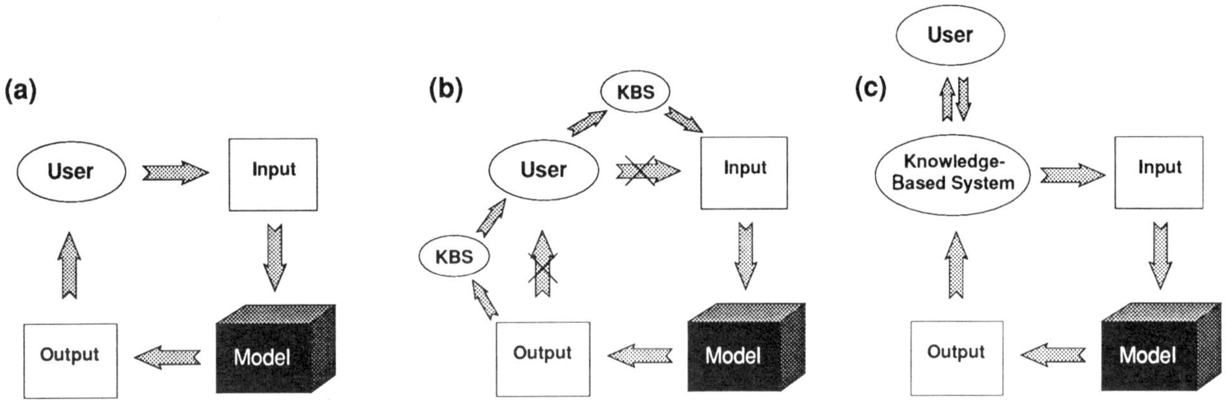

Figure 1.3 Incorporation of simulation models (black box) into knowledge-based systems. (*a*) The user interacts with a model. (*b*) The knowledge-based system is acting as an interface to the model. (*c*) The user interacts with an integrated knowledge-based system that runs the model and uses its output to support its reasoning.

mendation is made by setting up and running the model, FLIPSIM-V (Richardson and Nixon, 1985), and interpreting its output for the user (Helms et al., 1987). The simulation model's input file is constructed from an extensive farm database describing the finances and capital resources of the farm. At this level of integration, the user need not even know that a simulation model is being used or a database accessed. These components are embedded within the overall system.

Several researchers have linked simulation models to expert systems since 1985, mostly for soybean and cotton. Sometimes the goal is to make the model easier to use, and at other times it is to provide more information to a knowledge-based system (Jones et al., 1987; McClendon et al., 1987; Stone et al., 1987). Chapter 5 is dedicated to the latter problem.

1.4 Scope of Knowledge-Based Systems in Agriculture

The field of knowledge-based systems in crops spans several disciplines: computer science, agronomy, entomology, agricultural economics, and agricultural engineering. The problems being addressed in these disciplines are just as variable. Up to now, knowledge-based systems have proven their great potential as tools for the organization of research, as tools in research, as tactical advisers to farmers, and as whole-farm planners. Knowledge-based systems provide us with new ways to record our information and build models of systems.

Figure 1.4 shows the scope of knowledge-based systems categorized by problem type. At the core of knowledge-based systems are the fundamental problems of inference, knowledge representation, search, and pattern matching. Research in these areas is generally the purview of computer scientists and decision theorists. However, as we will see in later chapters, problems in agriculture are demanding fundamental solutions in these areas. The outer circle represents application types. At the beginning of each subsequent chapter, we use this figure to highlight the basic and applied issues covered.

1.4.1 Knowledge representation and inference

Artificial intelligence programs derive their power from their ability to manipulate nonnumeric information. The term artificial intelligence has even been defined as a nonnumerical science and engineering discipline (Clancy, 1989). In particular, knowledge-based systems derive their power from representations of human knowledge that are

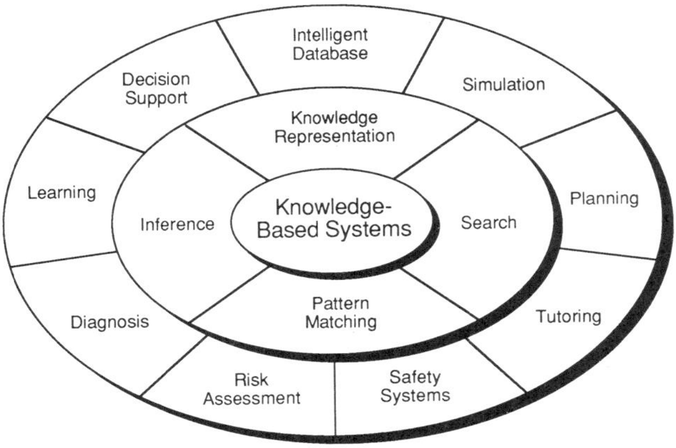

Figure 1.4 Fields of knowledge-based systems discussed in this book. Core areas and application areas of knowledge-based systems research are shown within the inner circle and outer circle, respectively.

typically recorded in written and spoken languages and in the human mind. The key to making computers better problem solvers is to mimic the way humans store, retrieve, and manipulate knowledge. Storage and retrieval is the problem of knowledge representation.

It is very difficult to distinguish common terms like *fact, information, knowledge,* and *understanding*. Our definition of knowledge is interpreted or interpretable information in context. For example, the fact guthion is an organophosphate is not knowledge to the average person. It is without an appropriate context. However, to toxicologists it is knowledge since it tells them about guthion's chemical structure and probable mode of action. Knowledge is a meaningless term without an associated domain.

The approaches developed by computer scientists to represent knowledge are varied and complicated by jargon. We are necessarily encumbered by much of the jargon. In this book, for instance, you will read about rule-based systems, frame-based systems, case-based systems, script-based systems, semantic nets, blackboards, and model-based systems. Each of these names refers to a type of knowledge representation. Although the diversity may seem daunting, each of these strategies for encoding knowledge is based on the same principle, adding associations to information.

The simplest systems associate truth values with facts or assertions. The statement, the soil is wet, can be true or false. Given this association of truth or falsehood, rules can be written to make inferences, e.g., *if* the soil is wet, *then*.... At the next level of complexity, objects of interest in the problem domain, like soil processes, are associated with attributes, each of which can have some truth value, e.g., the water capacity of the soil is low or the soil type of the soil is sandy loam. The attributes water capacity and soil type are first associated with the object soil. A third level of association would be to allow the objects in the knowledge domain to be linked. For example, soil is one component of a crop field, and corn field is a kind of crop field.

As more and more associations are added to the structure of the knowledge base, the systems become capable of more intelligent behavior. For instance, if a knowledge base has all three levels of association mentioned above, then by telling it that you have a problem with a corn field, you would automatically tell it that the corn field includes soil and that the soil has a moisture content and a soil type. We have discussed other associations that some knowledge-based systems implement. Certainty in an assertion is an association of a fact with a number. Explanation requires an association of a linked list of rules with an assertion.

As we will see, more complex knowledge representation schemes typically add new associations to information. Frames and semantic nets are two ways to add the same type of associations of attributes with objects and objects with other objects. Truth maintenance systems associate each assertion in the knowledge base with the justification for the assertion, i.e., a set of facts and rules already processed. This allows such systems to change their minds when new evidence contradicts earlier assumptions. Scripts are structures that associate objects or attributes of an object with an order; scripts are used to represent sequences of related actions like the operations needed to prepare a field for planting.

As knowledge representation schemes add more and more associations to each piece of information contained in the knowledge base, the inference engine must become more sophisticated to follow the chains of associations. Inference is the process of manipulating knowledge to form new knowledge. Rule-based systems, for example, use elementary logic to chain IF-THEN rules together to form new conclusions and to solve problems. Other systems rely on the association of activities with knowledge to determine what to do next. For example, a knowledge base entry describing soils might include an associated instruction to find out the soil type, slope of the field, and vegetative cover as soon as any information about soils is entered into or sought by the system. Chapters 2 and 8 discuss different types of knowledge representation and inference.

1.4.2 Pattern matching and search

Pattern matching and search are techniques used in almost any AI program, and they are also occasionally used as the main problem-solving strategy in a knowledge-based system. The inference engine uses pattern matching to determine whether two objects, assertions, frames, or any structures are the same. For example, when an inference engine examines a rule to see whether it applies to a problem, it compares the IF-side or THEN-side of the rule with those facts it knows about the problem and its goals. Suppose an expert system contains the rule

IF crop is corn AND
crop is-susceptible-to European-corn-borer AND
plant-height $\geq$ 22
THEN time-for-first-sample is true

Now suppose that the inference engine is trying to determine whether TIME-FOR-FIRST-SAMPLE is true or false. It must search through the rule base looking for statements of the form: TIME-FOR-FIRST-SAMPLE is ?X. In this statement ?X is a variable, which can take on any value. This form is a pattern, and when the inference engine finds the above rule, it recognizes that the conclusion of the rule matches the pattern. This elementary process is called pattern matching. If the inference engine wanted to determine that TIME-FOR-FIRST-SAMPLE was true, it would match the IF-side conditions to facts in its memory. In this case, it would need to find exact matches for CROP IS CORN and CROP IS-SUSCEPTIBLE-TO EUROPEAN-CORN-BORER. It would then look for match for PLANT-HEIGHT IS ?X AND ?X $\geq$ 22. Again, ?X can match any number, but the second pattern, a mathematical inequality, would eliminate all values less than 22.

Most rule-based expert systems have very simple pattern-matching algorithms to allow their inference engines to do this kind of rule selection. Pattern matching can also be a useful problem-solving tool when used to determine to which group or category a particular object belongs. In diagnosis, for example, a powerful pattern-matching algorithm could be used to eliminate unlikely diagnoses as more and more information is learned. Pattern matching has also been used in the STRIPS program, a planning system to determine discrepancies between the desired goal and the current state of a system (Fikes and Nilsson, 1971).

Adding efficiency to the reasoning process is a secondary goal of pattern matching, but it is the primary goal of a *search*. Search is the general term for how one gets from point A to point B. For example, since the simplest rule-based expert systems impose no structure on the rule base, there is no obvious way for the inference engine to find the next rule it needs to solve a problem. At each iteration, any of the remaining rules could be chosen. Assuming a rule base of just 50 rules, there are 50!/40! or 3.7×10^{16} ways the system could search for the first ten rules alone. In practice, inference engines use a search strategy to pick the next rule for evaluation. Many systems use a linear search of the rules in the order they were written or entered into the system. Other systems provide some structure to their rule bases, dividing them into modules or providing some discrimination rules that help the system select which rules to search through based on what has already been discovered.

1.4.3 Application areas

The outer circle in Fig. 1.4 shows application areas for knowledge-based systems in agriculture. We have already discussed diagnosis and simulation as areas in which knowledge-based systems have proven their worth. The general area of decision support covers systems that help farmers make tactical and strategic management decisions. These include systems that recommend pest control treatments and strategies, timing of planting and harvesting, cultivar selection, and marketing options, and that help farmers make policy decisions like whether or not to participate in the federal farm program.

Safety systems expand on the use of knowledge-based systems for diagnosis and decision support. These systems act as early warning systems, detecting failures before they become serious and recommending solutions while solutions are still inexpensive and are subtle modifications to a system. They rely on real-time monitoring of a system and model-based reasoning (see Chap. 11). Instead of waiting for the system to break down and then recommending a treatment, a safety system constantly compares the monitored system to its simulated system looking for unexpected or unwanted behavior. Diagnosis of faults and recommendation of treatments are then undertaken. Safety systems have been implemented mainly in the nuclear power industry but offer an interesting challenge for agriculture. As explored in Chap. 12, the goals of seasonal agricultural management are very similar to the goals of regulating a nuclear reactor.

The use of knowledge-based systems to add power to databases creates what is known as intelligent databases. We will discuss these applications only briefly in Chap. 12. The task of making databases intelligent is very similar to the problem of knowledge representation. The goal in both cases is to provide information to a user or an inference engine in context, i.e., when appropriate. The distinction is that in an intelligent database application, the representation scheme is already established in the database. The intelligence must be added as an interface. These systems are useful when a decision depends on large data sets that require high maintenance. Such data demand the power of a database management system for upkeep. Access to those data for analysis and decision making should not require all the power and complexity of a large database management system and can be streamlined with a knowledge-based system. Applications in agriculture will probably remain in the food processing, distribution, and financial businesses, rather than on the farm. Databases intended for the farm will more likely fit directly into knowledge representations directly used by inference engines.

We have also briefly mentioned planning as an application. Planning can be thought of as a problem of choosing a sequence of operations on a system to transform the state of that system from its current state to a goal state. The constraints on the plan are the set of allowable transformations. The classical planning problem comes from robotics and involves moving blocks on a table. As we will see in Chap. 9, even stacking and unstacking blocks present nontrivial problems. An agricultural example is deciding how to prepare a field for planting. The current state is usually a harvested field, with some remaining plant material from the previous crop. The goal state is a field planted with the next crop, and the rules of transition are the set of cultural operations possible, e.g., disking, burning, plowing, incorporating manure, planting into stubble, and ridge-tilling.

Learning is a self-modifying capability of some knowledge-based systems. Learning describes the ability of a knowledge-based system to incorporate new knowledge into its knowledge base during or after consultations. Learning systems are very attractive because they can be constructed initially with relatively little knowledge and can learn from experience, becoming more powerful with time. Many diagnostic systems and game-playing systems operate in a learning mode. Game-playing systems learn to avoid moves and positions that lead to a loss. In agriculture, this is an appealing capability. Decisions that lead to reduced profit could be avoided in the future. However, learning systems are also potentially dangerous. Often it is not clear why a bad result has occurred. Agricultural systems are unpredictable and variable. A farmer can make the right decision, which 9 times out of 10 will work better, and still end up losing money because of a fluke of nature. We would not want to change the system's logic that gave the proper advice in such a case.

Tutoring and risk assessment are applications made possible by knowledge-based systems' capabilities to explain their reasoning and to handle uncertain information. Tutoring systems transfer heuristics from the computer to the user. Effectively, they train inexperienced decision makers to act like experienced managers. Such systems require not only domain knowledge and explanation facilities but also heuristics about how students are learning and what approaches might help. Risk assessment is an application area based on the capability of knowledge-based systems to handle uncertain information and to reason under uncertainty. These areas are briefly covered in Chaps. 2 and 3.

1.4.4 Knowledge-based systems and the future

If our agriculture is to meet the challenges facing it today, then the issues of agricultural management must be addressed. How can farmers be expected to deal with diminishing profit margins, elimination of chemical pesticides, and restrictions on practices that promote erosion or pollute surface and groundwater? Agricultural researchers must face the fact that while answers from basic research may be forthcoming, they are coming slowly and not keeping pace with the new managerial complications.

Decision making in agriculture must become more objective and more based on up-to-date knowledge. Agricultural decision makers will be required more and more to justify and account for their actions. Seat-of-the-pants decision making simply will not do. Similar trends have been occurring in the management of other natural resources like forests and parks. The prospect of farm managers being brought to court to justify their decisions to apply pesticides to the environment is becoming increasingly ominous.

Knowledge-based systems offer the foundation for the science of agricultural management. It is intertwined with the fields of systems analysis and decision theory and with cognitive and computer science. For those agricultural researchers who wish to tackle management issues in the next decade and beyond, we strongly advocate a solid grounding in knowledge-based systems.

Chapter

2

The Diagnostic Expert System

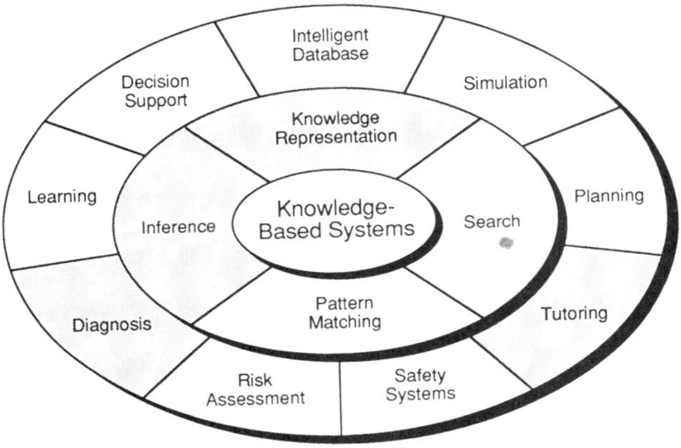

This chapter discusses the development of a basic diagnostic expert system. The problem to be solved is the following: given a set of mutually exclusive hypotheses, identify which one of them, if any, is correct. This problem could, of course, be generalized to include hypotheses that are not mutually exclusive, uncertainty as to which hypothesis is correct, and so forth. Topics such as these will be covered in later chapters. A good example of the type of problem that lends itself to solution by a diagnostic expert system is the taxonomic problem of identifying a biological specimen. This problem will be used as a motivation for the chapter, and for the next two chapters as well.

2.1 Introductory Problem: Taxonomic Identification

Taxonomy, or systematics, is the study of the relationships of different kinds of organisms. Abbott et al. (1985) describe taxonomy as "providing an information system for biology." The problem of the taxonomist is to fit different kinds of organisms into an orderly classification and to identify the class of particular organisms. Since all organisms are generally presumed to have evolved from a common ancestor, the most common means of classification is by their phylogeny, or evolutionary history. The standard classification system separates organisms into the familiar hierarchy of categories: kingdom, phylum, class, order, family, genus, and species.

We are concerned in this chapter with the problem of identifying, determining the correct classification of, a particular specimen. This problem has considerable application in agriculture, for example in the identification of insects and other arthropods found in a crop to determine their pest status and of weeds and plant pathogens. One of the taxonomist's tasks is the construction of identification aids for this problem. After discussing some representative examples of these identification aids, we shall describe how an expert system may provide an approach to the solution. In this way the similarities and differences between expert systems and more conventional identification aids will be clarified.

Classical identification aids are called *keys*. There are a number of different identification methods and correspondingly a number of different types of keys. Unfortunately, the nomenclature used to describe and compare methods is not standard. Moreover, from our perspective there are really two parallel sets of terminology, one used in taxonomy and the other used in artificial intelligence and expert systems theory. For example, in taxonomy the problem of identifying from a set of classes the particular class to which a given object belongs is generally called *identification,* while in expert systems theory the problem is called *classification.* Taxonomists use the word classification to mean the construction of the information system upon which the identification is based. In many cases the same term is used by different authors to mean different things, and sometimes the meanings of terms are exchanged. We cannot maintain a consistent and useful terminology that is loyal to one particular author, but we shall at least try to be consistent throughout the book. In many cases we shall use two synonymous terminologies, one more appropriate to taxonomy and another to artificial intelligence. We will attempt to be consistent with the systematics-based terminology given by Morse (1975). Other important references providing general information about taxonomy and identification are Pankhurst (1978) and Abbott et al. (1985).

Our discussion of the various forms of keys makes use of a hypothetical example that is similar to those given by Pankhurst (1970) and Duda and Gaschnig (1981). The example involves the identification of an animal from among the following possibilities: horse, tiger, lizard, sparrow, duck, and eagle. Table 2.1 shows the data on which this identification will be based. The table can be considered as a matrix, with the type of animal given by the row and the particular character by the column. The matrix forms the basic set of information that will be used to construct the identification aid. It is not necessarily a complete

TABLE 2.1 Classification of Animals

	Flight	Habitat	Birth method	Food	Coloring
Horse	No	Land	Live	Grass	Various
Tiger	No	Land	Live	Meat	Striped
Lizard	No	Land	Egg	Meat	Green
Duck	Yes	Water	Egg	Various	Brown
Eagle	Yes	Land	Egg	Meat	Brown
Sparrow	Yes	Land	Egg	Seeds	Gray

set, and indeed both classical identification keys and expert systems attempt to make use of information that may be known to the person constructing the key but not expressed in simple matrix form. For example, the key constructor might know that in Table 2.1 the value of the character *coloring* is easier to determine than the value of *birth method,* and might therefore construct the key so that the former character is tested before the latter.

There is sufficient information in the table to distinguish among the six possibilities if all of the information is available. Indeed, there is quite a bit of redundancy. For instance, the animal can be identified as a duck if its habitat is *water* or if its food is *various.* Similarly, an animal can be identified as a lizard if it does not fly and its birth method is *egg* or if its coloring is *green.* In this simple example redundancy is not a problem in using the table to make an identification, but in a real identification problem there may be dozens or even hundreds of alternatives from which to choose, making the use of a matrix such as that shown in Table 2.1 impossible. The information contained in the table must therefore be organized in some way. In Chap. 1 we made the distinction between *information* and *knowledge.* In this context the contents of Table 2.1 is the information, and the organizing and structuring process converts this information into knowledge.

Chapter 1 also described the (*class, attribute, value*) triple commonly used in artificial intelligence. In the systematics literature this is sometimes called a (*taxon, character, state*) triple (e.g., Dallwitz and Paine, 1986). Such a triple links the type of animal, e.g., *duck,* a particular character, e.g., *habitat,* and the nature of that character for the given type, e.g., *water.* Dallwitz and Paine (1986) distinguish the following four types of characters:

1. *Unordered multistate.* Characters have two or more states, with the states having no order relationship. Examples are color, e.g., *red, brown,* and *yellow,* and the boolean variables, *present* and *absent.*

2. *Ordered multistate.* Characters have more than two states, and

the states can be arranged in some natural order, e.g., *white, gray,* and *black,* or *does not fly, weak flier,* and *strong flier.*

3. *Integer numeric.* Characters have states that are always a whole number, e.g., number of legs.

4. *Real numeric.* Characters have states that can be any real number, e.g., *length.*

In the rule-based systems described in this chapter, the linkage between classes and the values of their attributes will be indirect, through rules. These rules will make use of attribute-value pairs. Since these pairs provide sufficient representative power, we will use them through Chap. 8, at which point the more complex structures will be introduced.

Two primary objectives in constructing an identification key are to maximize the likelihood that the user makes the correct identification and to minimize the amount of effort needed by the user to make the identification. These objectives are not necessarily in conflict. Constructing the key to enable the identification to be made using as few attributes as possible contributes to the first objective since it avoids unnecessary questions. It contributes to the second objective since each time the user is asked to provide the value corresponding to a given attribute, there is some chance that the information provided will be incorrect. The consequences of incorrect user information in making the correct identification range from annoying to disastrous, but by minimizing the amount of information requested of the user, the chance of error is minimized. Osborne (1963) discusses theoretical estimates of the probability of error in identification. While there is an obvious trade-off between minimizing the number of attributes used and maximizing the reliability of the identification, it is also true that reducing the number of unnecessary questions reduces the probability of error.

The purpose, then, of an identification key is to present data such as that represented in Table 2.1 in a way that makes the identification of a particular class, taxon, as easy and reliable as possible based on observations of the values, states, of particular attributes, characters, of that class. Some of the data in the table may not be used in the identification. For example, if the color of an animal is determined to be green before its habitat is known, there is no need to determine the habitat because the animal can be identified as a lizard. In selecting which attributes to present initially and which to eliminate or defer, the taxonomist must often use knowledge not represented in the table. There are several properties of attributes that may influence their selection. One property is the *consistency* of the attribute for different specimens of the same class. For example, in Table 2.1 the attribute *birth method* is very consistent; all instances of a given class are sure to have the same value for this attribute, i.e., all specimens are sure to give birth in the same way. The attribute *coloring,* on the other hand, is not very consistent for the class *duck,* since many ducks are not brown. A second property is *ease of observation.* For example, in Table 2.1 it might be easier to determine whether or not a specimen flies than the manner by which it gives birth. A third property is the *reliability* of the attribute, i.e., the degree to which it is diagnostic for a given class. For example, in Table 2.1 the attribute *food* is very reliable for the class *horse,* since only the horse eats *grass.* The attribute *flies,* on the other hand, has the same value, *yes,* for three different objects, so this attribute is not reliable for choosing among these three. The attribute *coloring* is very reliable (each instance except brown of a value corresponds to only one class) but is a poor choice for most identifications since it is not very consistent (many objects such as duck and lizard actually have several values). Considerable care must obviously be taken in selecting the attributes to be used in constructing the key, and this construction must take into account many factors of a heuristic nature not necessarily represented in a table of data. This is why keys are so difficult to construct.

The next sections describe several of the most common methods of constructing identification keys. The single most common form of key is the dichotomous key, described in the next section. Other methods including polyclaves, tabular keys, pattern matching, and on-line keys are described in the following section. Each of these methods has its strengths and weaknesses. Expert systems offer an alternative to the identification key, and the strengths of the expert system format often make it the method of choice for identification and classification. Expert systems were not developed as a taxonomic identification tool, and indeed the application of expert systems to this problem is still in its infancy. The rule-based expert system was developed, however, for a very similar problem, that of diagnosis. Both problems involve the selection from a number of alternatives of a correct hypothesis or hypotheses. Taxonomic identification is therefore a useful application for introducing many of the properties of rule-based expert systems.

2.1.1 Dichotomous keys

The dichotomous key is the oldest, simplest, and best known method of specimen identification. Pictures of dichotomous keys dating to the seventeenth century are given by Morse (1975). Figure 2.1*a* shows a dichotomous key constructed from the data of Table 2.1. Other examples of dichotomous keys, both humorous and real, are given by

1a	Can fly	2
1b	Cannot fly	
2a	Lives by water	*Duck*
2b	Lives on land	3
3a	Eats seeds	*Sparrow*
3b	Eats meat	*Eagle*
4a	Eats grass	*Horse*
4b	Eats meat	5
5a	Live birth	*Tiger*
5b	Lays eggs	*Lizard*

(a)

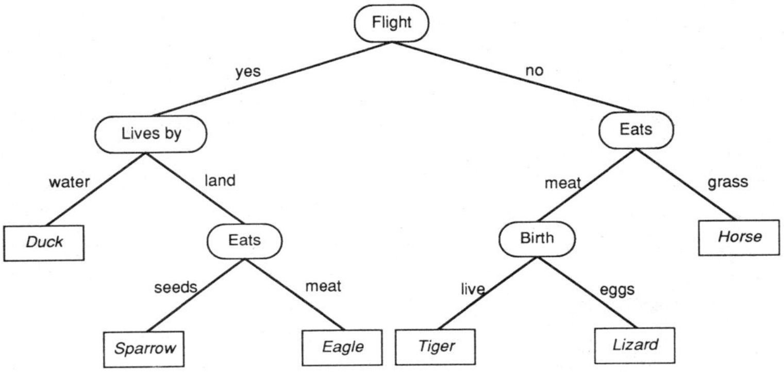

(b)

Figure 2.1 (*a*) A simple example of a dichotomous key constructed from the information in Table 2.1. (*b*) Decision tree constructed from (*a*). Decision nodes are rounded and terminal nodes are rectangular.

Metcalf (1954). A dichotomous key is a sequence of *couplets,* each couplet consisting of a pair of mutually exclusive choices for the values associated with some attribute or set of attributes. By progressing along the couplets, one is led ultimately to the appropriate class. Thus, a dichotomous key is simply a decision tree in which the user has at each node, i.e., at each decision opportunity, two possible branches, i.e., two options. There is no particular theoretical reason why the number of choices at each node must be limited to two, and indeed polychotomous keys, in which some or all nodes have more than two choices, are not uncommon. There is a feeling among some experts, however, that dichotomous keys are likely to be more reliable because they restrict the user's options (Pankhurst, 1970).

One of the principal advantages of the dichotomous key is its simplicity. A dichotomous key can be encoded on a sheet of paper or small pamphlet and taken directly into the field; there is no particular need for, or advantage gained by, encoding the key on a computer. If the specimens are difficult or impossible to bring back from the field this may be a real advantage, although the advent of the lap-top computer has reduced this advantage somewhat. No particular training in the concept of the dichotomous key is required to use one, and in principle it can be used reliably by persons with no prior experience. Moreover, even if the key contains a large number of objects it is possible to reach the correct class fairly quickly. The speed of identification can be increased if, as is usually the case in key construction, an effort is made to have the most common or most easily identifiable objects "key out" first.

The dichotomous key has, however, some considerable disadvantages. It is a monothetic procedure, that is, a procedure in which a single attribute or predefined combination of attributes is sufficient for identification. For example, in the key of Fig. 2.1*a* the combination of attributes sufficient for identification of an eagle is *flies:* yes; *lives by:* land; and *eats:* meat. Other precise combinations could also be chosen to lead to the conclusion that the specimen is an eagle, but it is difficult to incorporate imprecise information into a dichotomous key. A method in which identification is made based on overall similarity between a collection of attributes is called *polythetic.* The monothetic nature of the dichotomous key, together with the strict ordering of its options, means that to use the key successfully the user must be able to choose the correct branch at each node (cf., Fig. 2.1*b*). If some piece of data is missing, then the key comes to a dead end with no information at all, unless the user laboriously follows each possible remaining path to its conclusion. For example, in Fig. 2.1*a* if one gets to couplet 4 and does not know what the animal eats, then it is impossible to proceed. Another serious drawback of the dichotomous key is that a mistake made at any stage can have catastrophic results. For example, if on the first stage of the example key the user incorrectly chooses the "no" path from the node *Flight* at the first step and the specimen is an eagle, then it will be impossible to reach the correct conclusion. Typically the user proceeds a few more steps until it becomes clear by the inappropriateness of the choices that an error has been made. At this point, however, it is often impossible to determine which couplet led to the error. The property of gradually losing accuracy and precision as the quality of information declines is called *graceful degradation*; dichotomous keys do not degrade gracefully.

2.1.2 Other methods

Despite the attractive simplicity of dichotomous keys, the problems associated with their use have motivated the creation of a variety of other types of identification methods. Prominent among these are tabular keys, polyclaves, pattern-matching algorithms, and on-line keys.

These are listed in order of increasing reliance on the computer, with the last two being completely computer based.

Tabular keys. A tabular key (Newell, 1970) is basically a table of values of a class-attribute matrix such as that given in Table 2.1. To save space in displaying the table, each value is represented by a mnemonic code consisting of only one or a few characters. Since keys with more than about ten objects or attributes cannot be represented by a single table, Newell advocates the use of *key groups* for keys with large numbers of objects or attributes. A key group is a subset of the whole set of objects in which all members of the key group are related in some way. For example, in Table 2.1 the animals could be divided into two key groups: birds and nonbirds. When key groups are used, rows of the tabular key matrix represent whole groups. Each group then has its own tabular key matrix.

Polyclave methods. Computerized polyclaves are identification programs that employ a process of elimination. The user of a computerized polyclave sequentially enters attribute-value pairs. With each pair, the program eliminates all objects not matching that pair and provides the user with a list of remaining objects. A computerized polyclave can be made polythetic by using some comparison formula that allows for imperfect matching. These pattern-matching formulas are discussed below.

The polyclave has several advantages over the dichotomous key. The polyclave is reasonably easy for the novice to use and can be made polythetic, allowing for errors in observation. It displays graceful degradation in the sense that it automatically provides a list of remaining possibilities. The user can never be stymied by a missing piece of data. In the hands of an expert user who knows which are the crucial attribute-value pairs, a polyclave can be very efficient. The primary disadvantage of the polyclave is that the constructor has no control over the order in which attribute-value pairs are entered. This order is selected by the user, and, particularly with an inexperienced user, the selected order may be a very poor one. In this case the polyclave can be frustrating since few alternatives are eliminated and there is no way to determine the most efficient attribute-value pairs.

Attribute-set (pattern-matching) methods. Pattern-matching methods are the most polythetic of those discussed here. The theory of these methods is often called *numerical taxonomy*. Numerical taxonomy is most commonly used in the identification of objects whose prominent attributes take on numerical values. Examples of such attributes are length, wingspan, number of spots, and so forth. Numerical taxonomy is particularly common in bacteriology and microbiology, and much of the important literature originated there (e.g., Gyllenberg, 1963; Sokal and Sneath, 1963). Pankhurst (1978) and Abbott et al. (1985) provide excellent reviews of the field. Pattern-matching methods require numerical representations of attribute values and are best suited for objects whose attribute values can easily be placed on such a numerical scale.

After constructing the numerical scales, observations of a number of sample specimens are entered as data. A pattern-matching algorithm (e.g., Duda and Hart, 1973) is then used to locate clusters of entries corresponding to separate objects. These pattern-matching algorithms use some form of distance measure to isolate the clusters. Pattern-matching methods can also be used in conjunction with polyclave methods by basing the match on the number of attribute-value pairs that match those of the specimen. They share many of the advantages and disadvantages of the polyclave method, including the advantage of graceful degradation and the disadvantage of not having a logical order of entry of information.

On-line key methods. Like pattern-matching methods, but unlike other key-based methods, on-line keys can only be implemented on a computer. On-line keys are described by Wilson and Partridge (1986) and Dallwitz and Paine (1986). These keys are sufficiently similar in many of their features that a generic description broadly applicable to all of them is possible. The principal feature that distinguishes an on-line key is that it sequentially presents to the user questions or sets of questions, and that the order of presentation of these questions is selected dynamically based on the user's response to earlier questions and on an attempt to optimize the search for the correct answer. Optimality is determined by a mathematical formula taking into account one or more criteria such as the following : expected number of questions to be asked, evenness of split of the remaining taxa, commonness of the remaining taxa, and likelihood of a correct answer. Wilson and Partridge (1986) and Dallwitz and Paine (1986) provide mathematical formulas on which to base the optimization.

2.1.3 Comparison of identification methods

The discussion of the various key methods in this section is restricted to keys as implemented on a microcomputer and used by someone who is familiar with the basic use of the machine but not necessarily with the key. This restriction, while it may be regarded as unfair by advocates of simpler methods such as the dichotomous key, at least establishes a uniform frame of reference. The discussion in this section gives a list, adapted from Woolley and Stone (1987), of some of the most important desirable criteria of an identification system and a consumer's rating of the extent to which each of the methods discussed here meets each criterion.

Ease of use with minimal learning curve. If the method is not easy and intuitive to use, it probably will not be used very much at all. The dichotomous key, in which each couplet is simply presented to the user and the user responds with the appropriate choice, is perhaps the easiest method for a completely naive user. With all of the methods the ease of use is dependent on the construction of the user interface. Each of them, with the possible exception of the tabular key, can be made fairly easy to use with a well designed user interface.

Ease of construction and modification. Once it has been constructed, an identification key will be subject to frequent revision as new taxonomic data become available. Experienced taxonomists consistently describe the difficulty in constructing a dichotomous key by hand. The last decade has witnessed the development of numerous automated key construction programs such as DELTA (Dallwitz and Paine, 1986), which have almost completely eliminated the laborious aspects of constructing a computerized identification key. The use of such tools, however, restricts the form of the key to that produced by the tool. For instance, dichotomous keys developed automatically by DELTA have branches constructed using DELTA's optimization formula. Any modifications to the key to include taxon specific properties must still be done by hand.

Graceful degradation in face of missing data or erroneous input. The dichotomous key is a perfect example of an identification device that does *not* degrade gracefully. If the user does not know the response to a couplet, then the identification process stops. If the user makes a mistake, then something even worse happens. The identification process proceeds down the wrong path until a couplet is reached where the obvious incorrectness of both options indicates that an error has been made. At this point it is usually impossible to trace back along the path to determine where the error was made. Aside from the dichotomous key, all of the identification devices offer some degree of graceful degradation, with the polyclave being possibly the most graceful in the face of erroneous information and the on-line key the most graceful in the face of missing data.

Efficiency of use. The identification key should require as little time and effort on the part of the user as possible. Woolley and Stone (1987) identify two kinds of efficiency, *structural efficiency* and *dynamic effi-*

ciency. Structural efficiency is the property that only information pertinent to the identification is required of the user. Dynamic efficiency is the capability of determining, while the identification is in process, which characters will provide the most efficient progress toward identification. Most methods possess structural efficiency to some extent, but none, with the possible exception of some on-line keys, possess dynamic efficiency.

Provides explanation of reasoning. After the identification is made, the key should be able to provide the user with the logic that went into the decision. In this way the user can determine the credibility of the answer. The dichotomous key, of all the identification methods, provides the best trace of its logic because all the logic is included in the key's structure.

Able to deal with uncertainty explicitly. In western North America, if you see a bird with a red breast it is either a robin or a varied thrush. Of course the likelihood that it is a robin is, in general, much greater than the likelihood that it is a varied thrush. In most situations if an identification cannot be made with certainty, it can at least be narrowed to a few alternatives, and some of these alternatives will be more likely than others. The identification key should be capable of displaying alternatives and commenting on their relative likelihood. None of the keys has the latter property.

2.2 The Rule-Based Diagnostic Expert System

In Chap. 1 an expert system was described as a computer program meeting the following three criteria: able to solve real-world problems, able to interact with a nonexpert user, and easily incorporates new knowledge about the domain. These three criteria were interpreted in terms of the following three functional properties: the ability to process nonnumeric information, the automated ability to communicate with the user at the user's level and to explain its reasoning, and the separation of the inference mechanism from the domain-specific information. A wide variety of computer programs meeting these criteria have been constructed. By far the most widely used such programs, however, are the *rule-based* expert systems. The concept of the general purpose, rule-based expert system originated with the program EMYCIN (van Melle et al., 1984). A generic rule-based expert system is divided into three basic components: the knowledge base, the inference engine, and the dynamic memory (Fig. 2.2). In this chapter we shall only be concerned with simple rule-based expert systems, so we shall use the terms *knowledge base* and *rule base* synonymously.

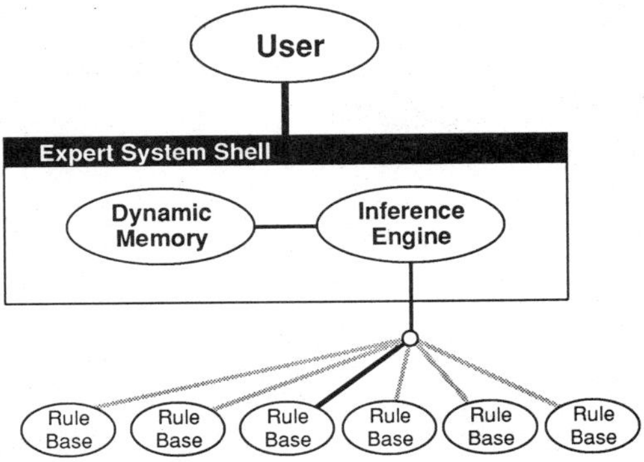

Figure 2.2 Schematic diagram showing the structure of a generic rule-based expert system.

The inference engine and dynamic memory taken together form the expert system *shell*. The rule base is maintained separately from the expert system shell. Consistent with the third criterion for an expert system, the same shell can access, one at a time, many different rule bases. This is represented in Fig. 2.2 by showing one rule base connected to the shell through a dark line while the others are connected through light lines. The expert system is in this sense analogous to a word processor that can access many different documents but allows only one document to be edited at a time.

The knowledge base, or rule base, contains all of the information specific to the problem at hand, e.g., the identification of animals. In a rule-based expert system this knowledge is expressed in the form of rules, which look something like: "If an animal flies, then it is a bird." The knowledge base may contain dozens or hundreds of these rules. The task of the inference engine is to piece together logical chains of reasoning based on these rules to verify a conclusion. For example, it might, under appropriate circumstances, chain the rules "If an animal flies, then it is a bird" and "If an animal is a bird and lives by the water, then it is a duck" to permit the conclusion that an animal that flies and lives by the water is a duck. The task of the dynamic memory is to store information gathered in the inference process for later use. For example, if the user is asked whether the animal flies and responds that it does, then this response is stored in dynamic memory. The following three sections present a more thorough description of the three basic components of the expert system, beginning with the knowledge base.

2.2.1 The knowledge base

Rules. The knowledge base represents the compendium of knowledge preprogrammed into the expert system. It is kept separate from the rest of the expert system, often in a separate file on the disk, so that the same expert system shell can process different knowledge bases. Knowledge bases vary considerably in the details of their construction. The knowledge bases of all rule-based expert systems, however, share the same most fundamental component, the production rule. The term *production rule* comes from the earliest work on systems called *production systems* (Post, 1943), which were the ancestors of today's expert systems. As mentioned in Chap. 1, this work, which predated by almost two decades the incorporation of artificial intelligence as a discipline, showed how computer programs could piece together chains of logical reasoning that are independent of a particular subject matter.

In the simplest case, a production rule has the form

IF < assertion 1 >
AND < assertion 2 >
AND . . .
AND < assertion m >
THEN < assertion $m + 1$ >
AND . . .
AND < assertion $m + n$ >

The m assertions in the IF part of the rule are called the *antecedents* and the n assertions in the THEN part are called the *consequents*. For example, in the dichotomous key of Fig. 2.1 we could combine the first half of couplet 1 with the first half of couplet 2 to write

RULE 1:
IF Animal flies AND
Animal lives by water
THEN Animal is a duck

Note, however, that we could *not* write, in the manner of a dichotomous key, the following:

RULE 1:
IF Animal flies
THEN Go to rule 2

RULE 2:
IF Animal lives by water
THEN Animal is a duck

Rules cannot contain direct references to other rules. Determining which rule to invoke as a logical consequence of another rule is a part of the inference process and as such is the responsibility of the inference engine. Thus, both the antecedents and the consequents of a rule must be assertions that contain no reference to any rule itself.

The form of the rule given in the preceding paragraph, in which the antecedents and consequents are each single statement assertions, or propositions, that can be true or false, implies that the expert system uses a form of logic called *propositional* logic. This is the simplest form of logic but is perfectly adequate for many of the tasks associated with expert systems. We shall discuss the different forms of logic in more detail in Chap. 8. The assertions in a rule may take any form, but for our animal identification problem they will ordinarily make use of a subset of the *class, attribute, value* system. Each assertion will be written in the form *ATTRIBUTE* IS *VALUE*. In this way references to objects will be removed, and the taxa that were formerly associated with objects will become values of an attribute in the consequent, the THEN part, of the rule. For example, the rule regarding ducks can be rephrased in a way that is less grammatical but shows this attribute-value relationship more clearly as follows:

RULE 1:
IF Flies is yes AND
Habitat is water
THEN Animal is duck

Here the attribute of the first assertion in the antecedents is FLIES and the value is YES. The attribute of the first assertion of the consequents is ANIMAL and the value is DUCK. In the previous terminology, DUCK would have been a class, but this level of association does not exist in simple rule-based systems.

It is often said that an expert system is nothing more than a dichotomous key in a different form. While we would disagree with this statement, and we hope you will, too, after reading this chapter, it is true that a dichotomous key can be directly translated into an expert system rule base. For example, we can translate the dichotomous key given in Fig 2.1 into a simple rule base as shown in Fig. 2.3.

While the rule base of Fig. 2.3 is certainly legitimate, it has a number of defects, not the least of which is that it makes no use of the reasoning power available from an expert system. A simple and obvious improvement is to break up the chain of reasoning as shown in Fig. 2.4.

The rule base shown in Fig 2.4 contains the same information as that of Fig. 2.3 in the sense that the same questions will be asked of the user and the user's responses will lead to the same ultimate conclusion. However, the rules of Fig. 2.3 have been broken up into smaller segments, and subsidiary conclusions have been inserted, for example that the animal is a bird, a mammal, and so forth. There are at least two advantages to this. One is that it begins to provide graceful degradation in the face of incomplete information as discussed previously. For example, if the user is only able to answer that the animal flies, then the rule base can tell the user that the animal is a bird. We shall see later in this chapter that the inclusion of uncertainty can

RULE 1:
IF Animal flies AND
Animal lives by water
THEN Animal is a duck

RULE 2:
IF Animal flies AND
Animal lives on land AND
Animal eats seeds
THEN Animal is a sparrow

RULE 3:
IF Animal flies AND
Animal lives on land AND
Animal eats meat
THEN Animal is an eagle

RULE 4:
IF NOT Animal flies AND
Animal eats grass
THEN Animal is a horse

RULE 5:
IF NOT Animal flies AND
Animal eats meat AND
Animal gives live birth
THEN Animal is a tiger

RULE 6:
IF NOT Animal flies AND
Animal eats meat AND
Animal lays eggs
THEN Animal is a lizard

Figure 2.3 A simple rule base consisting of a direct translation of the dichotomous key of Fig. 2.1.

provide still more graceful degradation. A second advantage of breaking up the rules as shown in Fig. 2.4 is that it makes modification of the rule base more straightforward and logical. For example, if the rule base contractor wants to add BEAVER to the identification capabilities, a rule of the following form can be written:

RULE 13:
IF Animal is a mammal AND
Animal lives by water
THEN Animal is a beaver

RULE 1:
IF Animal flies
THEN Animal is a bird

RULE 2:
IF Animal eats meat
THEN Animal is a carnivore

RULE 3:
IF NOT Animal eats meat
THEN Animal is a herbivore

RULE 4:
IF Animal gives live birth
THEN Animal is a mammal

RULE 5:
IF Animal is a bird AND
Animal lives by water
THEN Animal is a duck

RULE 6:
IF Animal is a bird AND
Animal is a carnivore
THEN Animal is an eagle

RULE 7:
IF Animal is a bird AND
Animal is a herbivore
THEN Animal is a sparrow

RULE 8:
IF Animal is a mammal AND
Animal is a herbivore
THEN Animal is a horse

RULE 9:
IF Animal is a mammal AND
Animal is a carnivore
THEN Animal is a tiger

RULE 10:
IF NOT Animal is a bird AND
NOT Animal is a mammal
THEN Animal is a lizard

Figure 2.4 A more complex rule base containing the information of the dichotomous key of Fig. 2.1.

In the rule base of Fig. 2.3, making a modification requires piecing together a long list of antecedents for each new animal. With the rule base of Fig. 2.4, however, an assertion that is further along the chain of reasoning can be used, e.g., the assertion that the animal is a mammal can be used without worrying about having to verify yet again that the animal is a mammal. The larger format is not without its disadvantages. One disadvantage is that it generally takes longer to process. That is, the inference engine ordinarily examines each rule in the rule base several times during the course of the inference process. The more rules there are, the more time it takes to examine them.

Prompts. In addition to rules, most expert systems require some form of prompt in requesting information from the user. For example, a knowledge base for the animal identification problem might include the following prompts:

PROMPT *Animal flies:* Does the animal fly?
PROMPT *Animal eats meat:* Does the animal eat meat?

and so forth. These are used to provide data to the inference engine so that it knows which rules to use in performing the identification. The way prompts are used depends on the type of inference engine. This will be discussed below.

2.2.2 Dynamic memory

Dynamic memory contains all of the information that the expert system has gathered in pursuit of the solution. In artificial intelligence jargon this collection of information is often called the *state of the world.* Information can be obtained in the following two ways: through the inference process itself and through direct interactions with the user. In its simplest form, dynamic memory usually contains information about two things: rules and assertions.

The information about rules that must be stored are the rule itself and whether it has already been used. Rules are tested when the truth value, i.e., whether it is true or false, of at least one of the antecedents is unknown. If all of the antecedents are known to be true, then the rule is invoked, the usual term for this is that the rule is *instantiated* or *fired,* and the validity of the consequents is established. If one or more of the antecedents of a rule is known to be false, then the rule cannot be instantiated. Once the truth values of the antecedents are determined and the rule is either instantiated or not, it would be a waste of time to test the rule again. Such a rule is therefore marked as "used."

The attributes of the assertions that need to be stored are the assertion itself and whether the assertion has been proven true or false. For example, in the rule base of Fig. 2.4 if the assertion ANIMAL LIVES BY WATER has been established to be false, then this assertion, together with the fact that its truth value is *false,* is stored in memory. If the inference engine comes to Rule 6, for example, it can retrieve this information from the dynamic memory as part of the test of the rule's antecedents. The *justification* of an assertion (most simply, the rule that leads to the assertion being verified) should be stored with each assertion. This is not strictly necessary to obtain a conclusion but it is necessary to provide an explanation. For example, if the conclusion is that the animal is a lizard, then Rules 1, 5, and 12 will be included in the justification chain of this conclusion. The manner in which this is implemented is to store the identifier associated with the rule that lead to an assertion's being verified. Explanations are discussed in more detail below.

2.2.3 The inference engine

The inference engine contains the control logic for manipulating the knowledge in the rule base to reach a conclusion or conclusions. The source of the rather poetic term *inference engine,* by the way, is the traditional use in artificial intelligence of the term *engine* to describe an isolated part of a computer program that performs some well-defined function. The part of the code that performs inference is called the inference engine. The process of inference as used in the inference engine is similar to solving a homework problem in mathematics. If John has 10 apples and he gives Mary 2 apples and gets 3 from Jane, how many does John have left? To solve this problem the rules of mathematics are applied until a solution is obtained. First, 2 is subtracted from 10 to get 8, and then 3 is added to 8 to get 11. While the process seems trivial in this case, it does result in the creation of new knowledge not existing before the exercise was completed. Complicated mathematical theorems that are far from obvious are proved in the same way. Moreover, the fact that the problem deals with John's apples is irrelevant to the inference process, which is carried out in abstract mathematical terms:

$$10 - 2 = 8$$
$$8 + 3 = 11$$

Forward chaining. In a rule-based expert system the process of inference is often called *chaining* since rules are chained together in the inference process. The two fundamental types of chaining are forward, or data-driven, chaining, and backward, or goal-driven, chaining. Forward chaining is simpler and we will begin with it. A forward-chaining inference engine starts with all of the facts and repeatedly works through the rule base adding new conclusions until no more can be added. Conclusions are reached by instantiating rules if and when all of their antecedents, the IF parts, have been verified.

In processing the animal rule base example of Fig. 2.4 the forward-chaining inference engine user would first ask if the animal flies, if it lives by the water, if it gives live birth, and if it eats meat. Suppose, for example that the answers are that the animal flies, does not live by the water, does not give live birth, and eats meat. The inference engine would first test Rule 1. Since the antecedent of this rule has been verified, the rule can be instantiated and its consequents added to the dynamic memory. In the case of Rule 1 there is only one consequent, that the animal is a bird. The inference engine next tests Rule 2. Since the antecedent of this rule is verified, the rule is instantiated and its consequent, that the animal is a carnivore, is added to memory. Rule 3 is next, and this rule cannot be instantiated since its antecedent, that the animal does *not* eat meat, cannot be verified. The inference engine proceeds in this way through the entire rule base, failing to instantiate Rules 4 and 5, instantiating Rule 6, and failing to instantiate Rules 7, 8, and 9.

Having come to the last rule, the inference engine starts over to see if the newly verified assertions permit any other rules to be instantiated. It retests each rule that has not been instantiated already. In the present example no new rules can be instantiated on the second pass through the rule base so the inference process is complete. The expert system can then announce all of its conclusions, i.e., ANIMAL IS A BIRD, ANIMAL IS A CARNIVORE, and ANIMAL IS AN EAGLE.

The order in which the rules are tested does not affect the final conclusions. For example, suppose that in Fig. 2.4 the locations of Rules 1 and 6 were interchanged. On the first pass through the rule base, Rule 6 would be encountered before the assertions ANIMAL IS A BIRD and ANIMAL IS A CARNIVORE had been verified by instantiation of Rules 1 and 2. It would therefore not be instantiated. However, Rules 1 and 2 would eventually be encountered and instantiated on this first pass through the rule base. On the second pass, the assertions ANIMAL IS A BIRD and ANIMAL IS A CARNIVORE would now be in memory and Rule 6 would be instantiated. This would force a third pass through the rule base during which nothing new would be instantiated.

Winston and Horn (1981) describe a simple but powerful inference engine for forward chaining. The inference engine consists of eight functions (or procedures): `Deduce`, `Stepforward`, `Tryrule`, `Usethen`, `Remember`, `Testif`, `Promptuser`, and `Recall`. The relationship among these procedures is shown in Fig. 2.5. In the figure,

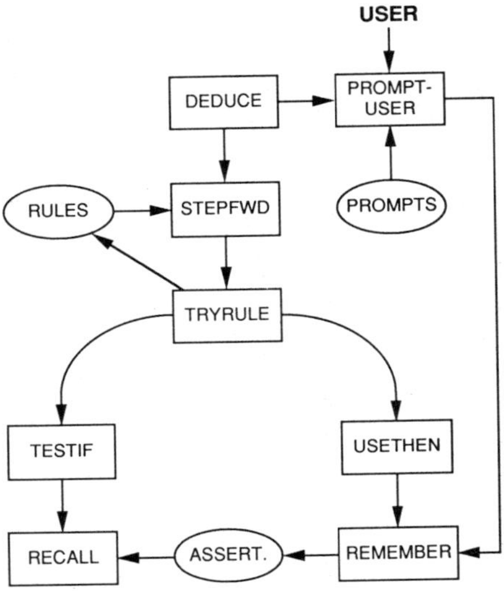

Figure 2.5 Flowchart for the procedures comprising a simple forward chaining inference engine.

the functions themselves are shown as rectangles, and the three components of dynamic memory, rules, assertions, and prompts, are shown as ovals. The function `Deduce` first repeatedly calls `Promptuser`, which presents each prompt to the user and determines and stores the truth value of the corresponding assertion based on the user's response. For example, if the user responds with a "Yes" to the prompt "Does the animal fly?" above, then `Promptuser` assigns a truth value of *true* to the assertion ANIMAL FLIES.

After calling `Promptuser`, `Deduce` repeatedly calls `Stepforward` until `Stepforward` signals that it should no longer be called and the inference process is complete. Each call of the function `Stepforward` causes it to work through the entire set of rules, passing them one at a time to the function `Tryrule` for examination. If the rule is marked "used," the function `Tryrule` ignores it. Otherwise, `Tryrule` passes the antecedents of the rule to the function `Testif`. `Testif` passes each antecedent one by one to function `Recall`, which examines the assertions stored in dynamic memory to see if any of them match the antecedent. If a match is found, `Recall` passes the truth value, true or false, back to `Testif`. In this way, `Testif` is able to tell `Tryrule` how to categorize the set of antecedents of the rule. The following three categories are possible: (1) All the antecedents are true; (2) at least one antecedent is false; and (3) no antecedent is false, but the truth value of at least one antecedent is unknown.

If the antecedent set of a rule falls into category (3), then there is not yet enough information to make a decision on the rule. `Tryrule` does nothing to it. If the antecedent set falls into category (2), then the rule cannot ever be instantiated because at least one antecedent is false. `Tryrule` marks the rule as used but does nothing else to it. If the antecedent set falls into category (1), then the rule is marked as used, and since all of the assertions are true, the rule is instantiated. In processing the rule base of Fig. 2.4, for example, suppose the user has answered "Yes" to the prompt "Does the animal fly?", "No" to the prompt "Does the animal live by the water?", and "Yes" to the prompt "Does the animal eat meat?", indicating that the animal is an eagle. `Stepforward` first passes Rule 1 to `Tryrule`, which passes the antecedent ANIMAL FLIES to `Testif`. `Testif` calls `Recall`, which checks the assertions in dynamic memory and finds that ANIMAL FLIES has been assigned a truth value of *true*. Since this is the only antecedent, `Testif` passes back to `Tryrule` the information that the antecedents of Rule 1 are all verified.

If a rule is to be instantiated, `Tryrule` passes the consequent set of the rule to the function `Usethen`. `Usethen`, together with `Remember`, forms a sort of mirror image to the `Testif` and `Recall` pair. `Usethen` takes each consequent of the rule and passes it to `Remember`, which stores it, together with its truth value as specified in the rule, in the *assertions* portion of memory. The identity of the rule itself is matched with the consequent in the *justifications* portion of memory. After the memory storage is complete, `Tryrule` tells `Stepforward` into which category the rule fell. For example, in instantiating Rule 1 `Tryrule` passes the consequent set, which consists of the single assertion ANIMAL IS A BIRD, to `Usethen`, which assigns a truth value of *true* to it and calls `Remember`. `Remember` places the assertion and its truth value in memory.

After it has competed its pass through each of the rules in the rule base, `Stepforward` sends a message back to `Deduce` telling it whether or not `Stepforward` was able to instantiate any new rules. If it was successful, then there are new assertions in memory that may in turn be useful in instantiating unused rules. `Deduce` therefore calls `Stepforward` again and the process is repeated. This continues until `Stepforward` signals `Deduce` that `Stepforward` has been unable to instantiate any new rules. At this point the process terminates and all conclusions, i.e., all consequents that have been determined to be true, are announced. In processing the rule base of Fig. 2.4, `Stepforward` on the first pass would instantiate Rules 1, 2, and 6, and on the second pass it would not instantiate any new rules, indicating that the processing is complete.

Backward chaining. Backward, or goal-directed, chaining starts with a set of hypotheses, or goals, and attempts to piece together chains of rules to reach one of the members of the set. The simple expert system described in this chapter must work with a predefined set of hypotheses. A more complex system might include the ability to match different values to attributes. For example, the values DUCK, HORSE, and so forth could be matched to the attribute ANIMAL. Such an expert system could then start with the goal attribute ANIMAL and work to find the appropriate value. This sort of system is discussed in Chaps. 5 and 8.

The discussion in this chapter will describe the situation in which the set of hypotheses is included in the knowledge base. We therefore include in our simple knowledge base the set of hypotheses consisting of the assertions describing the six types of animals, namely

Hypothesis 1: Animal is a duck
Hypothesis 2: Animal is a horse
Hypothesis 3: Animal is a tiger
Hypothesis 4: Animal is a lizard
Hypothesis 5: Animal is a eagle
Hypothesis 6: Animal is a sparrow

The hypotheses may be in any order. In this simple system it would make sense to put the most likely hypotheses first so that the inference engine can check them first and possibly eliminate some unnecessary questions.

The inference engine begins with the first hypothesis, that the animal is a duck. It looks for any rules having ANIMAL IS A DUCK as a consequent. Finding Rule 5 in Fig. 2.4, it begins its check of the antecedents of this rule, which are ANIMAL IS A BIRD and ANIMAL LIVES BY WATER. Rule 1 has ANIMAL IS A BIRD as a consequent. So if this rule could be instantiated, it would verify the assertion ANIMAL IS A BIRD, which would in turn permit the instantiation of Rule 5. The inference engine attempts to instantiate Rule 1 by testing the antecedent of this rule, the assertion ANIMAL FLIES. There is no rule having this assertion in its consequents, so the inference engine checks the prompts and finds a question it can ask the user. If the user answers "Yes" to the prompt "Does the animal fly?" then the assertion ANIMAL FLIES is assigned a value of *true,* and Rule 1 is instantiated. This verifies the assertion ANIMAL IS A BIRD, which is also placed in memory as true. If the user were to answer "No" to the prompt, then the assertion ANIMAL FLIES would be put in memory as *false,* Rule 1 would not be instantiated, and nothing would be done with the assertion ANIMAL IS A BIRD. In either case, Rule 1 is marked in memory as used so that the inference engine does not waste time trying to use it again.

If Rule 1 is instantiated and the assertion ANIMAL IS A BIRD is verified, then the inference engine moves to the assertion ANIMAL LIVES BY WATER and repeats the process, testing Rule 2. If both

Rules 1 and 2 are instantiated, then the inference engine has found a chain of valid assertions and instantiated rules leading to the hypothesis ANIMAL IS A DUCK and announces this conclusion. There is no reason to proceed in the case of animal identification since the hypotheses are mutually exclusive, so the program terminates. If the hypotheses were not mutually exclusive, then the inference engine could continue with another hypothesis until all the hypotheses had been checked.

Winston and Horn (1981) describe a simple backward-chaining inference engine that uses many of the same functions as the forward-chaining inference engine presented earlier (Fig. 2.6). Although this simple inference engine is used as an example in an introductory programming text, it is sufficiently powerful to have formed the foundation for at least two high-level, state-of-the-art expert systems, namely, REX (Gale, 1986) and CALEX (Plant, 1989). The functions `Tryrule`, `Testif`, `Recall`, `Usethen`, `Remember`, and `Promptuser`, and the dynamic memory of assertions, rules, and prompts are lifted intact from the forward-chaining system. The function `Infer` replaces `Deduce`, the function `Stepforward` is deleted, and new functions called `Verify` and `Inrule` are added.

In addition to the changes in functions, the process of *recursion* is added in the form of a recursive loop from the function `Testif` to the function `Verify`. That is, the function `Verify` is allowed to "call itself" by calling the function `Tryrule`, which calls `Testif`. `Testif` in turn calls `Verify`. Backward chaining is virtually impossible to im-

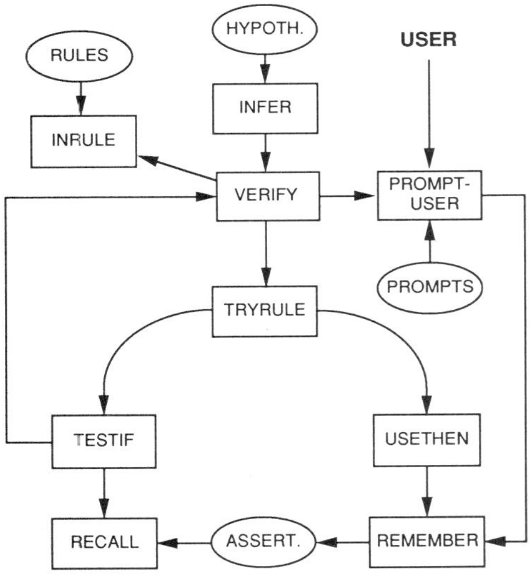

Figure 2.6 Flowchart for the procedures comprising a simple backward chaining inference engine.

plement without some form of recursion, so it is essential to understand how this process works. Dale and Orshalick (1983), in their introductory Pascal programming text, give a nice example to explain recursion, and we shall borrow from this in our own discussion. The example involves the computation of x^n, where x and n are integers, n being positive. We wish to write a computer program that will do this computation. The program will be based on a function called `power`, which has the two arguments x and n. For example, if we assigned the values 4 and 3 to x and n, respectively, then `power(4,3)` would return the value 4^3, which is 64. Figure 2.7 shows a way this function could be programmed using recursion. The function is written in correct Pascal syntax but should be understandable to those who do not know Pascal.

Let us see how this function works when it is called with the arguments $x = 4$ and $n = 3$. Since n is not equal to 1, the function is assigned the value `x * power(x,n - 1)`. This means that the function `power` is called with the arguments 4 and 2 (Fig. 2.8). The information contained in the call to `power(4,3)` is saved in memory in a location called the *stack*. The reason for this term is that the information from each successive, recursive call to the function will be placed adjacent to that from the last call, i.e., the process is like putting plates on a stack. The function `power(4,3)` calls the function `power(4,2)`, which in turn calls the function `power(4,1)`. The function `power(4,1)`, returns the value 4 to `power(4,2)`. Now the `power(4,2)` memory "plate," which contains the values $x = 4$ and $n = 2$, is retrieved from the stack. The value of x, which is 4, is multiplied by the just received value of `power(4,1)`, which is 4, and thus the value 16 is returned by `power(4,2)`. The function `power(4,3)` takes this value and the contents of its plate from the stack to return the answer `power(4,3)` = 64; the process is complete. The reason for the stack storage arrangement is that the successive calls to the function can be unstacked in the reverse order that they were originally placed on the stack.

Let us now examine the workings of the backward-chaining inference engine of Fig. 2.6. The inference process begins in the function

```
Function power (x,n: integer) : integer;
Begin
   if n = 1 then
     power : =  x;
   else
     power : =  x * power(x,n - 1)
End;
```

Figure 2.7 A simple Pascal function for recursively computing the value of x^n.

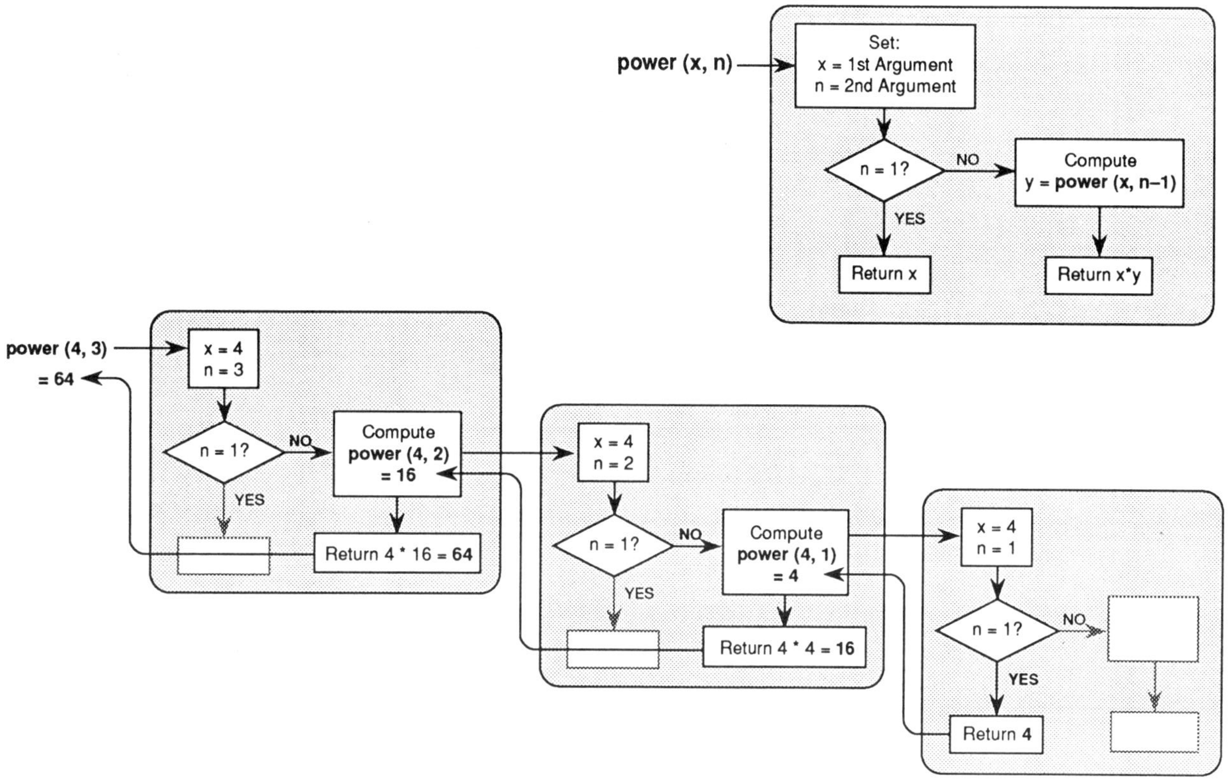

Figure 2.8 Schematic diagram of the recursive computation of the value of 4^3 using the function power of Fig. 2.7.

Infer, which has access to all of the possible hypotheses. Each hypothesis is selected, one at a time, for testing. In the case of our animal identification example, the first hypothesis is ANIMAL IS A DUCK. Infer passes this assertion to the function Verify, which attempts to verify it. First Verify calls Inrule, which checks the rules to see if any of them have the assertion in their list of consequents. In the present case, Inrule finds Rule 5 and returns this to Verify. Verify in turn passes this rule to Tryrule, whose job is the same as it was in the forward-chaining inference engine. As before, Tryrule calls Testif, which extracts the first antecedent of Rule 5, ANIMAL IS A BIRD, and calls Recall. Recall searches the list of already verified assertions and is unsuccessful. Recall notifies Testif that it was unsuccessful. At this point the recursive step is taken.

Testif calls Verify and asks it to verify the assertion ANIMAL IS A BIRD. The contents of the previous call to Verify, which consist of the assertion ANIMAL IS A DUCK together with Rule 5, are placed on the stack, and the function goes to work on the new assertion. It calls Inrule, which discovers that Rule 1 has ANIMAL IS A BIRD in the consequent. Verify passes Rule 1 to Tryrule, which seeks to verify the antecedent, ANIMAL FLIES. The function Recall is unsuccessful at verifying this assertion, so Testif, in another level of recursion, passes the assertion to Verify. Verify passes the assertion to Inrule, which is unable to find any rules with the assertion in the consequent. Verify therefore calls Promptuser, which finds a prompt associated with this assertion and presents it to the user. If the user answers "Yes," the assertion is verified, otherwise it is not.

Suppose the user's answer is "Yes." In this case Promptuser calls Remember to associate the value *true* with the assertion ANIMAL FLIES and save it. Promptuser returns a positive answer to Verify, which passes it to Testif. Testif in turn passes this to Tryrule. Since all of the antecedents of Rule 1 are verified, the rule is instantiated. Tryrule calls Usethen, which passes each assertion in the consequent set, in this case the single assertion ANIMAL IS A BIRD, to Remember for placement on the list of verified assertions.

After the function Remember does its task, Tryrule returns a positive answer to Verify, which at this level of recursion is working on Rule 5. Verify sets about testing the next assertion in the antecedent set of that rule, ANIMAL LIVES BY WATER. If this assertion is also verified through an affirmative user response, then Rule 5 is instantiated and the hypothesis ANIMAL IS A DUCK is verified. In this case the inference process ceases. If the assertion ANIMAL IS A DUCK cannot be verified, the function Infer moves to the next hypothesis and the process is repeated.

A practical comparison of forward and backward chaining. If an inference problem contains a set of well-defined, mutually exclusive set of hypotheses, as is the case in the simple animal identification problem presented in this chapter, then an inference engine should arrive at the same conclusion using either forward or backward chaining. In this situation, which may be regarded as a pure diagnostic problem, backward chaining is generally better. In the first place, it is more efficient. While forward chaining may spend a great deal of time on inferences that will not prove useful to gaining the final result, backward chaining minimizes this wasted inference since it works from goals. In the second place, a user generally finds it easier to interact with a backward-chaining inference engine. Most human experts probably reason at least in part by formulating hypotheses and then testing them; human experts almost certainly do not reason in a manner similar to forward chaining. The similarity of backward chaining to human reasoning is reflected in the order of prompts presented to the user. In a forward-chaining system these prompts have no particular order. In a backward-chaining system, however, each prompt is presented to the user as soon as the relevant assertion is tested, so the questioning appears to follow a discernible chain of reasoning.

Backward chaining is not, however, superior to forward chaining in all situations. Forward chaining is most useful when most of the data is known in advance but the hypotheses are not. It is also useful when the object of the inference process is to determine all, although possibly not mutually exclusive, true hypotheses, as opposed to the identification problem which focuses on a single hypothesis. For example, part of the task of a weed management expert system might be to identify all herbicides that control a specific weed or combination of weeds. Suppose that the hypotheses are the herbicides themselves. Then there will generally be an unknown number of them suitable for a given set of weeds, so it makes no sense to try to identify a single herbicide. Forward chaining is most suitable for any application which, like the weed control example, involves learning all that one can about a problem rather than making a single identification.

2.2.4 Explanation systems

One of the most widely touted features of rule-based expert systems is their capability of automatically providing explanation facilities. Rule-based expert systems make use of chains of rules to move logically from the user's input to the conclusions drawn from that input. Thus, by saving these rules and their connections in dynamic memory, the expert system can provide the user with information on the logic underlying decisions. The two most commonly provided pieces of information are answers to the *why* and *how* queries. The why query is invoked when the user wants to know why a particular question has been asked. The computer can respond by displaying the rule that is currently being tested. For example, if the user asks, "Why?" when prompted by "Does the animal fly?" the expert system can respond by indicating that it is trying to determine that the animal is a bird through the rule "If the animal flies, then it is a bird."

The how query is invoked when a conclusion is displayed. It means that the user wants to know how the system arrived at its conclusion. By tracing back through rules that led to a conclusion, the computer can reconstruct the chain of reasoning. For example, if the expert system concludes that the animal is a duck and the user invokes the how query, the expert system can respond that the expert system determined that the animal is a duck because it is a bird and lives by the water. It determined that the animal is a bird because it flies. Displaying this response makes the reasoning transparent to the user, and the user can decide if this reasoning is acceptable.

Although the display of a rule or chain of rules in response to user queries can provide a useful explanation device, in actual practice the usefulness of this device is sometimes limited. In most expert systems the rules are written in a syntax that is often unintelligible to someone who does not know this syntax. This problem can be ameliorated by the automatic display of English translations of the rule rather than the rule itself, and most expert system shells are capable of this.

Even if the user can make out the contents of the individual rules, it is sometimes difficult to determine by simple examination of these rules the reasoning and decision processes that underlie them. One problem is that in virtually all rule-based expert systems, rules are the only logical device. Because of this, rules are often called upon to play two roles, one being to increase the level of knowledge and the other being to control the flow of inference. The rules contained in the simple rule base of Fig. 2.4 are all examples of rules designed to increase the level of knowledge.

Rules intended to control the flow of logic are often used to arrange the order in which questions are asked, or to prevent seemingly foolish questions from being asked of the user. For example, in the animal identification system, the knowledge engineer might consider it useless to ask whether or not the animal lays eggs unless the user has seen females. The following rule might be included:

IF Females have been seen
THEN Can ask about egg laying

The assertion CAN ASK ABOUT EGG LAYING would then be one of the antecedents of any rules dealing with whether or not the animal lays eggs. The rule given above does not increase the level of knowledge about the identity of the animal, rather it controls the interaction of the expert system with the user. Rules are often used to perform other tasks involved more with control than with knowledge. To the expert system, however, rules are rules, and all relevant rules are displayed to the user upon a request for explanation. Explanations involving rules intended to control inference rather than to increase knowledge are often confusing or meaningless to the user.

A second problem is that rules simply represent a decision and often do not provide any insight into the thought processes that underlie that decision. For example, in the animal identification rule base of Fig. 2.4 there is a rule that says that if the animal flies, then it is a bird. This is, of course, not really correct in general, only in the context of the limited selection of animals under consideration. The user, in asking "Why?" might really be asking, "Why do you say that if the animal flies, then it is a bird?" This decision was, however, made by the person who wrote the knowledge base and is not evident in the rule itself.

Despite these problems, explanations can be a useful aspect of an expert system, and a well designed explanation system can play a crucial role in the acceptance of that system. Explanation systems are an active area of research, and this research has addressed these and other problems. Papers on the subject include Weiner (1980); Clancey (1983); Swartout (1983); Hasling et al. (1984); Scott et al. (1984); and Dhar and Pople (1987). One of the most promising means of increasing the explanatory power of an expert system is *model-based* reasoning. This is discussed in Chap. 11.

2.3 Discussion

Expert systems for taxonomic identification, similar in general concept to the one described here, have begun to appear in the literature (Forget et al., 1986; Atkinson and Gammerman, 1987; Woolley and Stone, 1987). Persuasive arguments can, however, be made that a format such as the on-line key is superior to that of the rule-based expert system for solving the taxonomic identification problem. As a matter of fact, medical diagnostic programs have been written using the on-line key format (Pankhurst, 1980). We must consider three questions when comparing a rule-based expert system of the kind described in this chapter to other types of computer programs that might perform a similar diagnostic function. First, how does the rule-based expert system differ from the other types of programs? Second, what are the advantages and limitations of the expert system format for general diagnostic applications? Third, how do rule-based expert systems compare with other forms of computerized diagnostic keys for taxonomic identification?

The primary function of a simple rule-based expert system of the kind described in this chapter is to select from a set of alternatives the hypothesis or hypotheses consistent with information provided by the user. A key feature of the expert system is the separation of the inference engine from the knowledge base. In writing the knowledge base one need not be concerned with the details of operation of the inference engine, only with how it processes the contents of the knowledge base. In this sense the rule base is a computer program within a computer program, similar to some of the familiar commercially available database and spreadsheet packages. These packages provide the user with a restricted, relatively simple language in which to write specialized spreadsheet or database applications. For example, a number of papers have been written (e.g., Carpenter, 1984) on the construction of simulation models using spreadsheets. Before the advent of the spreadsheet (and other specialized modeling software) such models would ordinarily have been written in Fortran or some other computer language. The spreadsheet language provides a simplified means to accomplish the same purpose. Similarly, commercially available database management programs provide a language for manipulating a database. In both the spreadsheet and the database programs the user need not know anything about how the host program is written, only how this program processes the "program within a program" written by the user.

The computer language provided by the rule-based expert system is the syntax in which the rule base is encoded, i.e., the precise form of the rules, prompts, hypotheses, and so forth are specified by the particular expert system shell. As with any language, this syntax must be followed with frustratingly complete precision or program "bugs" will result and the program will not run as expected or desired. Expert system rule bases, as well as spreadsheet and database programs, may be compared to classical computer programs written in Fortran or LISP as follows. The statements of a classical programming language as encoded by the programmer are called the *source code.* Programming languages are divided into two kinds, compiled and interpreted, depending on how they act on the source code. A compiled language, such as Fortran and C, comes with a compiler that compiles and links the source code and thereby converts it into an *executable file,* i.e., a separate file that can be executed independently. Most familiar applications such as word processors, communications packages, and spreadsheets are written in a compiled language and sold as executable files. An interpreted language, such as most forms of Basic and

LISP, comes with an interpreter that executes the source code one line at a time. The program is maintained as source code and cannot be distributed without the interpreter. Some computer games, shareware, and public domain products are distributed in this form, taking advantage of the inclusion of Basic in the MS/DOS operating system.

In this context the rule base, and a spreadsheet or database program as well, can be considered as the source code and the inference engine as the interpreter. Like any interpreter the inference engine processes the code of the rule base. An expert system rule base, however, is different in a very fundamental way from a computer program written in a language such as Fortran or Basic and from spreadsheet and database programs. A simple Basic program might appear as

```
read x
if x > 0 goto 5
y = 4.2 * x
goto 6
y = 5.6 * x + 3.0
write y
```

The Basic program consists of six statements in a particular order. The Basic interpreter works through the source code line by line, guided by the `goto` and other control statements. In a similar way, the inference engine works through the rule base one item at a time to arrive at its conclusion. The fundamental difference between the Basic interpreter and the inference engine is that, unlike the rule in a rule base, the statements in the Basic code above are not only logically dependent, they are also *ordinally* dependent. That is, the statements in the Basic code must be presented in a certain order to make sense. For example, if statements 3 and 4 are interchanged, the program will not function correctly. Moreover, Basic statements such as `y = 4.2 * x` in line 3 have no meaning by themselves. They only take on meaning in the context of the rest of the program. In the rule base of an expert system each rule has some meaning when taken by itself, and the order in which the rules are placed in the rule base can be altered with no effect on the functioning of the rule base itself. In this context, a traditional computer program is called *procedural,* or imperative, and a program such as an expert system rule base is called *declarative*. In a procedural programming language the interpreter follows the code line by line, or as directed by `goto` and other control statements, while in a declarative programming language the statements can be in any order and a part of the job of the interpreter is to determine which statement to execute next. Some programming languages, most notably Prolog, are declarative.

Both expert systems and traditional computer programs provide solutions to complex problems. For example, a statistics program and an expert statistician would provide similar solutions to statistical problems. In the same way, a financial adviser or budget program, possibly encoded using a commercially available spreadsheet package, provides expert level solutions to practical financial problems. One of the features that distinguishes an expert system from one of these traditional programs is the ability of the expert system to work with *symbolic* information, i.e., information that does not involve numerical values. Often this information can be expressed in terms of rules of thumb. Quantitative rules of thumb, e.g., if the g statistic is less than 0.2, do not do inverse regression, may be incorporated into a classical program, but descriptive rules, e.g., if a car will not turn over, it's usually either a dead battery or dirty terminals, are more difficult to incorporate.

However, an expert system is not the only form of computer program that can work with symbolic information. Database systems do this as well. How then does an expert system differ from a database system? One way to answer this question is by analogy. A problem in algebra is solved by beginning with some data, e.g., $x + y = 2$ and $x - y = 0$. Rules of arithmetic are applied to the data to arrive at a solution. The values of x and y in the solution represent an increase in our store of knowledge since they are information that was not previously known. We did not do research to arrive at this new knowledge; instead, we manipulated the information already available to us. In a similar way, database systems provide convenient means for storing and accessing information, such as $x + y = 2$ and $x - y = 0$, but they do not provide any means to increase the store of knowledge by manipulating this information. This is what expert systems do.

However, this distinction does become blurred at the edges. For example, most people would consider an on-line identification key as a database management tool, and yet it does increase the user's knowledge in the sense that it provides matches to taxonomic information given by the user. Perhaps it is better to set aside definitional questions and consider instead a functional comparison of expert systems with other taxonomic identification methods. Three commonly cited advantages of the expert system are the availability of an automated explanation facility, the capability of dealing precisely with uncertain information, and the unstructured format of the rule base. The automated explanation facility was described previously in this chapter. As discussed, this facility basically repeats rules or chains of rules invoked to determine a conclusion. The "explanations" provided are sometimes of questionable explanatory value to the user. Nevertheless, the display of a rule or chain of rules provides a useful check for the knowledgeable user. This is important in establishing credibility of the system in applications such as crop management. The display of rules is also an excellent debugging tool for the knowledge base developer. However, the terse logical statements making up the rules often do not provide the inexperienced user with tutorial information. Sometimes this information can be provided in the form of "boilerplate" verbiage displayed along with the rule.

Expert systems provide a means of precisely representing uncertainty that is unmatched by any other currently available identification system. The mechanics of this representation are discussed in the next chapter. Its potential utility is not hard to appreciate. For example, if a specimen can be identified as definitely belonging to one of two species, say A and B, it may be that the odds of its being an A are 10 times that of its being a B, or it may be that the odds are equal. Expert systems provide a ready means of computing and conveying this sort of information. The possibilities and limitations of the representation of uncertainty are discussed extensively in Chap. 3. At this point we will only say that in many practical applications, at least in agriculture, its utility has been questionable.

The potential of the unstructured knowledge base format, manifested in rule-based expert systems by the declarative format of the rule base, offers a clear potential advantage over traditional programming methods. The format, in principle, permits the knowledge base developer to encode knowledge locally without regard to the global structure of the knowledge base. For example, in the animal identification problem discussed in this chapter, rules describing one means of classifying an animal, e.g., for determining whether or not it is a mammal, exist independently of rules classifying it in other ways, e.g., whether or not it is a carnivore. These more fundamental identification rules are also independent of rules that make more precise identifications through the use of fundamental knowledge already acquired. For example, the inclusion of new rules for identifying an animal as a carnivore, e.g., the possession of forward pointing eyes, does not affect rules that have this classification in the antecedent, e.g., distinguishing between relatives of the dog and relatives of the cat. This declarative format permits, in principle, the development of a knowledge base to proceed in a more unstructured, "organic" manner, with new knowledge entered at any level as it becomes available.

As a practical matter the utility of the declarative structure of a knowledge base is currently limited by two factors. One is the lack of programmers, or "knowledge engineers," experienced in the declarative style of programming. It is difficult for someone who has spent years writing computer code in Fortran or C to shed old habits and adopt the new style of encoding knowledge. A more serious limitation, at least at the present time, is the need to provide users with a dependable, credible product that is free from logical gaps or flaws. An expert system with such gaps or flaws in the knowledge base is unlikely to win acceptance by an already skeptical public, particularly in a conservative field such as agriculture. The situation is analogous to that faced by a new county extension agent trying to win acceptance. The novice agent will probably spend extra effort to make sure that any given advice is correct, knowing that incorrect information will destroy the agent's credibility. A more experienced agent who has won acceptance by the farming community can afford to be a bit more relaxed. Expert systems currently face, even more than novice county agents, the problem of winning acceptance. This means that most knowledge bases, just like conventional programs, must be carefully structured and debugged. As with explanation facilities, there is active research in the development of automated knowledge base debugging and development code (e.g., Davis and Lenat, 1982; Nguyen et al., 1987), which in the future will alleviate this problem.

In summary, the automated explanation facility is of mixed usefulness, the explicit representation of uncertainty is of unproven utility, and the utility of the unstructured format is primarily unrealized as yet. What then are the practical advantages offered right now by the expert system format? What are its advantages and limitations compared both with conventional programming techniques and with other less conventional methods such as on-line keys? Based on our experience, we would assert that the principle, realized advantages of rule-based expert systems are the explanation facility described above, the separation of the inference mechanism from the knowledge base, and the modular, simple, and intuitive nature of the syntax of the rules.

The IF-THEN structure of the production rule has proven to be a simple and yet powerful construct for encoding knowledge. Because the components of the knowledge base have this simple, uniform structure, the logic of the knowledge base is generally easy to understand. Also, the simplicity of structure permits a generality of application since the structure of rules is appropriate for a wide variety of problem domains. The IF-THEN sequence and the rule of thumb concept are familiar and natural to nearly everyone. Therefore, in most cases the domain experts can readily comprehend what is desired in the rule base; frequently it is not hard for them to explicitly cast their decision-making procedures into the production rule format. The modularity of the rules makes it easier to work with these rules as individual entities, without getting caught up in the totality of the knowledge base. While the rules are logically connected within the rule base, each rule has a meaning of its own, and rules can be moved in and out more easily than sections of code in a conventional computer program.

The rule-based format for knowledge bases does, however, have some serious drawbacks. One of the most serious is the problem of organizing and maintaining a large rule base. A rule base has no structure or organization imposed on its rules except by their location in the rule base. The logical relationships among the rules is often not evident at all from inspection of the rule base. This can lead to serious bugs within the rule base and to the inadvertent introduction of bugs when the rule base is modified. The development of computer assisted rule base maintenance packages (e.g., Nguyen et al., 1987) may be expected to alleviate this problem somewhat. Another problem with the rule base format is that not all forms of knowledge are conveniently represented in the production rule format. Indeed, the remainder of this book, and particularly the last third of this book, is devoted to dealing with this problem. At this point it is sufficient to say that frequently complex knowledge is difficult to encode efficiently in production rule form.

How does the expert system format compare with other computer based taxonomic identification procedures? It should probably be stated right at the start that if the identification is being carried out by a trained expert who is in possession of a relatively complete set of data on the specimen, then some other method such as the on-line key method is probably superior. The on-line key format has much to recommend it. An on-line key, for example, can display to the user at each stage in the consultation process a list of all those species that remain as possibilities. This provides the graceful degradation property described earlier. Moreover, the presence of a database management program such as DELTA (Dallwitz and Paine, 1986) makes development, maintenance, and modification of the key's database, which is analogous to the expert system's rule base, relatively easy.

The chief advantage of the expert system format is that it provides greater flexibility than the on-line key format in how the information contained in the knowledge base is organized and presented to the user. For example, in their SYSTEX expert system for insect identification, Woolley and Stone (1987) organize their knowledge base so that information is gathered in a precise manner. Information about the host plant, geographical range, and anatomical features are collected in a prespecified order and alternative possible taxa are eliminated until only one remains. If possible, confirmatory information is collected to be sure of the correct identification. There are at least three advantages to this structure. First, and this is a relatively minor advantage, is that the order of questioning is probably less jarring to the user; questions are asked in a logical manner that the user can follow. Second, the manner of questioning is very efficient, possibly more efficient than that of the on-line key. The on-line key determines which question to ask next based on an attempted minimization of the number of remaining questions, but this minimization often involves a very coarse estimation of this number. Third, the expert system can easily provide important information to the user, besides simply the number of remaining possibilities, in the presence of incomplete information. This information can include relative likelihoods of the alternatives, undetermined characters most likely to result in a complete identification, and so forth. It is more difficult to accomplish this with an on-line key.

In summary, while expert systems are not a panacea, they do provide a number of useful features, particularly for diagnosis and decision support. The domain independence of the rule-based format provides a great deal of flexibility in the types of problems an expert system can address. The relative simplicity of their rule base syntax makes programming easier. The openness of the knowledge base facilitates automated explanations and easier understanding of the logic behind their conclusions. The separation of the knowledge base from the expert system shell permits the programmer to focus on the knowledge base without worrying about the user interface, the inference mechanism, and so forth. Many of the shortcomings of the rule-based format will be addressed in subsequent chapters of this book; some are the subject of active research.

Chapter 3

Dealing with Uncertainty

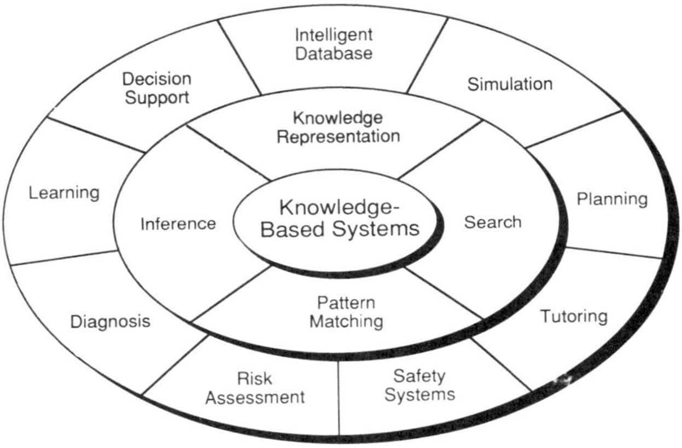

This chapter discusses reasoning under uncertainty in rule-based expert systems. There are a number of techniques for accomplishing this; we describe four of them in some detail. Two examples are given. The first is the animal identification problem carried over from Chap. 2. The second is the problem of deciding whether to apply an insecticide to an agricultural crop.

3.1 Introduction

In the simple animal identification example discussed in Chap. 2, we have little doubt when distinguishing a sparrow from a duck. In a real identification problem, however, there may be considerable uncertainty about the final identification, particularly if the specimen is damaged or has parts missing. It may be possible only to say, "The specimen is most likely this, but it could possibly be that or that." Rule-based expert systems have a built-in capability of dealing with uncertainty. If a rule is written abstractly in the form IF A THEN B, then the verification of A can imply a verification of B only to some degree of certainty. In addition, the antecedent A may be verified with less than complete certainty, and the degree of certainty in the conclusion B may be dependent of the level of certainty with which A is believed. This chapter describes some of the more prominent methods of dealing with uncertainty in rule-based expert systems.

There is no one universally accepted method of dealing with uncertainty; rather, there are about four or five competing methods, each having its own advocates. Indeed, the question of superiority of methods has been one of the most controversial and hotly debated issues in expert systems research. Perhaps because of the quantitative and theoretical nature of the question, it lends itself to academic research and has inspired innumerable articles in scientific journals and conference proceedings. The various methods have taken on almost religious qualities, with each method having its high priests and established dogma. The most prominent methods, and the ones on which we shall focus our attention, are bayesian probability theory, the Dempster-Shafer theory of belief, fuzzy set theory, and the theory of endorsements. We shall also discuss a simple, ad hoc method.

If the methods for handling uncertainty are viewed as religions, then bayesian theory is the "Established Church." This is somewhat ironic since among statisticians bayesian methods are controversial. In any case, the discussion in this chapter will not attempt to pass judgment on the various methods or to rank them. In fact, the choice of method probably depends on the nature of the problem. The discussion does, however, try to point out the strengths and weaknesses of the methods. Some of the methods involve very complex mathematics. The discussion does not go into any mathematical detail at all; interested readers can pursue the theories in the references. The motivational problem of this chapter is the same as that of Chap. 2, the taxonomic identification of a specimen. Most of the discussion of the chapter, however, is carried out in a more abstract vein. The taxonomic identification problem appears only at the end, in a practical comparison of the methods. We also give an example of the application of expert systems to a problem in crop management.

3.2 Uncertainty in Rule-Based Expert Systems

3.2.1 Types of uncertainty

Virtually any decision-making situation involves some degree of uncertainty. A farmer who must determine whether or not to apply a pesticide is uncertain about the intensity of the pest infestation, the effect of the pesticide, and the savings in yield for a given degree of control. The diagnostic or identification problems most suitable for simple expert systems frequently involve considerable uncertainty.

Since the form of the expert system and user interaction and of the structure of the knowledge base are quite constrained in the simple expert systems described in Chap. 2, it is possible to divide the forms of uncertainty into a relatively few categories. We will use the following three: user uncertainty, rule uncertainty, and vagueness.

User uncertainty occurs when the user is uncertain of the correct response to the prompt. For example, in the case of animal identification the user may be asked the color of a bird's beak. The user may not have gotten a good look at the specimen and may be fairly certain that the beak was a light color, possibly yellow, but on the other hand it might have been brown or red.

Rule uncertainty involves the introduction of uncertainty into a rule in the rule base. Rule uncertainty may be further subdivided into antecedent uncertainty and consequent uncertainty. *Antecedent uncertainty* occurs when one or more of the antecedents of a rule are not known with certainty. For example, suppose that a rule is invoked that has as an antecedent the color of the bird's beak as described in the previous paragraph. Since the user was unsure of this color, the antecedents of this rule are not known with certainty. Therefore, even if the rule involves the precise identification with complete certainty of the species of bird with a yellow beak, the fact that the antecedent, i.e., the beak color, is not known with certainty should prevent the identification from being made with certainty. *Consequent uncertainty* occurs when even precise knowledge of all the antecedents of a rule is insufficient evidence to conclude the consequent with complete certainty. For example, consider the rule: If the animal flies and is a vertebrate, then it is a bird. This rule cannot be stated with complete certainty since other vertebrates such as bats also fly.

Vagueness is a concept that is related to uncertainty but is not exactly the same. As an example of vagueness in an antecedent, consider the rule: If the animal has forward pointing eyes, then it is a carnivore. There are some animals such as eagles that have clearly forward pointing eyes, and some, such as rabbits, whose eyes clearly do not point to the front. Some animals, however, horses, for instance, have eyes that do not point clearly in either direction. An even more characteristic example is a rule such as: If the animal has sharp, pointed teeth, then it is a carnivore. Anyone who has been bitten by a rabbit will attest to the sharpness of its teeth, and what constitutes pointedness in teeth? Rules such as these two clearly beg the question of defining terms. Vagueness can also appear in the consequents. For instance, some insect and plant species are not clearly delineated, so that a specimen could in some instances be identified as belonging to more than one species.

3.2.2 A simple example rule base

In discussing and comparing the various methods, it is most useful to work with a simple rule base. The one we shall use is shown here. The precise validity of the rules is not important; their use is only illustrative. The rule base shown in Fig. 3.1 distinguishes between a duck and an eagle.

The words in parentheses, such as DEFINITELY and POSSIBLY, represent the degree to which the consequent is to be believed if the antecedent is known with certainty. Incorporating uncertainty into the inference process involves constructing a procedure or algorithm for determining certainty when rules are logically chained together. Such an algorithm is often called an *uncertainty calculus*. In many, but not all, uncertainty calculi the degree of uncertainty, which is represented verbally in the rule base above, is translated into a numerical form. Often the permitted numerical values are between 0 (indicating complete uncertainty) and 1 (indicating complete certainty). Sometimes negative numbers between 0 and −1 are also allowed, with −1 indicating complete certainty that the assertion is *not* true.

Our discussion of methods for dealing with uncertainty begins with a list of some of the factors that must be taken into account in a rule-based expert system. A simple, ad hoc uncertainty calculus is pre-

RULE 1:
IF The bird has forward pointing eyes AND
It has a sharp, curved beak
THEN The bird is a carnivore (definitely)

RULE 2:
IF The bird is a carnivore
THEN It is an eagle (probably)

RULE 3:
IF The bird lives by the water
THEN It is a duck (possibly)

RULE 4:
IF The bird quacks
THEN It is a duck (very likely)

RULE 5:
IF NOT The bird is a carnivore
THEN The bird is a duck (possibly)

Figure 3.1 A simple rule base including uncertainty.

sented initially. This simple algorithm illustrates how the degree of belief is propagated when rules are chained together. The simple calculus is based on a numerical representation of certainty, with the numbers taking on values between −1 and 1. The numerical representation is a translation of a verbal description in which only a limited number of values are distinguished. A certainty of 1 indicates that the assertion is *definitely* believed; values between 0.75 and 1 indicate that the assertion is *very likely*; values between 0.5 and 0.75 indicate that the assertion is *probably true,* and values between 0.25 and 0.5 indicate that the assertion is *possible.* Negative certainties indicate belief in the *negation* of the assertion. For example, a certainty of −1 indicates that the assertion is definitely not true, values between −0.75 and −1 indicate that the assertion is very unlikely, and so forth. A numerical value between 0.25 and −0.25 indicates *complete uncertainty.*

To characterize the uncertainty calculus, it is necessary to characterize how it deals with four specific types of uncertainty. These are described below.

Uncertainty in conjunction of antecedents. In Rule 1 above, if the assertion THE BIRD HAS FORWARD POINTING EYES is considered very likely, and the assertion THE BIRD HAS A SHARP, CURVED BEAK is considered possible, what is the certainty associated with the conjunction of these antecedents? The numerical representations of these certainties are 0.75 and 0.25, respectively. A simple but frequently used ad hoc formula is to take this value to be the minimum of the two certainty values, the justification being that "a chain of reasoning is only as strong as its weakest link." Thus, the certainty of the conjunction in Rule 1 would be 0.25. A different situation arises in the case of contradicting belief. For example, suppose the certainty of the assertion THE BIRD HAS FORWARD POINTING EYES is very unlikely, and the certainty of the assertion THE BIRD HAS A SHARP, CURVED BEAK is possible. In this case the numerical representations are −0.75 and 0.25, respectively. Various formulas are possible here, e.g., addition, but the simplest is to simply throw up one's hands and assign the conjunction a certainty of 0.

Effect of IF-certainty on THEN-certainty (propagation through a rule). Continuing with the above example, if the certainty of the conjunction is possible, how does that affect the certainty of the consequent THE BIRD IS A CARNIVORE? Once this certainty is known, how is it propagated to the certainty of the final conclusion THE BIRD IS AN

EAGLE? A simple procedure to use is to multiply the certainty in the consequent by the certainty in the antecedent. Thus, in Rule 1, if the certainty of the consequent is 1 and of the antecedent conjunction is 0.25, then the certainty of the consequent is 0.25, i.e., possible. This certainty is then propagated through Rule 3 to give a final certainty of 0.125 to the conclusion THE BIRD IS AN EAGLE. This indicates complete uncertainty about the conclusion. If, on the other hand, the certainty of the assertion THE BIRD HAS A SHARP, CURVED BEAK were 0.75 instead of 0.25, then the certainty of the conclusion THE BIRD IS AN EAGLE would be 0.375, or possible.

If the certainty of the assertion THE BIRD IS A CARNIVORE is 0.25, then what is the effect of Rule 5, in which the antecedent is the negative of this assertion? The simplest procedure is to not instantiate this rule at all, unless the certainty of the assertion THE BIRD IS A CARNIVORE takes on a negative value.

Uncertainty in combination of evidence. Moving on to the rules for ducks, suppose the certainty of the assertion THE BIRD LIVES BY THE WATER is definite (1.0) and that of THE BIRD QUACKS is probable (0.5). What, then, is the certainty of the conclusion THE BIRD IS A DUCK? The certainty of the conclusion due to Rule 3 alone is 1.0×0.25, or 0.25. The certainty due to Rule 5 alone is 0.5×0.75, or 0.375. One formula, called the *probabilistic sum,* for combining the evidence is to let the certainty equal $0.25 + 0.375 - 0.25 \times 0.375$, or 0.5. The reason for the term probabilistic sum is that this represents the probability of the conclusion if the certainties are regarded as independent probabilities. A second formula is to simply take the maximum of the two certainties, which in this case would be 0.375.

These simple ad hoc formulas provide a simple procedure for computing uncertainties in rule-based expert systems, and they produce results that are at least not totally implausible. They are not, however, axiomatic or substantiated by any theory whatsoever. A number of more theoretically sound, but also more computationally difficult, algorithms for computing and propagating uncertainty in rule-based expert systems have been developed.

3.2.3 Bayesian methods

Bayesian methods for dealing with uncertainty are based on the notion of subjective probability (Savage, 1954). This concept originated with a paper written by Reverend Thomas Bayes and published posthumously in 1763; it has since been reprinted (Bayes, 1958). Subjective probability attracted little interest until the late 1950s, but since that time it has been the object of intense study in the fields of statistics, operations research, and optimal control theory. In brief, the idea is as follows. The classical, or *frequentist,* interpretation of probability defines the probability $P(A)$ of an event A in terms of hypothetical experiments. Suppose an experiment whose possible outcomes are A and NOT A is repeated over and over again. For example, if the experiment is the tossing of a coin and A is the event that the coin lands heads, then the repetition would consist of tossing the coin over and over and recording the total number of tosses and the number of heads. As the experiments are repeated over and over again the fraction of times that the outcome A is observed approaches some number that is defined to be $P(A)$. The value of $P(A)$ must obviously lie between 0 and 1, with a higher value if A is more probable, i.e., the fraction of experiments is higher for A than for NOT A.

By contrast, the *subjective* interpretation of probability is that it is the value that a rational person would associate with the probability that A is true or that the outcome of a particular experiment will be A. Subjective probability is often defined in terms of betting odds. For example, the statement $P(A) = 0.6$ can mean that this individual would be willing to take odds of 6 to 4 in a bet that the outcome of a particular experiment will be A, i.e., to make a bet of \$6 that the outcome will be A against a bet of \$4 that the outcome will be NOT A.

Subjective probability works well with experiments or statements for which the concept of an infinite number of repetitions is either difficult or meaningless. For example, in some years lygus bugs are a serious pest in the Central Valley of California while in other years they have little impact. Consider the statement: Lygus bugs will be a serious problem next year. Next year the truth or falsehood of this statement will be known with certainty, but for now it is a matter of conjecture. However, at the present time a probability value can be assigned that represents an individual's personal degree of belief in the truth of the statement. An experienced farmer or entomologist might be able to make a more accurate assignation than someone who has never heard of lygus bugs, but anyone could nevertheless assign some number. Such a number must take on values between 0 and 1.

In the case of lygus bugs in the Central Valley, the concept of repeated experiments is difficult, since there is only one Central Valley and each year's lygus bug situation, unlike a coin toss, takes place in an entirely different set of circumstances. We could, however, interpret each year as an experiment and construct some sort of statistical forecast of the likely pest status of lygus bugs next year. Uncertainty need not be confined to events that have not yet occurred. Suppose, for example, that a farmer notices a distinct but inexplicable loss in yield from an irrigated crop in a particular year. Without further analysis the farmer could propose several explanations, e.g., nematodes, salinization, soil compaction, nutrient deficiency, soilborne pathogens, and so forth. A thorough analysis might indicate which of these is the culprit, but before this analysis the farmer could assign *a priori* probabilities to each. These probabilities could in principle be statistically estimated based, for instance, on historical data or on surveys of nearby fields. They could just as well, however, be pulled directly out of thin air and interpreted as subjective probabilities.

In each of the cases presented in the preceding paragraph, an interpretation of probability as a frequency, e.g., the fraction of years with a large number of lygus bugs or the fraction of fields affected by nematodes, made some sense, although a subjective interpretation was also possible. There are some circumstances, however, in which the interpretation of probability as a frequency is so nonsensical as to be meaningless. While we could probably think up an agricultural example of this, we can make the point more dramatically by using an example from outside the agricultural domain. Consider the following question: What caused the extinction of the dinosaurs?

A popular theory is that dinosaurs became extinct as a result, directly or indirectly, of the collision of an asteroid with the earth. This may or may not be true, but as with the case of lygus bug infestation, anyone could assign this theory a personal subjective probability. However, there is a major philosophical difference between predicting the status of lygus bugs and guessing about an event that occurred only once in the distant past. We can predict the status of lygus bugs using either a statistical or a subjective approach. In the case of the dinosaurs, only the second option is available. It is meaningless to think of repeated experiments in which dinosaurs on identical copies of the earth are allowed to become extinct, and the fraction is recorded of those that are caused by an asteroid. The event either did or did not happen as described, and a statement of probability can only be taken as a numerical representation of one's degree of belief in a theory.

The important connection between subjective probability and the more classical frequentist interpretation is that subjective probability is assumed to obey the same mathematical rules. For example, using the classical probability theory, if events A and B are independent and have respective probabilities of occurring of 0.5 and 0.4, then the probability that both A and B will occur is 0.2. Similarly, using the subjective probability theory, if our degree of belief that statement C is true is 0.5 and that of statement D is 0.4, and if the truth or falsehood of these statements is independent, then we would be compelled to say that our belief in the truth of both statements is represented by the number 0.2. It turns out that the entire classical theory of probability is based on only a few simple axioms. Subjective probabilists assume that subjective probability obeys exactly the same axioms. It turns out that this assumption can be justified mathematically, provided that the individual holding the subjective belief is completely rational (French, 1986).

Subjective probability is also called bayesian probability because Bayes' theorem is at its core. In the subjective interpretation of probability, Bayes' theorem tells individuals how to modify their beliefs in the light of new evidence. Suppose, for example, that we decide to let the number 0.6 represent our personal degree of belief in the statement that an asteroid killed the dinosaurs. Suppose further that we then read in the newspaper that investigators have found evidence of a giant crater in the earth with an estimated age of about 65 million years. Clearly this evidence should modify the number attached to our probability, but how? The answer is Bayes' theorem, which is written

$$P(C|E) = \frac{P(E|C)P(C)}{P(E)}$$

There are four quantities in Bayes' theorem, and, reading from left to right, they have the following meanings (e.g., French, 1986). The quantity $P(C|E)$, called the *posterior* probability, represents the probability that conclusion C is true given the weight of evidence E, where in our example C represents the hypothesis that an asteroid killed the dinosaurs and E represents the apparent 65 million year old crater. We use the symbol C to represent *conclusion* rather than H for *hypothesis* because Bayes' rule is applied at all stages of the inference process, to intermediate assertions as well as to hypotheses. The quantity $P(E|C)$, called the *likelihood*, represents the probability that evidence E would be available given that conclusion C were true. In our example, this is the probability that we would find evidence of a 65 million year old crater given that the dinosaurs had been killed by an asteroid. The quantity $P(C)$, called the *prior* probability, represents our degree of belief prior to learning of the evidence E. Finally, the quantity $P(E)$, called the *marginal* probability, represents the probability that evidence E would be observed, independent of whether or not conclusion C is true. In our case, this is the probability that the evidence for a 65 million year old crater would be found, independent of what killed the dinosaurs.

There are three quantities on the right side of the equals sign and we only have a value, $P(C) = 0.6$, for one of them. Actually, the value of $P(E)$ is computed from the equation

$$P(E) = P(E|C)P(C) + P(E|C')P(C')$$

where C' represents the statement that C is *not* true, $P(C') = 1 - P(C)$, and $P(E|C')$ represents the probability that E would be observed if C were not true. This still leaves two quantities, $P(E|C)$ and $P(E|C')$, that must be assigned values if we are to proceed. To pick a value of the likelihood $P(E|C)$ we reason that if an asteroid of the magnitude described in dinosaur extinction theories has crashed into the earth, then there should be some evidence of it, even if the evidence is at the bottom of the sea. Therefore we set $P(E|C)$ to 0.75. We set $P(E|C')$, the probability of finding evidence of a 65 million year old asteroid impact given that the dinosaurs were *not* killed by an asteroid, at 0.1. Thus, $P(E) = 0.75 \times 0.6 + 0.1 \times 0.4$, that is, $P(E) = 0.49$. Substituting this together with the values for $P(C)$ and $P(E|C)$ into Bayes' theorem gives us $P(C|E) = 0.918$. The evidence of the crater has thus increased our measure of belief in the impact theory, if we are bayesians, from 60 percent to just over 90 percent.

To implement a bayesian scheme for dealing with uncertainty in a rule-based expert system such as the one we have developed, we must associate a probability with each assertion and make each instantiation of a rule an implementation of Bayes' theorem. In actual practice most bayesian rule-based systems work with the *odds* associated with an assertion, rather than directly with probabilities. Suppose assertion A has probability $P(A)$, so that A' has probability $1 - P(A)$. The odds on an assertion A are defined as

$$O(A) = \frac{P(A)}{P(A')}$$

For example, if $P(A) = 2/3$, then $P(A') = 1/3$ and $O(A) = 2$, i.e., the odds that A will occur are 2 to 1. Bayes' theorem can be written in terms of odds, in which case it has the form

$$O(C|E) = \lambda(E|C)O(C)$$

where $\lambda(E|C)$ is called the *likelihood ratio,* and is defined to be

$$\lambda(E|C) = \frac{P(E|C)}{P(E|C')}$$

Thus, to implement a bayesian scheme we need the following data: a prior probability for each hypothesis, or, in some schemes, for each assertion, before we start and a likelihood ratio associated with each IF-THEN statement. If a rule is written abstractly as "If E, then C with likelihood ratio λ," then if that rule is instantiated, the current odds attached to C are multiplied by λ. Because the odds are obtained from probabilities that must sum up to one, they must be renormalized after each computation. A number of procedures have been developed for this. Details of these procedures are given, e.g., by Duda et al.

(1976), Spiegelhalter et al. (1984), and Pearl (1982, 1985). We shall not discuss these methods here; suffice it to say that they can become quite complicated.

Provided that the individual holding the beliefs adheres to certain precisely defined axioms of rationality, it can be shown mathematically that the individual's beliefs *must* obey the laws of probability. This is a strong statement, and it is the reason that bayesians are so fundamentalist in their belief in the accuracy of Bayes' theory as a model for human belief. Other theories are based on some perceived inadequacy of subjective probability as a model for human uncertainty. Each of the three theories to be discussed emphasizes a particular inadequacy. The Dempster-Shafer theory focuses on the inadequacy of a single numerical value as a representation for uncertainty; the fuzzy set theory focuses on the failure of probability to represent *vagueness* as opposed to uncertainty; and the theory of endorsements focuses on the inadequacy of any numerical representation for uncertainty.

3.2.4 Dempster-Shafer theory

The Dempster-Shafer theory of evidence, originally developed by Arthur Dempster (1968) and later extended by his student Glenn Shafer (1976), provides a mathematical model for the human process of the accumulation of evidence in attempting to verify a hypothesis. The theory attaches measures of belief to sets of hypotheses. For example, consider a bird identification problem. Suppose there are only three possibilities—the bird is an eagle, denoted E, a duck, D, or a sparrow, S—and it must be one of these three. Probability theory would maintain probabilities $P(\mathrm{E})$, $P(\mathrm{D})$, and $P(\mathrm{S})$ associated with each of these and $P(\mathrm{E}) + P(\mathrm{D}) + P(\mathrm{S}) = 1$; it would use Bayes' theorem to accumulate evidence in an attempt to increase the probabilities of one of them and decrease the probabilities of the others. Dempster-Shafer theory, on the other hand, permits a currently held measure of belief in each of the 2^3 subsets of the set of all possible conclusions. That is, one may associate a number with each of the combinations (E,D,S), (E,D), (E,S), (S,D), (E), (S), (D), and (∅), the last being the null set, which represents the belief that none is true. This last element may be eliminated by the assumption that the bird must be one of these three kinds. This number represents a measure of belief and is computed as follows.

Suppose that pieces of evidence are obtained in the following order: First, evidence is obtained that tends to support the possibility that the bird is a duck. Second, evidence is obtained that tends to support the possibility that the bird is either an eagle or a duck. Third, the evidence tends to support the hypothesis that the bird is not an eagle. The evidence that the bird is not an eagle is interpreted as supporting the belief that the bird is something else, i.e., that it is one of the other possible birds, without specifying which one. Thus, the third piece of evidence disconfirming the belief that the bird is an eagle is interpreted as confirming belief in the set (S,D), i.e., that it is either a sparrow or a duck. The first piece of evidence is interpreted as supporting belief in the set (D), and the second as supporting belief in the set (E,D).

Intuitively it is clear that the first two pieces of evidence provide support for the belief that the bird is a duck. The Dempster-Shafer theory provides a mathematical representation for this intuition. This representation is in terms of two functions, a *belief function Bel* and a *basic probability assignment m*. The function $m(A)$ represents the fraction of belief assigned exactly to the subset A; e.g., m(S,D) represents the fraction of belief assigned to (S,D). Since all belief must be assigned somewhere, the functions m must sum in value to one. Also, since the hypotheses are assumed to be exhaustive, $m(\varnothing)$ must be zero. The function $Bel(A)$ represents the total amount of belief assigned to A through A itself or any of its subsets. For example, $Bel(\mathrm{S,D}) = m(\mathrm{S,D}) + m(\mathrm{S}) + m(\mathrm{D})$.

The process of combining beliefs is done according to a rule called Dempster's rule, which is a generalization of the probabilistic sum rule for combination of evidence given above. A mathematical statement of Dempster's rule, together with algorithms for performing the computations, is given by Barnett (1981) and by Gordon and Shortliffe (1984). In our example the three pieces of evidence would be combined as follows. Suppose the first piece of evidence supported (D) to a degree of 0.4. Then m(D) would be assigned the value 0.4 and the remainder of the belief, being unassigned, would go to the set (E,S,D) of all possibilities. The value of m(E,S,D) would therefore be set at 0.6. The quantity associated with the second piece of evidence would be partitioned out among the sets (E,D), (E), and (D) according to Dempster's rule. The quantity associated with the third rule would be partitioned out among the sets (S,D), (S), and (D). Since the functions m must always sum to one, this last apportionment would reduce the value of m(E). The final result of all the apportionments would be to weight the evidence in favor of (D), the duck hypothesis. Dempster-Shafer theory actually provides two numbers that measure belief about the hypothesis (D). One is the belief function Bel(D), which measures the degree of positive belief about (D). The other, denoted F(D), is a function derived from the belief function that measures the lack of reason to doubt (D). It turns out that F(D) is always at least as large as

Bel(D), i.e., an individual is more willing to not disbelieve something than to actively believe it.

The interval between the values of *Bel*(*A*) and *F*(*A*) represents the level of confidence of the individual in his or her degree of belief in the assertion *A*. For example, if *Bel*(*A*) = 0.1 and *F*(*A*) = 0.7 then the degree of belief in *A* is 0.1 and the degree of not disbelieving in *A* is 0.7. The halfway point between these is 0.4, so the individual feels that *A* is about a 40 to 60 proposition. The large interval between *Bel*(*A*) and *F*(*A*) indicates that the person does not have much faith in placing the odds at 40 to 60; they could be much higher or lower. If, on the other hand, *Bel*(*A*) = 0.39 and *F*(*A*) = 0.41, then the person has the same degree of belief in the truth of *A* but much more confidence in the degree of belief. Typically as one accumulates evidence the range between *Bel*(*A*) and *F*(*A*) decreases. Subjective probability, on the other hand, does not distinguish between a strongly held belief that *P*(*A*) = 0.4 and a weakly held belief. Actually, this is not strictly true since an entire distribution can be used to represent a subjective probability, in which case the variance of the distribution represents the strength of belief. This method, however, introduces still more complexities into the probability calculations.

The taxonomic identification system EXPERT KEY (Atkinson and Gammerman, 1987) uses Dempster-Shafer theory to assess degree of belief in a particular classification. The objective of the system in using this theory is to gradually eliminate alternative hypotheses in a reasonable and mathematically rigorous way. The system performs this elimination in the course of a classification session and displays confidence intervals for its most strongly believed conclusions.

3.2.5 Fuzzy logic

Fuzzy logic is an outgrowth of the theory of fuzzy sets, developed by Lofti Zadeh (1965). The original purpose of fuzzy set theory was to model *vagueness,* as opposed to uncertainty. The two concepts are slightly different. As an example of a vague statement, consider "John is tall." Another, more mathematical statement of this assertion is "John is a member of the set of tall men." If John's height is 7 feet, then this statement is clearly true, while if John's height is 5 feet, it is clearly false. Suppose, however, that John is 6 feet tall. In this case he has some degree of membership in the set of tall men, but not a strong one. Zadeh proposed that the set of tall men be a *fuzzy* set without clearly defined boundaries. Each man would have a measure μ between 0 and 1 that denotes his degree of membership in the set. Seven-foot-tall men would have $\mu = 1$, 5-foot-tall men would have

$\mu = 0$, and 6-foot-tall men would have a μ somewhere in between, say, $\mu = 0.5$. Zadeh developed a set theory, i.e., a means of taking unions, intersections, etc., for fuzzy sets.

Since mathematics in general and logic in particular can be based on set theory, one can develop a fuzzy mathematics and a fuzzy logic associated with fuzzy set theory. This logic technically deals with vagueness. For example, one could imagine a rule saying something like: If a bird has a sharp beak, then it is a fast flier. Here *sharp* and *fast* are vague concepts. It is a small conceptual step, however, from vagueness to uncertainty. Indeed, consider the rule: If a bird is a carnivore, then it is probably an eagle. Zadeh (1983) suggests that rules such as this can be interpreted in the context of fuzzy set theory by interpreting the word *probably* as a fuzzy quantifier. He suggests that "fuzzy logic provides a natural conceptual framework for knowledge representation and inference from knowledge bases which are incomplete, imprecise, or not totally reliable" (Zadeh, 1983).

This opinion is not shared by all. In a stinging diatribe against all things nonbayesian, Lindley (1985, p. 13) said, "The literature on fuzzy logic is vast, complicated, and somewhat obscure." A review of some of this literature reveals that he does have a point. However, Zadeh (1983) provides a thorough discussion of a means for applying fuzzy logic to the handling of uncertainty in expert systems. Zadeh's method involves the use of "possibility theory," which is the fuzzy logic analog of probability theory (Zadeh, 1978; Dubois and Prade, 1980). Zadeh considers a number of different cases somewhat analogous to our enumeration of operations in the calculus of uncertainty discussed at the beginning of the chapter. For example, one case, analogous to our discussion of the effect of IF-certainty on THEN-certainty, may be written abstractly as the pair of statements "X is in F^*" and "If X is in F, then Y is in G." This is called *generalized modus ponens* (the term modus ponens is defined in Chap. 8). A verbal example is "The bird's beak is fairly sharp" and "If the bird's beak is sharp, then the bird is a carnivore," or more precisely, "If the bird's beak is in the set of sharp beaks, then the bird is in the set of carnivores." The notation F^* is intended to indicate fuzzy membership in the set F, e.g., the set of sharp-beaked birds.

The mathematical computations of uncertainty involve the use of possibility distributions. One must compute the possibility distribution associated with the statement "X is in F^*" and the conditional possibility distribution associated with the possibility that Y is in G given that X is in F. There are rules for combining possibilities analogous to those for combining probabilities; for example, there is a rule involving the combination of membership in several fuzzy sets in which the minimum of the individual fuzzy membership values becomes the combined fuzzy membership value. This is somewhat analogous to the rule that the probability of two independent events both occurring is the product of the probabilities of each event. By using the fuzzy combining rules, which involve minimization and maximization rather than multiplication and division, the representation of the uncertainty associated with generalized modus ponens can be computed. Because the operations involve minimization and maximization, the actual computation is quite complex and must be done iteratively.

3.2.6 The theory of endorsements

The theory of endorsements, developed by Paul Cohen (1985), is a completely nonnumeric representation scheme for uncertainty. The numerical certainty factors found in other methods of dealing with uncertainty are replaced by nonnumeric *endorsements.* Calculations of uncertainty are replaced by a weighing of the endorsements. Endorsements may be positive, indicating support for the conclusion, or negative, indicating opposition. For example, an endorsement might be "maybe-too-general," indicating that the conclusion of the rule is too general, and that, given the available evidence, the user should arrive at a weaker conclusion. Such a situation would be true, for example, of the rule "If an animal flies, then it is a bird," since bats and insects can also fly. Cohen develops a scheme of ledgers for various types of evidence and rules applied to hypotheses. These ledgers replace the rules of combination of evidence found in numerical representations of uncertainty. Different rules, when instantiated with different types of uncertainty, are used to arrive at a ledger of endorsements for a given hypothesis.

If the theory of endorsements is being used, then the measure of uncertainty associated with individual rules will play a role in the order of selection of the rules for testing; this is not true for most numerical representation schemes. The inference engine attempts to weigh the value of the information associated with each conclusion and the degree and type of uncertainty before selecting a rule. At the end of the inference process the endorsements of each hypothesis are compared to determine the one that is most likely to be true. For example, suppose in using the simple rule base for deciding whether a bird is a duck or an eagle, the user indicated that the bird had forward pointing eyes and quacked. The user, however, could not give any response to the other questions. The ledger for deciding that the bird is a duck would then look like this:

Hypothesis: bird is a duck.

Evidence in favor	Evidence against
Bird quacks	Bird has forward pointing eyes

As is often the case in real situations, the evidence is contradictory. The endorsement examiner built into the expert system has a device for weighing the evidence based on a number of factors such as its reliability. In this case the expert system might determine that hearing a bird's call is more reliable than seeing the location of its eyes and reach the tentative conclusion that the bird is a duck.

At least one agricultural expert system, the SIRATAC cotton management system developed in Australia (Colomb et al., 1988), uses the theory of endorsements to manage uncertainty. The stated reason for using this method is that it makes the system more flexible and permits the generation of better explanations. The system is used to generate recommendations for chemical pesticides. Each chemical is judged on the basis of several factors, such as efficacy and cost, and endorsements, in the form of weighted votes, for and against each chemical are used to decide which chemical to recommend.

3.3 Practical Comparison of the Methods

Theorists, zealous in their belief in one of these methods, display an astounding lack of tolerance for the other methods. The bayesians are perhaps the most arrogant; those who favor the bayesian approach tend to dismiss any other method of dealing with uncertainty out of hand (e.g., Lindley, 1985; Pearl, 1985; Cheeseman, 1986). Lindley (1985, p. 1) said, "*the only satisfactory description of uncertainty is probability*. By this is meant that every uncertainty statement must be in the form of a probability; that several uncertainties must be combined using the rules of probability; and that the calculus of probabilities is adequate to handle *all* situations involving uncertainty. In particular, alternative descriptions of uncertainty are unnecessary." (his italics) Zadeh provides similar criticisms of probability theory (Zadeh, 1983) and Dempster-Shafer theory (Zadeh, 1979). There is seldom any admittance by a member of one school that the theory of any other school could ever be useful. Dempster-Shafer advocates are a frequent exception, many of them, including Shafer himself (Shafer and Tversky, 1985), will readily admit that there may be circumstances in which Bayes' theorem is superior. The debate provides plenty of raw material for making a comparison. Much of it is, however, highly mathematical in nature. Many of the arguments are mathematically compelling but seem to lose their force when confronted with reality. We shall begin by considering the merits and faults of each method individually and end by attempting to say something general, though not certain, about the problem of dealing with uncertainty.

We begin with Bayes' theory. The first question that must be addressed is whether probabilities are suitable for the description of subjective uncertainty. In the case of the classical, frequency-based definition of probability a person can do experiments involving the drawing of colored balls from an urn or the tossing of a coin and experimentally show that the axioms of probability correspond well with observations. Is this statement true for subjective probability as well? That is, do people, when asked for example to assign numbers to their degrees of belief in statements and then combine them in ways that can be modeled probabilistically, arrive without the aid of calculation at the appropriate probability? There is plenty of evidence that they do not. To substantiate this claim we present three well-known examples of ways in which human subjects, confronted with the need to accumulate evidence about an uncertain proposition, do so in a way that disobeys the laws of probability. Further discussion may be found in Hink (1987) and Hogarth (1980).

1. The first example is a famous paradox called the paradox of the ravens (Buchanan and Shortliffe, 1984, Chap. 10). Consider the statement "All ravens are black." Suppose you see a green vase. How does this affect your belief in the statement? Most people would say that seeing a green vase does not affect this belief at all. This, however, violates the laws of probability in the following way. Logically, saying "All ravens are black" is the same as saying "All things that are not black are not ravens." This is known as the *contrapositive*. Seeing a thing that is not black, the green vase, and is not a raven, provides evidence that should increase our subjective probability that all nonblack things are nonravens. By the logical equivalence of the two statements, subjective probability in the statement "All ravens are black" is also increased.

2. Suppose each of two bags has 1000 poker chips. One bag contains 700 red chips and 300 blue chips; the other contains 300 red chips and 700 blue chips. One of the bags is picked at random, and 12 poker chips are selected from it with the proviso that each selected chip is returned to the bag before the next selection. Suppose that of the 12 chips selected in this way, 8 are red and 4 are blue. What is the probability that the bag that was picked is the one with 700 red chips? Most people, when confronted with this problem, give a probability of about 0.7 or 0.8. The correct answer, which can be computed from Bayes' theorem, is 0.97. This failure to correctly adjust subjective probabilities in the face of accumulated evidence is called *conservatism* (Kahneman et al., 1982).

3. Two cab companies, blue and green, operate in a city where 85 percent of the cabs are blue and 15 percent are green. A cab involved in a hit-and-run accident at night is identified as green. The ability of the witness to identify the color of a taxicab at night is tested, and the

witness is found to correctly identify the color 80 percent of time. What is the probability that the cab involved in the accident was green? Most people give an answer of about 0.7. The correct answer is 0.41. This is called the *base rate fallacy* (Kahneman et al., 1982) because people ignore the data, the witness's ability to distinguish colors, and base their estimate too strongly on the base rate, the portion of green cabs.

In the face of this demonstrated unbayesian behavior by actual people, why is the subjective theory of probability to be taken as anything more than a mathematical curiosity? The reason is that mathematicians have shown that while people may not adjust their beliefs according to the rules of probability, if they were completely rational they *should* modify their beliefs in this way. That is, mathematicians have defined a set of seemingly reasonable axioms that rational people should obey in making a decision (for references see French, 1986). If they do not, the expected outcome of a decision they make involving uncertain outcomes will be less than optimal. There are a number of arguments of this type in the literature (for references see Lindley, 1971). One of the easiest to understand in principle is the "money pump" argument (French, 1986). To state the argument precisely would involve a discussion of the axioms of probability themselves, but the gist of the argument is as follows. Suppose two people are engaged in the purchase and selling of an object at a negotiated price. One person updates his or her beliefs about the value of the object according to the laws of probability while the other uses a different method. Then the situation can be arranged so that the nonprobabilist is forever buying the object from the probabilist at a high price and reselling it at a low price, providing a "money pump" to the wiser probabilist. Other examples of suboptimal behavior actually observed by real human subjects involve the buying and selling of lottery tickets (e.g., Merkhofer, 1987). Mathematicians have also shown that if people obey the axioms of probability in updating their beliefs then they are *automatically* acting in a bayesian manner. The mathematical certitude of these arguments is perhaps at the root of the bayesians' contempt for nonprobabilistic methods of dealing with uncertainty.

What negative comments about probability can possibly be voiced in the face of these mathematical proofs? Surprisingly, subjective probability is under attack from the right, occupied by statisticians who favor the classical frequentist interpretation of probability, as well as the left, occupied by artificial intelligence theorists who favor representations of uncertainty that do not involve probability theory at all. Objections to bayesianism may generally be divided into two categories. The first category contains theoretical questions of the ability of probability theory to adequately represent the process of decision making. The second contains practical questions of the possibility of successfully implementing a bayesian scheme to solve a real problem. We shall look at these objections in some detail, and then determine whether and to what extent the other competing representation schemes overcome these difficulties.

Many of the objections of the classical statisticians to bayesian methodology are summarized in a scathing article written by Feinstein (1977), a noted biostatistician and an authority on medical diagnosis. Feinstein makes a number of points, primarily addressed to the process of clinical diagnosis but having relevance to other uses of subjective probability. On the issue of representing diagnostic reasoning, he says of the bayesian approach, "Anyone who has ever practiced clinical medicine will recognize that the approach just cited does not resemble even a weird parody of clinical reasoning." Feinstein argues that clinical diagnosis is a decision-making process in which tests are selected from a number of alternatives and used to provide direct evidence for the diagnosis. Moreover, "it ignores the scientific demand for diagnostic explanations of clinical evidence and it provides no pattern of logic for seeking scientific explanations." Finally, Feinstein asserts that "numerical probabilities are too vague and uncertain for modern diagnostic precision." He concludes that "The diagnostic importance of probabilities in diagnostic reasoning is obscured rather than clarified by a Procrustean model that converts direct observation and reasoning into a convoluted morass of mathematical indirections."

On a somewhat more theoretical plane, Shafer and Tversky (1985) analyze two methodologies, Bayes' theory and Dempster-Shafer theory, for their representational adequacy. They consider the reasoning applied to case study decision problems and ask whether these theories offer sufficient power to represent the human reasoning that goes into these decisions. One decision involves predicting the winner of a swimming race based on the current positions of the competitors. Another involves selecting from among a number of competing hypotheses for the number of species of hominids present on earth approximately 2 million years ago. Shafer and Tversky show how both of the methodologies can be used to represent arguments as made by humans and conclude that both provide sufficient representational power. The interested reader should peruse the original paper to form an opinion. One uncharitable interpretation of their work, however, is that it actually demonstrates that the methods require highly skilled theoreticians to twist and turn and properly shape the equations in such a way as to appear to mimic the human thought process.

At this point a bayesian might interject that the inability of the method to mimic the human thought process is no detriment. Humans have consistently been shown, both through mathematical reasoning and through psychological experimentation, to be poor decision makers in matters involving uncertainty. Indeed, Buchanan and Shortliffe (1984, Chap. 10) point out that their initial discussions of uncertainty were centered on this very point. Should an expert system mimic what a human expert actually does, or should it do what the expert *should* do?

The argument of human inadequacy merits closer scrutiny. Many experiments purport to show that human beings make suboptimal decisions involving uncertainty because they do not reason probabilistically. However, in all of these experiments the subjects must make decisions about matters that can be represented as games of chance such as lotteries. For example, the experiments described above demonstrating the phenomena of conservatism and base rate fallacy each involve precisely defined numerical probabilities. It is natural that such numerical experiments would be performed because they provide for precise measurement of results. However, the results do not demonstrate that humans are poor at dealing with risk and uncertainty. Rather, they demonstrate that untrained humans are poor at dealing with risk and uncertainty when that risk and uncertainty is presented in a numerical context. One may fairly ask how frequently, outside the realms of gambling and high finance, such numerical situations are encountered in the real world. Indeed, one may turn the probabilists' argument on its head and assert that evolution favors individuals that are able to successfully deal with the rigors of living in an uncertain environment, and that the adaptive success of the human species is an indication of its skill at this endeavor. The failure of numerical schemes to conform to human reasoning may be interpreted as an indication of the weakness of the schemes to represent the real world, not of the humans to reason in the real world.

This argument is relevant because it unites the first and second objections to bayesian uncertainty described above, that it does not model how humans think and that it cannot be implemented in a real system. How would such an implementation take place? Prior probabilities $P(A)$ and likelihoods $P(E|A)$ must be assigned to all of the assertions. The prior probability is the subjective probability, before any evidence is collected, that the assertion A is true. The likelihood is the probability that the assertion A is true given that evidence E is observed. Many have pointed out that this requires a phenomenal amount of data for any real problem. Bayesians reply that the use of odds $O(A)$ and likelihood ratios $\lambda(E|A)$ reduce the amount of required data, but an examination of the mathematical equations reveals that the reduction is proportionally a small one (e.g., Duda et al., 1976). Another problem is that, since the data involve probabilities, they must obey the rules of probability. For example, probabilities of mutually exclusive assertions, e.g., that the animal is warm blooded and that it is cold blooded, must sum to 1. In a complicated maze of reasoning involving many assertions whose relationships are incompletely understood, ensuring this summation may prove an impossible task.

Even in a relatively simple domain in which adherence to the rules of probability can be enforced, the task of acquiring the proper values for the priors and likelihood ratios may prove impossible. These values come from one of two sources, human opinion and data. In asking a human expert for an opinion about probabilities, we are asking the expert to translate feelings about uncertainty into a number. Moreover, many of these probabilities, or odds, are virtually impossible to assess. In the animal identification problem, what is the prior probability, before any information is assessed, that the specimen is a carnivore? The bayesians themselves argue that probability assessment is a task at which humans are notoriously weak. While considerable research has been done by decision theorists to improve the extraction of probabilities from human experts, anyone who has ever actually participated in such a process will agree that the numbers acquired are anything but precise. Using a precision tool such as probability theory to manipulate subjective probabilities is like using a micrometer to measure the width of a bomb crater.

In summary, there appear to be two primary arguments against the use of bayesian methods in dealing with uncertainty in expert systems. The first is a theoretical one: Bayes' theorem cannot adequately be made to represent the human reasoning process except possibly in the hands of a skilled practitioner. The second is a practical one: even if it could represent the human reasoning process, there is generally no way to acquire from human experts numbers of sufficient precision to justify its use. What of the other methods? Do they overcome either or both of these objections? The answer appears to be no. The Dempster-Shafer theory, while requiring a bit less data and placing fewer restrictions on the form of the data than does Bayes' theory, nevertheless cannot address the theoretical issue of representation of the reasoning process any more adequately than can Bayes' theory. In addition, the Dempster-Shafer theory suffers from the same dependence on numerical data of possibly suspect origin. Fuzzy logic appears to be worse; not only does it suffer from the same defects, it is a theory about vagueness that is being applied to problems in uncertainty. The theory of endorsements overcomes the second, practical objection. However, it does not seem to address the first objection, and can be a difficult method to apply in practice.

In the face of this discouraging situation, how are we to deal with uncertainty? There appear to be three alternatives. The first appears to be the most attractive for many situations. This is to attempt to develop the knowledge base in such a way as to mimic the actual process of human reasoning under the conditions of the actual situation. For example, consider the case of animal identification. How does an actual systematist attempt to identify a specimen? The systematist almost surely will not assign prior probabilities, or numbers of any kind, to any stage of the identification process. In this sense, the bayesian format is as weird a parody of taxonomic reasoning as it is of diagnostic reasoning. The designer of the expert system knowledge base can attempt to arrange the logic of the rules to match the approach taken by the systematist in reasoning about the identity of the specimen. Such an approach was taken, for example, by Woolley and Stone (1987) in their development of a taxonomic knowledge base.

The second alternative, which appears to be simpler than the first and may be appropriate for situations in which the resources available to complete the task of constructing the knowledge base do not adequately meet the rather demanding requirements of the first alternative, is to go ahead with a numerical representation of uncertainty, but to use simple ad hoc formulas such as those given at the start of this chapter. Consider for example the development of a knowledge base for specimen identification. In many applications the most appropriate output from such an expert system is simply a listing of all the possibilities together with some coarse classification of the likelihood of each. For example, the expert system may say that some taxa are very likely, some are reasonably likely, and some are possible. The user is thereby provided with an initial screening of the possible solutions to the identification problem, which may be all that a computer program can reasonably be expected to provide. In such a case, coarse numerical representations of uncertainty together with simple ad hoc formulas for combining evidence may be sufficient.

The third alternative is to actually use a formal system such as Bayes' theory. Such an approach is reasonable if the reasoning process unavoidably involves numerically represented uncertainty, as would be the case, for example, in gambling, if adequate data are available to provide dependable values for the priors and likelihoods and if precisely delimited certainties are useful to the user. In the case of animal identification this situation might occur if the diagnostic attributes have numerical values, e.g., dry weight and wingspan, and if considerable statistical data on these values were available. In other biological domains, these requirements might be met, for example, in dealing with meteorological data.

The ability of rule-based expert systems to deal effectively with uncertainty, which is often cited as a major strong point of expert systems, appears to have been heavily oversold. While methods for dealing with uncertainty exist, their implementation faces serious theoretical and practical difficulties. The quantitative nature of the problem of representing uncertainty in expert systems has made it a popular problem for academic research, but much of this research is of a theoretical nature and is unlikely to have much practical impact. That is the bad news. The good news is that under many circumstances human beings are quite successful at decision making under uncertainty even when they do not possess the ability to represent that uncertainty quantitatively. People who develop expert systems can often deal very successfully with uncertainty by emulating the manner in which the human expert deals with it.

3.4 Uncertainty in Agricultural Management: An Example

The management of an agricultural crop or commodity is a problem characterized by an unusually high level of uncertainty in almost all decisions. The weather, the level of pest infestation, the price of the crop, the availability of nutrients, and many other factors are ordinarily known only with great imprecision. In this section we describe one of the most important decisions confronting the manager of an agricultural crop: whether or not to apply an insecticide to a suspected pest infestation. How should an expert decision support system deal with such a problem?

The theory and practice of decision making in insect pest management is often called integrated pest management (IPM). A key concept of IPM is the *economic threshold* (Stern et al., 1959). As used in practice, this concept is very simple. It says that insecticide should be applied only when the level of infestation is such that a substantial yield loss would be sustained if insecticide were not applied. This is obviously a vague statement. It can be made considerably more precise in theory (Headly, 1972), but because of the many sources of uncertainty associated with pesticide application, this theoretical precision usually cannot have much effect on actual practice (Plant, 1986).

An example of recommendations involving the economic threshold is the published recommendations of the Texas Agricultural Extension Service for bollworms in cotton. These recommendations include (Robinson and Stewart, 1987):

> SCOUTING PRIOR TO INITIAL CHEMICAL APPLICATION. Fields should be scouted at least once a week (in certain instances, twice weekly). The fields should be divided into four quadrants and 25 green squares (1/2 grown or larger) and terminals should be selected at random in each quadrant.... If numbers of beneficials are low and an average of 12 or fewer small bollworms is found per 100 terminals, consider using one of the microbial insecticides... Otherwise, before bloom the economic threshold is reached when 15 to 25 percent of the green squares are worm damaged. After bolls are present, the economic threshold has been reached when 8 to 10 percent of the green squares have been worm damaged...
>
> AFTER INITIATION OF INSECTICIDE APPLICATIONS. Fields should be inspected closely 2 to 3 days after each insecticide application. Examine the terminal buds and a total of 100 green squares and small bolls taken from at least four different areas in the field. The economic threshold has been reached when bollworm eggs and four or five small worms are found per 100 terminals and 5 percent of the squares and small bolls have been injured by small bollworms.

This recommendation is still an oversimplification of the problem. It is sometimes vague about specific thresholds and is clearly meant to leave room for interpretations. An expert system for this application has been developed (Stone and Schaub, 1990); bollworm management is the topic of Chap. 7 and will be discussed in detail there. There is, however, one important point that can be made at this juncture about expert systems for management applications.

An expert decision support system must act in a considerably different way than an expert identification or classification system. It is perfectly reasonable, and possibly even desirable, for an expert identification system to list several possible conclusions, i.e., possible species to which a specimen may belong, if no conclusion can be reached with certainty. Such is often not the case, however, for an expert decision support system. In the bollworm economic threshold example we could translate the statement "If numbers of beneficials are low and an average of 12 or fewer small bollworms is found per 100 terminals, consider using one of the microbial insecticides" into a rule of the form

IF Beneficials is low AND
  Number of small bollworms per 100 terminals ≥ 0.12
THEN Apply microbial insecticide (certainty = 0.7)

Most farmers would not particularly appreciate a statement from an expert system that it was 70 percent sure that a pesticide should be applied. A simple decision guideline, i.e., apply or do not apply, would ordinarily be preferred. One might argue that the more precise display afforded by the enumeration of percentages is actually more informative to the farmer, but it is also likely to be more irritating and more likely to cause the farmer to stop using the system.

If it is accepted that the expert decision support system should only display a single management guideline for each action, does this mean that uncertainty should not enter into the decision-making process? Not necessarily. A simple means of incorporating uncertainty is to simply do the calculations as usual and only display the most likely output, with no accompanying certainty factors. However, this is a rather impoverished approach and does not really use the full power of the expert system.

In making a management decision the farmer is in a similar situation to that of a physician making a diagnosis. Both must not only determine the cause of the observed symptoms but must also determine the most appropriate response. P. Cohen (personal communication; see also Feinstein, 1977) has pointed out that real physicians in this situation respond to uncertainty by a form of planning. That is, if the physician is incapable of making a correct diagnosis, and if further tests are available that would assist in identifying the causal agent, then the physician will call for those tests. The farmer faced with a management decision may not have the luxury of seeking such additional information. Like the physician, however, the farmer's decision must ultimately take the form of an action or actions in response to the observed symptoms. Also in common with the physician, the farmer is aware that any decision is made under uncertainty, and this uncertainty can therefore be factored into the decision process.

If, as may be the case with weather data for example, a reasonably precise quantification of uncertainty is available. and if the quantification of uncertainty can be useful in contributing to the decision-making process, then such quantification should be used. In cases where precise quantification is not available, it may be misleading to attach a numerical value to the degree of uncertainty; such a number may imply a precision that does not in fact exist. It is perhaps better to weigh the factors involved in the decision, taking into account the risks involved, but to avoid attaching a representation of certainty to the decision once it is made.

Chapter 4

Building a Knowledge Base

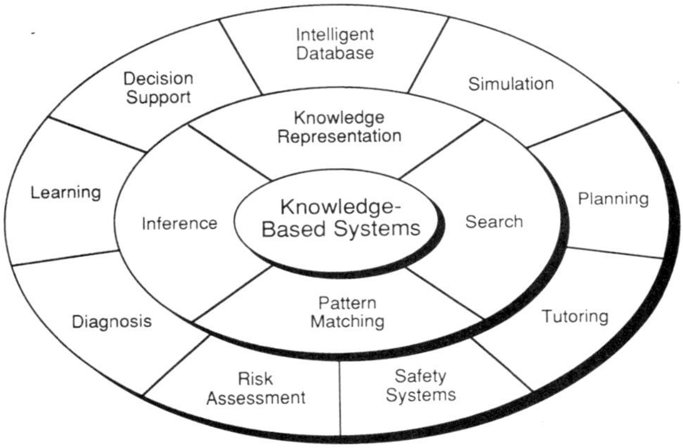

This chapter introduces the process of *knowledge acquisition,* the actual programming of the knowledge base. Most of the discussion in this chapter is devoted to the manual construction of the knowledge base. Computer-aided methods exist for knowledge acquisition; these will be discussed in Chap. 8.

4.1 Introduction

The previous chapters have been devoted to technical issues in expert systems. Understanding these issues, the inference engine, the calculation of uncertainty, and so forth, are necessary if one is to construct expert systems. However, they are subsidiary in importance to the most difficult part of the development of an expert system: the development of the knowledge base. Artificial intelligence research has given this process the high sounding name *knowledge engineering.* Essentially, however, it is simply another form of computer programming. The knowledge base is basically a computer program, albeit one written in a nontraditional computer language and frequently dealing primarily in descriptive rather than numerical information.

There is a difference of opinion over the most appropriate individual to actually program the knowledge base. The expert system textbooks are almost unanimous in asserting that the knowledge engineer should not be the same person as the domain expert. Most of the theory of knowledge engineering is based on this principle. For example, several papers (e.g., Buchanan et al., 1983) deal with the proper process for interviewing the expert. In agriculture, however, funding and other constraints have often forced a violation of this separation. In the experience of the authors and others involved in the creation of expert systems, there are some advantages to having the knowledge engineer be the domain expert.

The techniques used to figure out what to put into a knowledge base are collectively called *knowledge acquisition*. This chapter discusses knowledge acquisition at both the theoretical and practical levels. As might be expected in a discipline of computer science, much of the theory of knowledge acquisition involves the development of computer programs to automate the process. However, little of this theory has reached the level of practicality, and much of the practical information about the process can be found in references outside of computer science and in less technical magazines. We will attempt to draw a balance between computer-based techniques, which are interesting primarily as an indication of what the future may hold, and manual techniques, which are more mundane but of greater immediate use.

4.2 Developing a Rule Base

4.2.1 Stages in developing a rule base

Buchanan et al. (1983) have identified a number of stages in the development of a knowledge base. We have modified their interpretation of these stages and their order somewhat to make the stages more appropriate for the agricultural context. The four stages that we will describe then are problem identification, knowledge base formalization, testing and prototype revision, and implementation. As with any such prescription, these stages, while providing a useful guideline for correct behavior, should not be obeyed to the point of stultification.

Identification. It is implicit in the inception of development of a knowledge base that the problem the knowledge base needs to solve has been identified, at least to some degree. However, the expert system format is not appropriate for every type of problem. Indeed, as more experience is gained in the use of expert systems and in their successes and failures, it is becoming clear that the range of problems appropriate for the classical rule-based expert system is somewhat restricted. The first question that the would-be knowledge base developer must ask is whether the problem falls within this range.

Two criteria must be considered in assessing the appropriateness of an expert system for making a contribution to a problem. One criterion involves the nature of the knowledge involved in the solution of the problem, and the other involves the nature of the problem itself.

The nature of the knowledge. Artificial intelligence theory recognizes a fundamental dichotomy between types of knowledge. The two fundamental types are called *procedural knowledge* and *declarative knowledge*. The easiest way to describe and distinguish them is by giving an example of each.

Procedural knowledge is the structured sort of knowledge involved in carrying out a step-by-step process, such as doing a linear regression or calibrating a spray rig. A simple example of this process is the conversion of a temperature measurement from Celcius to Farenheit: First, multiply the number by 1.8, then add 32. In formulating a computer program that involves procedural knowledge, a classical procedural programming language such as Fortran or Pascal is most appropriate.

Declarative knowledge is a more unstructured kind of knowledge that is involved in identifying a type of animal or picking out the most suitable piece of machinery for a particular farming operation, e.g., use narrow equipment on uneven land. Declarative knowledge is difficult to incorporate into a classical computer program, and it is no surprise that problems involving declarative knowledge are most appropriate for expert systems.

Even if the knowledge involved in the solution of a problem is primarily declarative, there are still other considerations. These relate primarily to the size and ease of programming of a rule base. Rule bases are easiest to understand, and therefore to program and debug, if the rules are not too tightly coupled. Loose coupling occurs when the knowledge used in creating the rule base has several independent or partially independent states, each of which is relatively simple. Problems occur when it is difficult to separate the knowledge into such states. For example, in the case of the diagnosis of a crop problem, it might be the case that the symptoms can first be subdivided by where they occur, e.g., on the leaves, the stem, or the fruit, and then subdivided further by the nature of the damage, e.g., chewing or other insect damage, chlorosis, or wilting. After making these initial subdivisions, it may be relatively simple to determine the specific cause from among the few agents that cause that symptom on that organ. Knowledge such as this is well suited for an expert system. On the other hand, suppose that there are many disorders that caused the same symptoms, and that the symptoms are not specific to any particular organ. This situation would lead to a large, unstructured rule base that would be difficult to program.

A second consideration is the size of the knowledge base. If the size is too small, so that there are only a few rules, then it may be more appropriate to simply write these rules on a sheet of paper. There is no sense using a sledge hammer to pound a tack. A good rule of thumb for determining whether a compendium of knowledge is well suited for encoding into a rule-based system is this: If the rule base could be well described by formulating a flowchart, but the flowchart would be too big and complicated to be practical as a diagnostic or decision-making tool, then the problem is appropriate for an expert system.

The nature of the problem. Even if the body of knowledge is well suited for incorporating into an expert system, there are a number of criteria that must be satisfied in order for the endeavor to have a reasonable chance of success. These include the following:

1. *A formalized core of knowledge must exist.* It is impossible to construct a rule base if the necessary body of knowledge does not exist. This may seem obvious, but it is surprising how many knowledge base building programs are undertaken, and how far into these programs the participants go, before the realization occurs that the knowledge needed to formulate the rule base is simply unavailable.

2. *There should be a knowledge bottleneck.* The term *knowledge bottleneck* is one of many that have come out of expert systems and artificial intelligence. It refers to the situation where there are a few people who have the knowledge needed to solve a problem and many people who need to solve this problem. For example, in weed control it is often necessary to identify weeds at the seedling stage. Distinguishing weed species at this stage may require a highly trained weed taxonomist. If there were few such persons available, or if the nearest weed specialist with the necessary skills was inconveniently far away, then there would be a knowledge bottleneck. This is the ideal situation for an expert system. On the other hand, if the knowledge is widely distributed and readily available, then even if it were well suited for an expert system, no one would be motivated to use that system.

3. *The experts must be available to the knowledge engineer.* Assuming that the knowledge engineer is not the individual with the expertise, the knowledge engineer must at least have ready access to these individuals. It is most helpful if the knowledge is concentrated among only a few experts. In the first place, this makes it logistically easier for the knowledge engineer. In the second place, substantial disagreements among the experts may be less likely if there are fewer of them.

4. *The experts must be motivated to aid the expert system development.* Expert systems have had considerable success in large corpo-

rations (e.g., Austin, 1990). One reason for this may be that the experts who are employed by the same company as the expert system developers can be ordered to cooperate in the development. This cooperation is often time consuming and annoying for the expert, who usually gets little in the way of direct reward if the expert system is developed since this person has no need to use it. If the expert will not cooperate, then the project cannot proceed.

5. *Potential users must be motivated to use the system once it is developed.* Even if the expert can be persuaded to cooperate and a functioning, well-designed expert system can be developed, there is no guarantee that the intended users will actually use it (unless, again, they are also employees of the same organization and can be ordered to do so). Many people are intimidated by computers, and many more will be skeptical of any information that does not come from a human expert.

Knowledge base formalization: some good practices in writing a rule base. The same knowledge may often be encoded into rules in a variety of ways. In the case of the animal identification problem, for example, we saw that rules could be either directly translated from the dichotomous key (Fig. 2.3) or broken up into a more logical structure (Fig 2.4). Other encodings of the same knowledge are obviously possible. However, the fact that the encodings represent the same knowledge does not mean that they are equivalent. As with any computer program, the style of encoding a knowledge base greatly influences the ease with which that knowledge base can be understood, debugged, and modified. Programming a knowledge base is an art, and any two practitioners of the art will differ somewhat in their opinions as to what constitutes good practice. What follows are a few general principles that seem universally accepted. Many of these principles involve significant trade-offs so that a well-programmed knowledge base becomes a compromise between competing objectives. With this caveat, here are some rules of thumb:

1. *Keep the rule base as shallow as possible.* The "depth" of a rule base refers to the number of levels of reasoning necessary to go from the user's input to the final conclusion. For example, the rule base of Fig. 2.3 is one level deep since one goes directly from user input to conclusion. The rule base of Fig 2.4 is two to three levels deep. There are two general problems associated with a very deep knowledge base. The first is that depth generally implies a high degree of interdependence among the rules and an increasing chance that the modification of a rule at one level of the knowledge base will have unintended consequences on the logic of the knowledge base several levels away. The

second problem is that in a backward-chaining system, each level of depth of the knowledge base requires one more level of recursion by the inference engine. In some systems this can quickly lead to a growth in the stack size to the point where memory becomes limiting on a microcomputer.

This rule of thumb should not be pushed too far. A rule base that is only one level deep is almost certainly a poorly designed rule base. In the first place it is usually hard to modify since each new situation calls for the restatement of several antecedents, and in the second place the explanations that it provides to the user are almost never very satisfying.

2. *Avoid tight logic in relating rules.* The more tightly reasoned an argument is, the more difficult it is to understand. In the construction of the rule base there are two possible extremes. One is to set the rules down with no thought of their interrelationships, a sort of "stream of consciousness" from the experts. The other extreme is to fit the rules carefully together so that the knowledge base becomes logically little different from an ordinary computer program. The first extreme is often regarded as the ideal in knowledge engineering, but many actual rule bases come closer to the second extreme. This is not necessarily a bad thing, but a highly structured knowledge base or one involving very convoluted reasoning is often more difficult to understand and maintain. One problem with rule bases that are actually intended for use by the general public is that they usually must be reasonably "bulletproof," that is, resistant to crashing or giving ridiculous answers in response to unanticipated input. Making any computer program, including a knowledge base, bulletproof usually means tightening the program's logic.

3. *Organize the rules into independent subcomponents.* One way of maintaining independence among the rules is to organize them into semi-independent "modules" of increasing specificity. For example, in the animal identification problem the modules could subdivide the system into rules about birds, reptiles, and mammals, and then within each group into carnivores and herbivores, and so forth. This organization promotes graceful degradation by moving in a logical way from the general to the specific. Moreover, with some inference engines it is possible to reduce the size of the knowledge base by discarding blocks of rules in different modules during the course of the inference process. This can often greatly speed up the inference process by reducing the number of rules and assertions that must be considered at each stage.

4. *Develop the rule base one module at a time.* There are two possible ways to develop a rule base. One is to introduce rules in a

breadth-first manner, and the other is to introduce them depth first. For example, in the animal identification problem one could first write all the rules in the first layer, subdividing the categories into carnivore, herbivore, mammal, bird, and so forth, and then fill in the deeper rules more or less at random. Alternatively, one could first write all the rules for carnivorous mammals, then for carnivorous birds, and so forth. There is an almost unanimous consensus among experienced expert system developers that the latter method is better. It promotes modularity and avoids complex logical tangles.

Testing and prototype revision

Verification. Testing of a knowledge base involves verification and validation (Marcot, 1987; O'Keefe et al., 1987). These two activities reflect a dichotomy common in artificial intelligence (and, more generally, in all of computer programming), between *syntax,* or grammatical structure, and *semantics,* or meaning. A knowledge base must be syntactically correct, i.e., it must not contain any "bugs" such as mistyped rules or logical fallacies; it must mean what the programmer thinks it means. Checking that this is the case is called *verification.* The knowledge base must also be semantically correct, i.e., it must provide the correct information; what the programmer thinks it means should be correct. This is called *validation.* Of these two activities, verification is by far the easier since there is an absolute standard of correctness against which the program can be compared. We will discuss this process first.

By the standards of computer languages, the syntax of most expert system rule base languages is relatively simple and restricted. On the other hand, the nonlinear, declarative nature of rule bases often makes a thorough debugging very difficult. Because of the restricted syntax of rule base languages, most expert system bugs fall into one of a few categories. This has motivated the development of a number of computer programs for knowledge base verification (e.g., Suwa et al., 1982; Nguyen et al., 1987). These programs look for specific, well-defined types of errors, many of which are hard to spot by eye. As an example, consider the rule base shown in Fig. 4.1, adapted from Ostergard (1990). The rule base, if it is to be used with a backward-chaining inference engine, has seven errors. Can you find them?

Errors in a rule base may be divided into two categories: inconsistency and incompleteness (Nguyen et al., 1987). The five errors in the inconsistency category are conflict, redundancy, subsumption, unnecessary antecedents, and circularity. The two errors of incompleteness are unreachable consequents and dead ends. We will describe and demonstrate each of these seven types of errors.

Rule 1:
IF The animal lays eggs
THEN The animal is a bird

Rule 2:
IF The animal has hair
THEN The animal is a mammal

Rule 3:
IF The animal lays eggs
THEN The animal is a reptile

Rule 4:
IF The animal is a mammal
THEN The animal gives live birth

Rule 5:
IF The animal lays eggs AND
The animal flies
THEN The animal is a bird

Rule 6:
IF The animal has feathers AND
NOT The animal flies
THEN The animal is a bird

Rule 7:
IF The animal gives live birth
THEN The animal is a mammal

Rule 8:
IF The animal has hair
THEN The animal gives live birth

Rule 9:
IF The animal has feathers AND
The animal flies
THEN The animal is a bird

Rule 10:
IF The animal has scales
THEN The animal is a lizard

Hypothesis: The animal is a bird
Hypothesis: The animal is a reptile
Hypothesis: The animal is a mammal

Prompt: Does the animal fly?
Prompt: Does the animal have scales?
Prompt: Does the animal lay eggs?
Prompt: Does the animal have feathers?

Figure 4.1 A simple rule base with seven errors.

Inconsistency:

Conflict. Two rules are in conflict if they have the same antecedents and different conclusions. For example, the rules

RULE 1:
IF The animal lays eggs
THEN The animal is a bird

and

RULE 3:
IF The animal lays eggs
THEN The animal is a reptile

are in conflict in a knowledge base in which certainty factors are excluded. If uncertainty is permitted then these rules would not be in conflict unless both were concluded with certainty.

Redundancy. Rules are redundant if they succeed in the same situations and lead to the same conclusions. For example, the set of rules

RULE 2:
IF The animal has hair
THEN The animal is a mammal

RULE 4:
IF The animal is a mammal
THEN The animal gives live birth

RULE 8:
IF The animal has hair
THEN The animal gives live birth

is redundant.

Subsumption. A rule is subsumed by another rule if the two rules have the same conclusions and the second rule requires fewer antecedents. For example, the rule

RULE 5:
IF The animal lays eggs AND
The animal flies
THEN The animal is a bird

is subsumed by the rule

RULE 1:
IF The animal lays eggs
THEN The animal is a bird

If uncertainty is permitted, then subsumption does not occur if the conclusion of the rule that would have been subsumed has a lower certainty.

Unnecessary antecedents. Two rules contain unnecessary antecedents if the rules have the same consequents, an antecedent in one rule is in conflict with an antecedent in another rule, and all other antecedents in the two rules are equivalent. For example, in the two rules

RULE 6:
IF The animal has feathers AND
NOT The animal flies
THEN The animal is a bird

RULE 9:
IF The animal has feathers AND
The animal flies
THEN The animal is a bird

the antecedents "The animal flies" and "NOT The animal flies" are unnecessary since the two rules taken together imply that if the animal has feathers then it is a bird whether or not it flies. If certainty factors are permitted, then the antecedents are unnecessary if the certainty of the conclusions is the same.

Circularity. A set of rules is circular if the chaining of these rules forms a cycle. For example, the rules

RULE 4:
IF The animal is a mammal
THEN The animal gives live birth

RULE 7:
IF The animal gives live birth
THEN The animal is a mammal

are circular. If uncertainty is permitted, then the rules are only logically circular if the conclusions are made with certainty. However, because of the way chaining is done in the inference engine, the rules may produce an infinite loop even if they include uncertainty in their conclusions. Therefore, circularity should always be reported as a potential error.

Incompleteness:

Unreachable consequents. A consequent in a backward-chaining system is unreachable if it does not match either a hypothesis or the antecedent of another rule. For example, the rule

RULE 10:
IF The animal has scales
THEN The animal is a lizard

has an unreachable conclusion since the assertion "The animal is a lizard" is neither a hypothesis nor the antecedent of another rule. Unreachable conclusions often occur because of a typographical error or change in terminology. For example, the term *lizard* may have been inadvertently substituted for *reptile*. Unreachable hypotheses are not affected by certainty factors and are not necessarily errors (or unreachable) in a forward-chaining system.

Dead ends. An assertion must either be in the consequents of a rule or have an associated prompt, or the assertion will never be verified. Such an assertion is called a dead end. The assertion in Rule 2 "The animal has hair" is a dead end in the rule base of Fig. 4.1.

Validation. While verification involves the well-defined process of cleaning out bugs, validation is the often murky process of ensuring that the syntactically correct knowledge base is also "scientifically" correct. The word scientifically is placed in quotes because its meaning in this context is different from its usual usage. If one is constructing a computer simulation model, e.g., a plant growth model, then validation involves testing the model against the crop in a variety of conditions to determine whether the solution to the equations of the model matches with sufficient accuracy the observed behavior of the crop. If it does, then the model is scientifically correct, or, more accurately, it is not invalid at the desired level of precision. Thus, for example, a computer model is validated if it correctly predicts the response of the crop to severe moisture stress. In the case of an expert system, the situation is somewhat different.

An expert system is a simulation of a human expert, not of a physical or biological system. Thus, one can argue that the expert system is validated if it displays under a variety of conditions the response to input that the human expert would display. The distinction here is that the expert may be wrong. If that is the case, then the expert system would, if it is valid, be wrong, too. Scientific colleagues of expert system developers often fail to appreciate this distinction, and this can cause friction and hard feelings. Another source of friction between expert system developers and experts involves the validation process itself, particularly in the case of validating decision support systems. Scientific colleagues, and sometimes reviewers, are used to seeing hypotheses verified, and models tested, with statistical precision. Often they will ask whether a decision support system has been tested for increased productivity. What they have in mind is usually a field test in which two sets of plots are managed, one set using the expert system and one using a human manager, and the yields are compared statistically to determine if the expert system provides a significant improvement. There are two problems with this procedure. The first, and most obvious, is the selection of the human managers. These must somehow be representative of farmers in general and should not be strictly farm advisers or progressive, well-informed growers. The second problem, which is more subtle but potentially more difficult, is ensuring that the independence criteria for statistical validity are satisfied. That is, the human manager must make decisions that are independent of both the knowledge that a competition is involved and the knowledge of the decisions being made by the expert system. This may not be an impossible task, but no one has yet proved that it is possible.

Performance. Besides verification and validation, another factor that must be tested during the development stage is performance. The speed, reasonableness of interaction, completeness, and so forth of the expert system must satisfy the user. Testing performance is a tedious task involving repeated interaction with the system using a variety of inputs. It is perhaps the least enjoyable and most important part of the development process. There are a number of performance criteria that should be considered. Marcot (1987) provides a complete list of twenty criteria in all, listed alphabetically from accuracy to wholeness. Much of what he says is common sense, but a perusal of his article is worthwhile and recommended.

Implementation. Implementation of the expert system, that is, distribution to its ultimate users and maintenance of the program, offers special problems in agriculture. Many of these problems relate to the structure of land grant institutions and the reward system for researchers at such institutions. An expert system is a device for the transfer rather than increase of knowledge. By its very nature, this knowledge transfer is often classified as an "extension" function, and many of those who participate in expert system development are extension personnel. The distribution of expert systems has many differences, however, from the distribution of leaflets and books more common in extension work. Very few expert systems have been successfully implemented at the field level, and many of the questions surrounding the development and implementation of expert systems are researchable issues. Moreover, the development of an expert system is still a sufficiently novel task that few extension personnel are qualified to complete it. As such, the completion of an expert system is a major research task, often involving nonextension personnel whose primary interest is in research and academic publication, and whose interest in following a project through to field implementation may be secondary. This can naturally lead to difficulties in pushing an expert system through to complete field implementation.

A second problem involves the nature of computer software. As a piece of software, an expert system generally requires more in the way of technical support than a comparable item that is printed as a leaflet or bulletin. Many commercial software distributors provide toll-free numbers for users' troubleshooting, and all provide some level of technical support. Universities and other research institutions, however, do not typically have such facilities in place for software distributed by their faculty or staff. Moreover, the funding that supports expert system development is generally available only for a relatively short term. Once the project is completed, this funding is generally no longer available. In the case of computer software made for distribution, however, this is precisely the time when funding is needed to provide user support. This is especially true for novel software such as expert systems, which may be expected to generate considerable need for support among users who are not particularly computer literate.

In summary, there are at least two institutional roadblocks standing in the way of the successful field implementation of an agricultural expert system. These are the nature of the reward system, which in many cases focuses on academic publication of results rather than successful field implementation of these results, and the nature of the academic funding system, in which funds are made available for short-term projects involving development, rather than long-term projects involving maintenance. We have little to offer in the way of solutions to these problems; rather, our purpose is to warn potential developers of the difficulties they may face.

4.3 Knowledge Acquisition

One of the purposes of an expert system is to relieve a knowledge bottleneck, that is, a situation in which there are many people who need access to knowledge possessed by only a few individuals. It is somewhat ironic that the development of an expert system itself sometimes creates a knowledge bottleneck. Developing an expert system may not relieve the knowledge bottleneck but rather simply move it to a different place, that of *knowledge acquisition,* the encoding of the experts' knowledge into the expert system knowledge base. The bottleneck is the shortage of individuals skilled at knowledge acquisition.

The translation of the knowledge possessed by the experts into a knowledge base is the bottleneck in the process of knowledge acquisition. There are two problems associated with this translation. The first is that it is sometimes difficult for experts themselves to understand and describe how they do what they do in their field of expertise. A great deal of what we call "expertise" is unverbalized intuition developed through long experience; if the knowledge were readily transmitted, then the expert could write a book or manual for others to read. A second problem is that the experts themselves generally do not know how to program an expert system knowledge base. Therefore, a skilled knowledge base programmer, the knowledge engineer, is often called upon to interview the experts and construct the knowledge base. Particularly among agricultural and resource management organizations there is often a shortage of such individuals and of the funds necessary to employ them. If this is the case, then there is no person who can get the experts to communicate their knowledge to the computer.

These two problems are actually the same problem because the knowledge engineer, as a result of intuition and skill gained through experience in interviewing experts and transcribing their expertise, becomes an expert too. Thus, a new knowledge bottleneck is created: a shortage of expert knowledge engineers. It would seem that nothing has been gained! However, this pessimistic assessment is not quite true. While it is indeed true that the knowledge bottleneck has been shifted from one location to another rather than having been opened, this shift in location has been very significant. For one thing, since the previous location of the bottleneck had been with the experts themselves, the knowledge that was "bottled up" was widely varying, depending on the subject of expertise. As a result of the shift, this bottleneck has been opened. The newly bottled up knowledge is of a much more specific kind: the transcription of expertise on a subject into a knowledge base. For another thing, the subject of the newly bottled up knowledge is knowledge acquisition, which is of direct interest to the individuals who wish to develop the expert system in the first place. Thus, while the expert system developer cannot do much to relieve the knowledge bottleneck caused by a lack of access to experts in every subject, the developer can hope to help relieve the knowledge bottleneck caused by a lack of access to experts in the specific subject of knowledge engineering simply by gaining experience in it.

Before going on, it is well to note explicitly two ways in which the shared experience of expert system developers in agriculture has apparently differed from that of expert systems developers in other areas. The first has already been mentioned: the necessity of individual experts to program expert systems themselves rather than separating the knowledge engineer from the expert as suggested in the textbooks. The second point of departure from the norm is that it is apparently the case in many nonagricultural areas of expert system application that the expertise to be encoded is often a single individual. This is rarely true in agriculture. More often the expertise is shared among a group of people, frequently extension specialists and agents, who must pool their knowledge to create the knowledge base. This can cause special problems since the experts must arrive at a consensus, possibly suppressing great differences of opinion and/or personality clashes. In considering the adoption of knowledge base development aids for agricultural applications, these special features should be taken into account.

4.3.1 Manual knowledge acquisition

In this book we have frequently said that a knowledge base is essentially a computer program and that many of the problems associated with the construction of a knowledge base are programming problems. However, this is not the whole story. A conventional computer program encodes a procedure or set of procedures. A knowledge base encodes knowledge. The difference is that conventional computer programs are based on computational algorithms or formulas that can be expressed in a precise, mathematical form. This is true whether the program involved is a statistical software package, a word processing program, or whatever. Since computational procedures can be written with this precision, the task of the computer programmer is to translate the formula or algorithm involved into a functioning computer program. The translation, however, is a direct one, and the resulting computer program is a linear, ordered set of procedural statements that can precisely express the intended algorithm.

The situation in programming a knowledge base is much more complex, and perhaps it is for this reason that the more pretentious term knowledge engineer is warranted. A knowledge base is not a precise statement of the experts' knowledge; rather, it is an imprecise attempt to represent that knowledge within the constraints of the assigned knowledge base format and syntax. The knowledge engineer thus has the twofold task of interpreting the experts' knowledge and then translating it into the knowledge base form. The process of interpretation is knowledge acquisition. The most common knowledge base format, and the only one we have discussed so far, is the rule base. In this context knowledge acquisition is the process of extracting the domain knowledge from the expert or experts and translating it into a set of rules.

The principal problem associated with knowledge acquisition is that the expert is not always aware of the thought processes used in solving a problem. Intuition and unselfconscious interpretation of events are common in everyday experience as well as in expert behavior. For example, any individual, when greeting a close personal friend or family member, can usually discern that person's mood almost immediately. However, trying to figure out and express in words how this is done usually produces an unsatisfactory approximation to the actual process. In the same way, an expert farmer can frequently husband the crops based on intuition and experience. Trying to capture this in a format that can be transmitted to others or encoded in a rule base can be a daunting task. A number of specialized methods and a variety of computer programs have been developed to aid in this task. However, before presenting these we will discuss some general principles that are given in the literature and provide a few more based on our own experience.

Waterman (1986) provides a useful summary of some of the primary principles of knowledge acquisition. These include:

1. Observe the expert solving real problems on the job.
2. Discuss the kinds of data, knowledge, and procedures required for particular types of problems.
3. Develop descriptions with the expert of prototypical problems associated with each category of answers in the domain.
4. Have the expert solve a series of realistic problems aloud; ask for the rationale behind each decision step.
5. Solve problems provided by the expert using rules developed from the interviews.
6. Have the expert review and criticize the rules and control structure.
7. Have outside experts solve problems already presented to the prototypic system and expert.

Jones et al. (1986) discuss an attempt to implement these principles in an agricultural context.

One common problem with the encoding of expert knowledge is that, no matter what the knowledge representation method, the knowledge base is limited to an essentially verbal representation of the expert's knowledge. However, many experts make use of nonverbal information in analyzing a situation in which their expertise is being applied. For example, an agronomist in diagnosing a crop disorder may walk the field and take in visual information that plays a key role in the diagnosis. The expert may make use of other senses, such as feel, taste, sound, or smell, as well. The expert may not even be aware of this information, and even if the individual is aware of it, may be unable to characterize the information verbally. An ingenious attempt to solve this problem was used by Flynn et al. (1989) in the design of an expert system for the automatic application of nutrients in a greenhouse. The expert system was developed by growing plants in the greenhouse setting and asking the expert, who was in this case a single individual, to respond to needs as they arose. The expert was, however, never permitted inside the greenhouse. Instead, all the consultations took place over the telephone. This not only relieved the expert of the need to set aside time for visits, but, more importantly, it forced the knowledge engineer and the expert to characterize everything verbally in a way that could be more easily encoded in a rule base.

A number of methods, both manual and computer based, have been developed or adapted to aid in knowledge acquisition. In the remainder of this section we discuss a few of these.

Flowchart-related methods. A knowledge base is, as we have said, basically a computer program, and like other computer programs, its contents and logic may be represented with a flowchart. A flowchart for the simple rule base of Fig. 2.4 is shown in Fig. 4.2. Most real knowledge bases, however, cannot be represented by as explicit and complete a flowchart as that shown in Fig. 4.2. Indeed, if the contents of a knowledge base can be represented adequately by a simple flowchart on a page or two, there is ordinarily little reason to construct the knowledge base; it is better to simply work with the flowchart. In actual practice, constructing a flowchart for a knowledge base is the same as constructing a flowchart for any other complex computer program; the boxed items in the chart refer not to individual statements but to blocks of code. In the case of a knowledge base, these blocks of

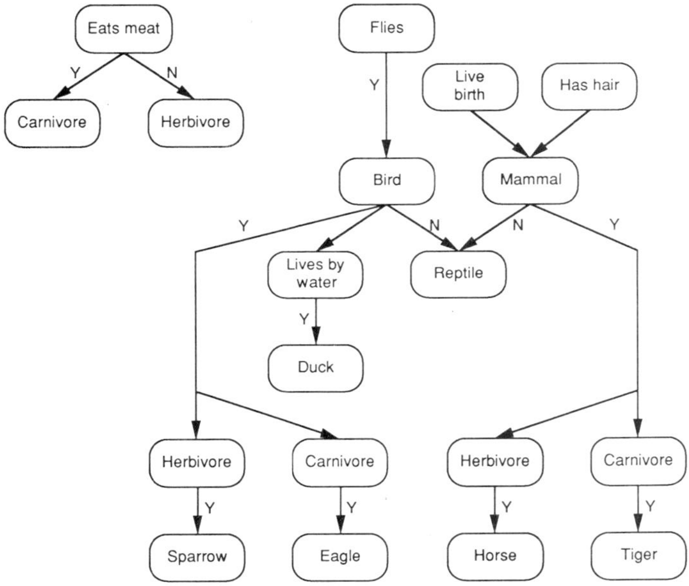

Figure 4.2 A flowchart representation of the simple animal identification rule base shown in Fig. 2.4.

code represent groups of rules. When dealing with a real expert system, one would not ordinarily represent the entire knowledge base with a single flowchart. Instead, segments of the rule base are represented as boxes in the overall flowchart. In the case of a traditional computer program these segments usually consist of subroutines or collections of subroutines. In the case of a rule base, the segments would ordinarily consist of conceptually related blocks of rules. The individual segments can then be broken up and described using flowcharts that do display blocks representing individual rules or small collections of rules.

An alternative to the flowchart that has proven useful in representing rule bases is the *dependency network*. The dependency network can be traced to the idea of the "network of supported concepts" described by Winston (1984). Figure 4.3 shows a dependency network for the same rule base as that represented by a flowchart in Fig. 4.3. Stated somewhat vaguely, a flowchart is more procedural while a dependency network is more declarative. That is, the flowchart explicitly shows each step that the program would take in arriving at a conclusion but leaves the logical relationships implicit. The dependency network does just the opposite: it explicitly shows the logical relationships among the elements of the program but leaves the flow of control implicit. Since this corresponds with the declarative nature of a rule base, many workers have found the dependency network more useful than the flowchart in representing a rule base. The two concepts are not, of course, mutually exclusive. Indeed, they are complementary. It is often useful to develop both a flowchart and a dependency network for the same rule base, with the dependency network emphasizing the logical relationships and the flowchart emphasizing the flow of control. Woolley and Stone (1987) provide an example of such a complementary use of a flowchart and a dependency network in representing a complex rule base, the SYSTEX rule base for insect identification.

Structured analysis. Building a knowledge-based system can be accomplished in a very unstructured way. In fact, many of the commercially available expert system development tools encourage the knowledge engineer to plunge right into the process of writing rules and building a system prototype. Such proof-of-concept prototypes are very useful to a knowledge engineer because ideas can be tested quickly, and the cost of failure is fairly small. Knowledge bases that do not work well are thrown out, and a new knowledge base using a different representation scheme is built. However, when building larger and more complex expert systems, or trying to get systems to the application stage without a great deal of wasted coding, some or-

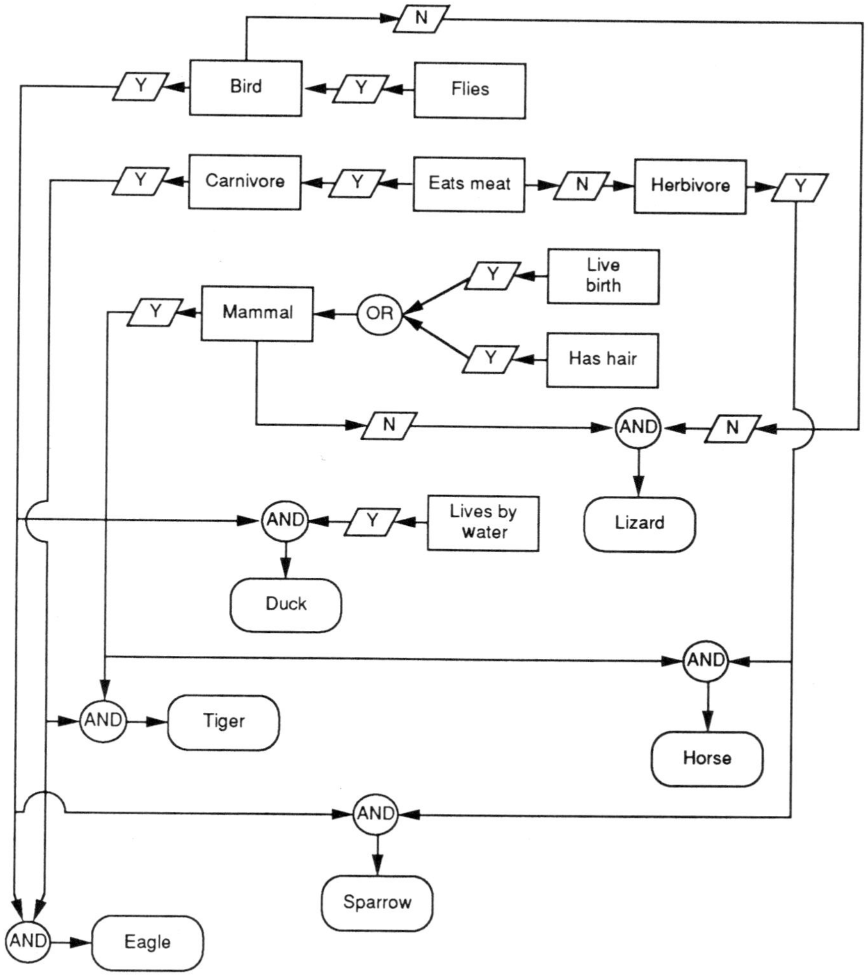

Figure 4.3 A dependency network representation of the simple animal identification rule base shown in Fig. 2.4.

ganizational work at the outset will greatly improve the efficiency of development.

For example, it would be helpful to know even before choosing an expert system shell that to solve a problem, the expert typically consults a database or has to perform a complex calculation. Also, it would be much better to know at the beginning that some actions of the expert must be performed in some explicit order or that one aspect of the problem cannot be addressed until another has been successfully concluded. These characteristics of the expert's problem-solving behavior can often suggest an appropriate knowledge engineering strategy, a system architecture, or a knowledge acquisition method that would otherwise be missed. Structured approaches to knowledge base development are designed to enable knowledge engineers to solve these organizational problems without developing the system first. The difficulty is that there are few well-worn rules for developing knowledge-based systems, and structuring the process too much can hinder rather than help the effort.

The best way to approach the overall problem of developing a knowledge-based system is to view the expert as a system that must be modeled rather than as a means to discover and transcribe some hidden set of rules. A system has inputs, outputs, and processes. Inputs to an expert are information about specific problems, outputs are the recommendations given or solutions to the problem, and processes are the tasks or actions performed by the expert during the problem-solving exercise. It is the knowledge engineer's task to build a computer model of this exercise. The knowledge engineer is therefore very much like a systems analyst preparing to build a model of a physical or biological system, and most of the structured methods of knowledge engineering are taken directly or with some modification from general systems theory (Lucas, 1981; Hart, 1986; McGraw and Harbison-Briggs, 1989).

Agricultural researchers were very quick to make the connection between knowledge engineering and systems analysis, probably because of the intensive use of systems analysis to study large and complex crop systems (Huffaker, 1980; Frisbie and Adkisson, 1986). Here we will discuss the basic steps involved in systems analysis and their relationship to knowledge engineering as discussed in Stone et al. (1987): problem definition, systems decomposition, synthesis, and evaluation. We will not describe systems theory in any detail here. Interested readers can refer to Schroderbek (1971) or Law and Kelton (1982). Likewise, more in-depth discussions of structured knowledge acquisition can be found in Hart (1986) and McGraw and Harbison-Briggs (1989).

Problem definition and domain decomposition. Setting the bounds of a system is fundamental to successful systems analysis and knowledge-based system development. Just as a systems analyst setting about to build a crop plant model must decide at the outset whether the model will incorporate edaphic factors and soil-root interactions, a knowledge engineer building a system to diagnose viral diseases in a crop must decide whether the system will also include knowledge about insect and mite damage that could cause similar looking symptoms. Either choice has important implications for the system that will be developed, both in the system's construction and in its use.

One point that may appear obvious is that the knowledge engineer should focus immediately on what the knowledge-based system will be expected to produce. That is, it is not enough to say that one wants to build an expert system for corn pest management. One must answer the question, what information, advice, or recommendations will the system be expected to provide? Will it give general information about how to grow corn and to control its pests? Will it have to help users identify pests? Will it recommend specific sampling plans and make control recommendations? The key to successful knowledge engineering is to describe the system's bounds and the exact types of output it will provide.

Once a knowledge domain has been defined, the next step is to decompose the domain. Breaking up a cognitive system into its component parts is not nearly as straightforward as decomposing a physical system. Nevertheless, the solution to a complex problem can often be subdivided into the solutions of many smaller problems. For example, the problem of how to control weeds in a crop system can be broken down into at least two problems: identifying the weed or weeds and then controlling each specific weed. The problem of weed identification can then be divided into tasks like determining the weed's general type, e.g., grass versus broadleaf, looking for specific diagnostic characters, e.g., the presence of serrate or peltate leaves, and finally making a species determination. In general, there will be many possible ways to subdivide such a system, and the knowledge engineer needs some way to compare possible decompositions without investing too much time in each.

One of the best ways to compare system decompositions is by drawing rough models of the decision system. Figure 4.4 shows a generalized decision system model after Stone et al. (1987). The key components are decisions, represented as boxes. Each box contains the possible outputs from that decision, and each box is connected by arrows to the information needed in making the decision. From this rough model of the decision system, the knowledge engineer can (1) verify with the domain expert or experts that the overall structure is correct, (2) identify the types of information and knowledge that will be needed to complete the system, (3) plan the knowledge acquisition sessions based on the breakdown of the overall problem into discrete subproblems, and (4) identify gaps in our understanding and information, i.e., decisions that are dependent on unavailable information or poorly understood relationships.

This diagram of the overall system structure is neither a flowchart nor a dependency network. It shows relationships of information flow.

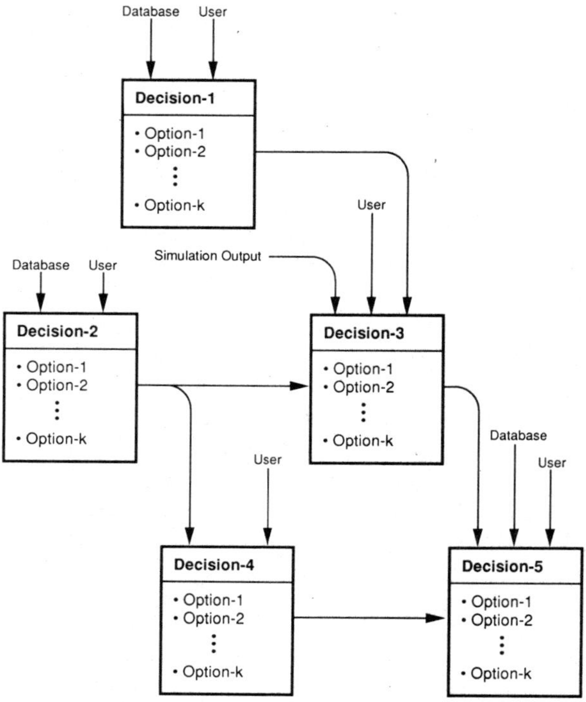

Figure 4.4 Decomposition of a decision system. Boxes are deductions and the possible choices. Arrows represent information flow. Sources of information include the user, databases, simulation models and analysis, and previous choices.

For example, Fig. 4.5 shows a possible information flow diagram for the animal identification system we have been discussing. The final recommendation of the system is based on the previous determination of the animal's class, i.e., mammal, bird, or reptile, and the animal's food and habitat.

Notice that this diagram does not tell us anything about what the actual rules will be. It only defines the relationships among sets of rules. That is, there will be some rules to determine whether an animal is a mammal based on its body covering and birth method.

McGraw and Harbison-Briggs (1989) suggest some additional types of structured analysis that can help to structure the knowledge base and the knowledge acquisition process, including: functional analysis, information flow analysis, and operational sequence analysis. Each is similar to the decision-flow diagram above, but may be more appropriate for specific domains.

Functional analysis concentrates, as one might expect, on the functions performed by the various subsystems. An expert system that di-

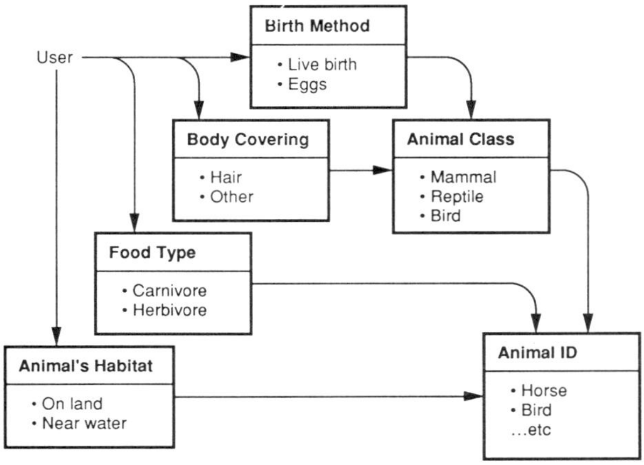

Figure 4.5 A decision-information flow diagram for the animal identification system.

agnoses automobile engine malfunctions, for example, might need to be broken up into subsystems that deal with the fuel system and the electrical system. If these separate modules are to work properly, the knowledge engineer needs to know how the proper functioning of one component can affect the diagnosis of faults in another component. For example, if the ignition system is not working properly, it will be very difficult to diagnose a problem with the carburetor. A functional model of the system would (1) represent the major components of each subsystem of the engine with a box, (2) describe the major functions of that subsystem, e.g., to diagnose problems with the carburetor, (3) draw linkages to other components or subsystems upon which each subsystem depends, and (4) show the inputs and outputs from each component as in Fig. 4.5.

Information flow analysis is very much like a classical programmer's flowchart of system operation. It shows how a user of the system provides information to the system and how the system returns information to the user. In effect, the only difference between information flow analysis and a classical flowchart is that internal systems functions, such as the calculation of variables or even the inference process itself, are not explicitly modeled. Only the input requirements of the user are shown, not what the system does with that information. Figure 4.6 shows an information flow diagram for our animal identification example, showing the consultation process and the ability of the user to request explanation.

Operational sequence analysis diagrams show the relationships left out of information flow analysis. That is, they concentrate on the processing tasks of the system and how these relate to one another, to the

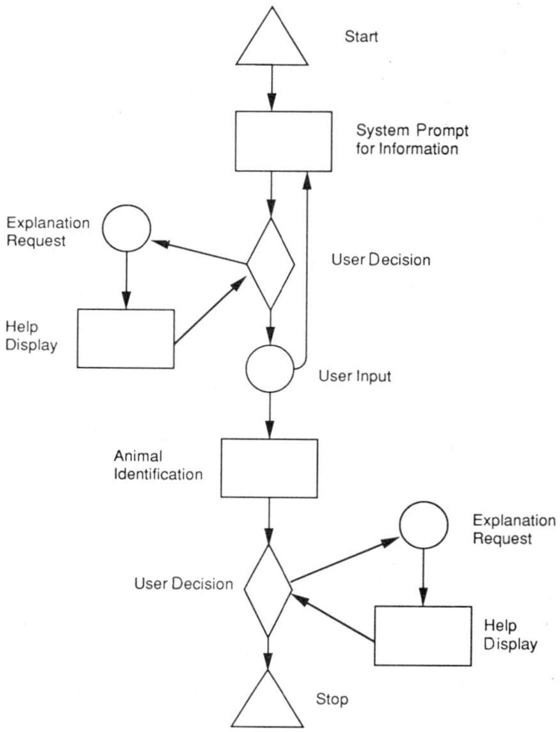

Figure 4.6 A very general information flow diagram for the animal identification system.

tasks performed by the system, and to the inputs of the user. A full system may require several of these analyses, each one describing a different type of interaction with the system. Once again, Fig. 4.7 shows an operational sequence analysis for the elementary animal identification system.

In general, the idea of structuring the process of knowledge engineering is very much like the idea of making an outline before writing a thesis. Often, one can greatly improve the efficiency of the development process by building holistic models of system interdependencies, flow, and function. The methodology is very flexible. There are few rules to help a knowledge engineer build the types of diagrammatic models presented here. However, some general advice is worth noting.

First, be consistent. Use like symbols to represent like items in a diagram, and make sure that all such diagrams use the same symbolic syntax. Second, be very clear about the purpose of the diagram. Do not mix model types. For example, do not include time-based flow infor-

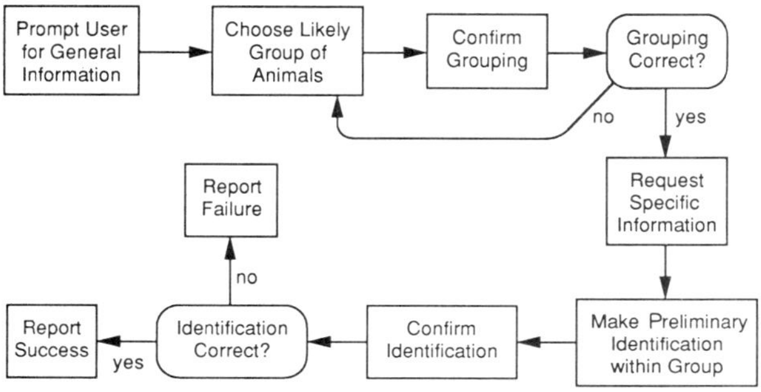

Figure 4.7 An operational sequence analysis for a more complex animal identification system that makes preliminary identifications and then attempts to verify them.

mation in a decision flow diagram. Third, use the developed models and diagrams as tools to interact with the expert or experts. These tools are more useful the more they point out areas of misunderstanding and potential conflicts.

4.3.2 Automated knowledge acquisition

This section describes three methods of computer-based knowledge acquisition, in which a computer program constructs the knowledge base "directly" from some basic form of information. The three methods, induction, personal construct theory, and natural language understanding, are by no means the only possible methods of automated knowledge acquisition. They are, however, among the most promising. The three methods may best be understood and compared if one recognizes that each method corresponds to a different way in which people learn things. The first part of the title of a paper by Hass and Hendrix (1983) on natural language understanding, "Learning by being told," emphasizes this comparison. One way to learn is through direct instruction, by being told. This corresponds to natural language understanding described in the natural language understanding section below. Another way to learn is by example; this corresponds to the inductive methods of knowledge acquisition described in the section on induction below. A third way is by observation, corresponding to the personal construct theory approach to knowledge acquisition described in the section below on the personal construct theory. The volume edited by Michalski et al. (1983), in which the paper cited above is located, is highly recommended for further reading on this topic. More recent information may be found in vol. 2 of this book (Michalski et al., 1986).

Induction. The basic form of the production rule states that IF something is true, THEN something else is true, e.g., IF an animal has hair, THEN it is a mammal. With such rules, the truth of the antecedent, the IF part, guarantees the truth of the consequent, the THEN part. This is called *modus ponens* and is discussed more fully in Chap. 8. It is not always the case with such rules that the truth of the THEN part guarantees the truth of the IF part. In other words, the fact that an animal is a mammal does not guarantee that it has hair. Nevertheless, people often learn things by acting as if this is the case. For example, an individual might see several examples of mammals and, noting that each one has hair, concludes that all mammals have hair. This kind of reasoning is called *induction*; sometimes it is called *abduction* by artificial intelligence workers and logicians to distinguish it from mathematical induction, which is a different process. While this form of induction is not guaranteed to produce correct results, it is nonetheless a very powerful learning tool. Indeed, most scientific discoveries are motivated by a process of induction in which the scientist observes examples of some relationship and sets about conducting experiments to determine whether this relationship always holds true.

Inductive learning was for a long time asserted to be strictly a human phenomenon, one that machines could never hope to duplicate. Not surprisingly, artificial intelligence researchers took such an opinion as a thrown gauntlet, and much effort was devoted to the analysis of the inductive process and the development of computer programs that could simulate this process. One of the best known workers in this area is Ryszard S. Michalski, who developed a number of highly successful inductive learning programs (e.g., Michalski, 1980). It happens that plant pathology was one of the domains that Michalski chose to apply his programs and methods. He participated in the development of PLANT/ds, an expert system for the diagnosis of soybean diseases (Michalski et al., 1985). This was one of the first agricultural expert systems. Michalski's primary motivation in the construction of this system was to test his computer program AQ11 for inductively generating rule bases. Specifically, he developed two rule bases for soybean disease diagnosis, one based on expert-generated rules and one based on rules inductively derived from case studies, and compared the performance of the two. Somewhat surprisingly, he found that the inductively derived rule base on average performed better (Michalski and Chilausky, 1980).

The precise method of induction used in Michalski's AQ11 program is rather technical; it is summarized by Michalski and Chilausky (1980). Basically, the method begins with data consisting of a list of attribute values and the corresponding classes, not unlike the matrix shown in Table 2.1. In the present instance, the table might be regarded as a series of field observations made by one or more experts. A significant difference between Michalski's table, which is called a *description space,* and Table 2.1 is that uncertainty is permitted in the associations of the description space table, so that, for example, a value might be associated with more than one class with differing degrees of certainty. The inductive program constructs a rule base from this description space. It does this by an iterative process in which antecedents and consequents are added and subtracted until a complete rule base is obtained.

Michalski and Chilausky express surprise, no doubt mixed with pleasure, that the inductively derived rule base consistently outperform the expert derived ones. They give four possible explanations for this phenomenon (Michalski and Chilausky, 1980, p. 79):

1. The information obtained from consultation with the experts was incomplete.
2. The knowledge representation scheme is inadequate.
3. The inference methods were inadequate.
4. Experts in making diagnoses are not necessarily experts in explaining the process of diagnosis.

Personal construct theory. Personal construct theory is a theory of human epistemology, i.e., the acquisition of knowledge, developed by the psychotherapist G. A. Kelly (1955). The theory lends itself to quantitative analysis and computation, and as such it has received considerable attention from computer scientists and artificial intelligence researchers. According to the personal construct theory, every person acts as a "personal scientist" in learning about his or her world (Shaw, 1980), i.e., each person as described by Boose (1985) "seeks to predict and control events by forming theories, testing hypotheses, and weighting evidence." Kelly hypothesized that each individual forms theories and learns things through the formation of *constructs.* A construct is a mental image or description of something. A key feature of Kelly's theory is that every construct is a dichotomous entity. That is, any description of what something *is* must a priori include a description of what it is *not.* For example, a person is described as *good* by placing that person on the good side of a dichotomous good-bad construct. An extension of the dichotomy is to think of constructs as consisting of scales with the dichotomies representing polar ends of the scale. For example, a person could be placed in an intermediate position on a scale between the extremes of bad and good.

Kelly used his personal construct theory to assist him in developing techniques for psychoanalysis. One of the quantitative techniques is the *repertory grid,* which is a device designed to elicit an individual's personal construct system. There are a wide variety of repertory grids (Fransella and Bannister, 1977); each type is used in some way to determine constructs that are alike in some way and that can be distinguished from other constructs. Boose and his coworkers have developed a well-known knowledge acquisition system based on personal construct theory that applies the principal of the repertory grid to knowledge acquisition. The initial version of the system was called ETS (Boose, 1985); a subsequent version is called AQUINAS (Boose and Bradshaw, 1987). The basic function of these systems is to perform the analytical function of the knowledge engineer, i.e., to interact with the expert and develop a rule base from the expert's knowledge. The systems permit the expert to perform a guided introspection, transferring unverbalized knowledge into verbal form.

The basic idea in the elicitation process is to attempt to determine the expert's personal constructs and translate these into a rule-based form (Boose, 1985). A simple example based on the animal identification problem will illustrate the process. The expert is asked to form a list of the animals to be considered. Suppose this list is the same as that of Chap. 2, namely, horse, tiger, lizard, duck, eagle, and sparrow. The system might set up the following dialog with the expert.

System: What is the name of one animal under consideration?
Expert: Horse
System: What is the name of another animal?
Expert: Tiger
System: What is the name of a third animal?
Expert: Lizard
System: What is an important characteristic that two of these share but the other does not?
Expert: Mammal
System: What is the opposite of that characteristic as it applies in this case?
Expert: Reptile
System: Please rate each of the three items as to Mammal or Reptile:
Expert: Horse = Mammal; Tiger = Mammal; Lizard = Reptile
System: What is another animal that is different from Lizard in some attribute?
Expert: Eagle

The consultation proceeds in this manner, with the expert being asked to identify animals in sets of three and provide attributes that distinguish two of the three from the third. The responses are tabulated in a structure called an *entailment graph,* and statistical cluster analysis techniques are used to sort them out and prepare them for translation into rules. The AQUINAS system extends the ETS system in providing facilities for hierarchical organization of the knowledge base, integration of knowledge from several experts, and so forth.

Natural language understanding. A knowledge engineer, in interviewing a domain expert, converses with the expert in a natural language such as English, transcribes the information provided by the expert, and translates it into a form that can be read and logically processed by the inference engine. Hass and Hendrix (1983) describe a preliminary effort to develop a computer program that can accomplish this same task. Hass and Hendrix (1983, p. 405) define a knowledge learning and using system (KLAUS) as a system that can "hold a conversation with a user in English about his specific domain of interest, subsequently retrieve and display information conveyed by the user, and apply various types of external software systems to solve user problems." Nothing close to such a system presently exists, but they describe the development of a preliminary program called NANOKLAUS that can process natural language, apply logical procedures and tests to English statements, and derive relationships among objects described in the statements.

A computer program for understanding natural language must accomplish two tasks. First, it must be able to parse sentences and phrases, i.e., it must recognize the various elements of the phrase, e.g., subject and verb, and how they relate to the meaning of the sentence. Second, it must have a vocabulary so that it can understand the meaning of the individual words, taken in context. Hass and Hendrix point out that while the sentence structures used in a knowledge acquisition session between a computer and an expert may be fairly simple in nature, the vocabulary used in each session might be completely different and highly specialized. Thus a major part of the knowledge acquisition session might consist of acquiring jargon words and understanding how the concepts described by these jargon words relate to each other. Hass and Hendrix employ concepts from the theory of sets to determine these relationships. For instance, NANOKLAUS might be told that a bird is a kind of animal, i.e., a subset of the set of animals, that a robin is a kind of bird, that a property of a robin is that it flies, and so forth.

NANOKLAUS is preprogrammed with a set of rules governing the structure of certain types of simple English sentences. Given a sentence in English, it attempts to match that sentence with one of a number of patterns and, if successful, it extracts new terminology from the sentence and derives the relationship of the new term to terms already known. For example, if it already had established a definition of the word "bird" and it was presented with the sentence "An eagle is a bird," it would recognize that "eagle" is a new term and that the set of all eagles is a subset of the set of all birds.

NANOKLAUS is also preprogrammed with a set of "seed concepts," which represent the meanings of various common English words, such as physical object, person, measure, etc. Hass and Hendrix emphasize that seed concepts are not primitives, i.e., new concepts do not have to be defined in terms of already existing seed concepts. Instead, new concepts are related by facts and constraints to already existing concepts. For example, in the case of an eagle, NANOKLAUS might progressively learn that an eagle is a member of the set of birds, that it is brown with a white head, that it is larger than a duck, and so forth. It is never provided with a definition of the concept "eagle," rather with a set of relationships to other concepts. NANOKLAUS is very limited in the sorts of English statements that it can decode, but it provides a starting point for the development of KLAUS programs.

Hass and Hendrix provide an example of a NANOKLAUS training session involving aircraft carriers. (It is often easy to infer the funding source of artificial intelligence projects.) From this example we can create a simulation of how a part of a session with NANOKLAUS involving animal identification might appear. Here it is:

User: What do you know about?
N: I know about PHYSICAL OBJECTS, PEOPLE, MEASURES, and THINGS.
User: A length is a measure.
N: OK, now I understand LENGTH.
User: A meter is a unit of length.
N: OK, now I understand METER.
User: A physical object has a length.
N: Must a PHYSICAL OBJECT have exactly one length?
User: Yes.
User: A bird is a physical object.
N: Is PERSON a proper subset of BIRD?
User: No.
N: Is PERSON necessarily composed of entirely different members from BIRD?
User: Yes.
N: Do PERSON and BIRD span the set of all PHYSICAL OBJECTS?
User: No.
N: OK, now I understand BIRD.
User: A bird can fly.
N: Give the -ed and -en forms of FLY.
User: flew, flown.
N: If a BIRD can FLY something, type in the most general description of that thing, otherwise type <carriage return >
User: <carriage return >
N: OK, now I understand FLY.

The session proceeds in this way. During the course of the interaction, NANOKLAUS develops a vocabulary of terms and their relationships that enables it to generate a hierarchy of these relationships resembling an identification key.

4.4 Maintaining an Established Knowledge Base

Once a rule base is developed, it must be maintained. New knowledge is always becoming available, and, like any computer program, the rule base must be modified to take this new knowledge into account. The new knowledge may come from original research and contain information that was not known to anyone at the time the rule base was developed, but it may also be derived from information that was available at the time the rule base was created but, either intentionally or through oversight, was not included. This latter possibility is especially important because it often manifests itself in the expert system's failure to reach a conclusion that it "should know about." Tracking down the reasons for the expert system's poor performance and correcting the problem can be considered a part of the validation process described earlier in this chapter. There is, however, a subtle difference of emphasis, which is that the validation process focuses on a rule base under construction while the maintenance process focuses on an existing rule base that has been through the validation process but now needs some "post graduate instruction."

One of the most sophisticated early programs for knowledge base maintenance was the program TEIRESIAS, created by Randall Davis (Davis and Lenat, 1982) for the specific purpose of debugging and maintaining the expert system MYCIN. MYCIN already had an established knowledge base at the time TEIRESIAS was created, and this heavily influenced the form of TEIRESIAS. Unlike the knowledge-based verification programs described earlier, TEIRESIAS and related programs actually execute the expert system as a part of the maintenance process and therefore require a working knowledge base. The TEIRESIAS knowledge base management system is a very complex program (actually, a very complex subprogram of an *extremely* complex program), and it would be impossible to do it justice in a paragraph or two. The best we can do is to give a basic idea of how it works.

The program makes use of *rule models* to allow a knowledge base developer to examine a knowledge base interactively and make changes if necessary. A rule model is a representation of the generic features of a rule. It contains a description of the rule, examples of such rules, and indicators of more general and more specific rules. For example, an animal identification system might have rules for identifying whether an animal is a carnivore, a herbivore, or an omnivore. Some of these might be represented using predicate logic as

RULE 1
IF The diet is meat
THEN The animal is a carnivore

RULE 2
IF The teeth are pointed
THEN The animal is a carnivore

RULE 3
IF The eye location is on the side of the head
THEN The animal is a herbivore

Given the above rules, a rule model for classifying the animal according to diet might be

EXAMPLES:	1, 2, 3
ANTECEDENT ATTRIBUTES	Diet, teeth shape, eye location
CONSEQUENT ATTRIBUTES	Carnivore, herbivore

Rule models are constructed automatically by TEIRESIAS; they do not have to be entered by the user. They are used in debugging sessions in connection with a natural language interface to aid in locating missing or incorrect rules.

Suppose, for example, that an expert is working with the animal identification knowledge base using a program like TEIRESIAS. The expert wants to test the system's ability to identify raptors; the knowledge base contains rules for identifying eagles, hawks, and owls. Suppose further that in the course of the test consultation it has already been established that the animal is a nocturnal bird with eyes at the front of the head, but the expert system quits without making an identification when the expert says that the diet of the animal is not known. The expert system "should have known" that the animal is a carnivore based on its eye location. Suppose the bug in the rule base is that it is missing a rule of the form, in English, "If the animal has eyes at the front of its head, then it is a carnivore." The expert knows that, given the three raptors in the knowledge base, the expert system should have been able to conclude without knowing the diet, based on eye location and nocturnal habit, that the animal is an owl. However, the expert does not yet know about the missing rule and therefore enters a command indicating that the answer is incorrect.

The program asks what the answer should have been and the expert answers "owl." The program then presents an English translation of all of the rules having "owl" in the conclusion. It also provides a list of all of the attributes whose values were determined and what those values are. The expert sees that one of the rules for concluding that the animal is an owl is "If the animal is nocturnal and a carnivore, then it is an owl," and further, that the system knows that the animal is nocturnal but not that it is a carnivore. This leads the expert to suspect that a rule is missing. The expert asks to see all of the rules with carnivore in the conclusion and immediately sees that a rule is missing. The expert indicates this to the computer, and the computer asks what the rule should say. The expert provides the answer "If the animal has eyes at the front of its head, then it is a carnivore." The program translates this into the rule base language and the expert reruns the program. This time the program correctly identifies owl as the answer.

The programs SEEK (Politakis and Weiss, 1984) and SEEK2 (Ginsberg et al., 1988) take a different approach to knowledge base refinement. In comparing their system to TEIRESIAS, Politakis and Weiss note that the primary function of the latter is to permit the expert to locate the need for new rules and to place these rules in the knowledge base. Politakis and Weiss (1984, p. 24) note that TEIRESIAS "does not, however, directly determine the impact of changes to the knowledge base on other cases previously processed by the consultation program." The SEEK system focuses on rules already in place and modifies these rules to bring the performance of the system in line with that of the experts. SEEK demonstrates the effect of such modifications and SEEK2 actually makes the modifications themselves. The programs work by comparing the performance of the knowledge base with a database of *cases,* which are files of medical cases containing the disease symptoms and the experts' conclusions about the diagnosis. As usual, we will translate the domain of the original paper, internal medicine, to the animal identification problem.

In this context each case would be a record consisting of a list of the attributes and their values and of the identity of the animal, or the possible identities and their degrees of certainty. For each animal, the SEEK system would compile statistics from the case file to determine the attributes and values characteristic of that animal. It would then compare this list with the rule base. Any discrepancy would be an indication that the rule base did not correspond to the experts' opinion and would signal the need for modifications.

The algorithm used in SEEK and SEEK2 to generate appropriate modifications to the rule base makes use of the observation that most modifications to an individual rule either make the rule more general, i.e., easier for the rule's consequents to be accepted, or make it more specialized, i.e., harder for the rule's consequents to be accepted. A rule is made more general by decreasing the number of its antecedents, decreasing the certainty with which they are required to be known, or increasing the certainty of its consequents. A rule is made more specialized by increasing the number of its antecedents, increasing the certainty with which they are required to be known, or decreasing the certainty of its consequents. SEEK2 basically tries to generalize rules that, according to comparison with the case studies, should have been triggered but were not; it tries to specialize rules that were instantiated and should not have been.

4.5 Discussion

One of the underlying themes of this book is that expert systems can help solve the knowledge bottleneck problem by shifting it and focusing it. There are many fields in which there is a demand for expertise and a shortage of experts. Expert systems technology attempts to solve this problem by providing computer programs to take the place of the unavailable experts. However, the problem is merely shifted, because there is a shortage of experts in the construction of expert systems. The gain is that this is only one field, and so people can be trained for it more readily. Moreover, since it is a field of computer science, researchers in this area naturally try to create computer programs to solve or at least ameliorate the problem.

All of this computer software has created a brave, new world of knowledge engineering. Little of it has, however, reached the desk of the average expert system developer, particularly if that developer is working in agriculture. This leaves the developer facing an arduous task with little in the way of shared practical knowledge or experience. However, among those who have some experience there are several points of general agreement. Many of the most important of these can be summarized by comparing the programming of a knowledge base with the programming of an ordinary computer program. To a novice knowledge engineer who is an experienced programmer the first and striking difference is the manner of thinking required to program in the declarative style of an expert system rule base. Programming in Fortran, or any other procedural programming language, enforces a linear, step-by-step approach to controlling the flow of logic in a program. Each step of the program is precisely sequenced, and each step either physically follows the next or contains a well defined jumping off point such as a GO TO statement or a subroutine call. Programming a rule base, with its declarative style, is an entirely different matter, and many old habits must be unlearned. The natural tendency is to approach the construction of a rule base as if it were a Fortran program and to encode a sequence of logical steps by using the rules as if they were IF-THEN statements in Fortran or Pascal.

Many rule bases can, to some extent, be encoded in this way, but the resulting rule base is almost always poorly written and awkward. One must adopt a more unstructured approach to write meaningful rule bases; this ability only comes with experience, or inexperience at traditional programming!

The dissimilarity between programming a rule base and traditional programming is a source of problems, and the greatest similarity between rule base programming and traditional programming is, unfortunately, also a problem. This is the amount of time that must be spent in verification, or, more prosaically, debugging. An early dream of expert system developers was that the unstructured nature of the rule base would permit rules to be encoded more or less in free form, as if the experts could enter the rules in a stream of consciousness. The rules need not form a logical chain as entered into the rule base; the inference engine could take care of the formation of logical chains.

While this idea seems reasonable in theory, in practice it produces rule bases that often cannot reach any conclusion or that have logical bugs of the type discussed earlier in this chapter in the section on verification. If it is to be released to the general public, and particularly if it is to be directed at the agricultural sector of the general public, it must not give egregious or annoying results. This restriction imposes a severe limitation on the free form nature of the rule base; indeed, experience indicates that, at least at present, practical rule bases intended for public use must be debugged and tested as thoroughly as ordinary computer programs.

What, then, is the advantage of using a knowledge-based approach over a traditional programming approach? After all, there are several clear disadvantages. Duda (personal communication) expresses these very succinctly: expert systems are large, slow, and strange. That is, the overhead associated with the inference engine and the separate knowledge base causes an expert system to take up more computer memory and run more slowly than a traditional program that accomplishes the same thing, and the programming techniques associated with rule-based programming are unfamiliar to most programmers. However, the expert system appears to have at least two clear advantages. One is the automatic availability of an explanation system. The presence of such a system can provide, if it is structured correctly and presents explanations that the user can understand, an "open" computer program that can more easily gain credibility with the user. Another is that the rule-based language is more restricted than a typical programming language and is often easier for persons unfamiliar with programming to understand and learn. On balance, the use of expert systems seems justified simply for these programmatic reasons even if there are no other advantages over traditional programs.

Chapter

5

Demons in Expert Systems: A Simulation Example

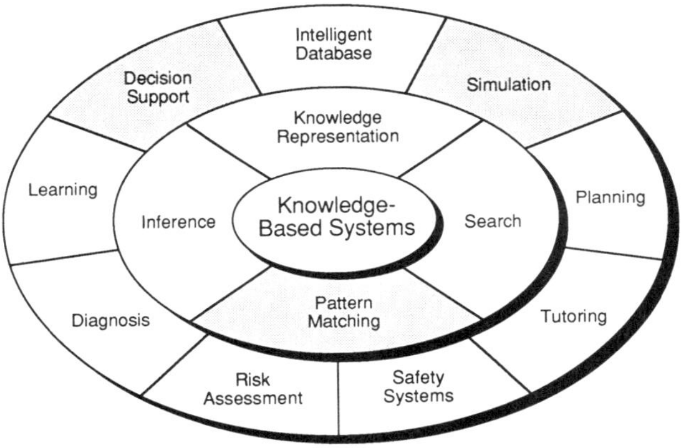

Expert systems can interface with simulation models; the knowledge-based approach often integrates a simulation model with an expert system, using the model as a decision support tool. Modeling a decision network is the first step in developing a knowledge-based system with multiple information sources. In addition, this chapter will discuss the elementary use of other demons. Some of the limitations, and possible solutions, of this type of linkage are discussed. The problem is explored using a predictive model of boll weevil emergence on the rolling plains of Texas to help determine coordinated planting dates for cotton in the area.

This chapter deals with the combination of qualitative or heuristic knowledge with quantitative knowledge. In particular, we step through in some detail the process of linking a knowledge-based system with a simulation model. Those of us who build simulation models in agriculture constantly have to deal with some commonly held misconceptions about the purpose, potential, and payoff of mathematical and computer models. Most people view models as complex black boxes that are used to predict the future. Many would probably add that the models do not do a very good job. As we saw in Chap. 1, this is a confused and simplistic impression of models; however, it is the prevailing impression, and the prevailing response from modelers is just as confused. The standard modeler's retort is that models are useful in and of themselves as research tools because they identify gaps in our knowledge and understanding. In other words, it is all right to build a model that has no potential practical application because it can help us plan research and synthesize our ideas. This defense is unconvincing to most people, and the result is that while modelers often build models without any field application in mind, funding agencies, scientific peers, and administrators have always perceived and promoted modeling as a way to solve complex problems. When models show no utility in solving field problems, their funding base often vanishes and their supporters quietly disperse.

In the late 1970s and early 1980s, with the emergence of artificial intelligence and particularly expert systems, modelers began to see a way to save their funding: Provide simple and user friendly artificial intelligence interfaces to complex and user unfriendly simulation models. An expert system interface, for example, could be constructed to determine when to run a simulation model, how to initialize it, and how to interpret its results. In such a scenario, even a model requiring considerable input could be repackaged in such a way that the expert system constructs its input files, replacing missing or unavailable information with expert guesses or appropriate default values. It has even been suggested that the confidence values assigned to the output of simulation models could be adjusted by an expert system interface based on the reliability and availability of model input data.

The truth is that if models are not designed with a practical application in mind, it is extremely difficult if not impossible to generate information from that model that will be useful in solving problems. Nevertheless, it is also true that many models developed with a field application in mind have never been used simply because the models are too complicated or too difficult to interpret. In virtually all cases, the model's output provides only pieces of information that must be interpreted in conjunction with other information to come to a proper conclusion.* For such models, integration with an expert system interface is a legitimate and productive marriage.

In relation to knowledge-based systems, simulation models are termed *demons,* external functions or programs called by the inference engine to perform a specific function. A demon is thus best un-

---

*Considerable research has focused on making optimal decisions based on mathematical and/or computer models of agricultural systems. The field of operations research (OR) is devoted to this endeavor. While OR applications have been very successful in other disciplines, in agriculture OR techniques have been very disappointing (cf., Chap. 10).

derstood as an "invisible helper" rather than as something diabolical. Inference engines must be specifically coded to call an external function. Many commercially available inference engines or expert system shells are provided with functions that allow the knowledge engineer to call C-language programs or to access specific software packages like spreadsheets or database management systems. In this chapter, we make the convenient assumption that the inference engine is coded in LISP, allowing the knowledge engineer to write demons in the form of LISP functions.

5.1 Introductory Problem: Boll Weevil Management

We will discuss one aspect of boll weevil management in cotton and see how the integration of a knowledge-based system with a demon, i.e., a simulation model, is more powerful than either system alone. Simpler demons that make the knowledge-based system work with the output from the simulation model are used. Our example is the boll weevil control program in the high plains and rolling plains of Texas, where boll weevil is controlled mainly by cultural practices. In this area, a strategy called delayed uniform planting is used. Cotton is planted late in the spring and as contemporaneously as possible. The idea is to prevent the boll weevils that are emerging from their winter hibernation, diapause, from finding any cotton suitable for their development. Unless the emerging adult weevils find cotton plants with flower buds, cotton *squares,* on which to feed, they will die without successfully reproducing. The delayed uniform planting program in that area saves farmers from having to control the boll weevil with insecticides throughout the growing period. When insecticide sprays during the early season can be eliminated, the community of natural predators and parasites of other cotton pests can be maintained, controlling such later season pests as the tobacco budworm and cotton bollworm (Sterling et al., 1989).

There are two trade-offs from the delayed planting strategy. One is reduced yield potential. By planting late, farmers in the area decrease the length of the growing season, thus reducing their yields. The second drawback is that planting late sometimes means growing cotton into the late summer and fall. This not only exposes the farmers to the possibility of bad weather, which can delay harvesting and reduce lint quality, it also provides a late-season food source for boll weevils that have developed over the summer and that need to feed on cotton fruiting structures before they go into diapause for the winter. If cotton could be harvested early and the fields cleared of squares and cotton bolls, the lint-producing fruit, the fall populations of boll weevils would not be able to store up the nutritional reserves needed for them to survive the winter. This danger of allowing weevils to go into winter diapause in a healthy state is usually outweighed by the benefit of starving the emerging spring populations. No matter how many weevils come out of diapause in the spring, if there is no food for them, they will die.

However, there is another risk due to the unpredictability of boll weevil spring emergence. In some years, boll weevils emerge late or over an extended period. Despite the farmers' delayed planting, significant numbers of these insect pests find cotton that has begun squaring, lay eggs in the developing squares, and build up considerable populations later in the season, thus requiring insecticide treatments. Therefore, in some seasons it would benefit farmers to plant earlier to maximize their yield potential and to control the weevil with chemicals, since they would have to use insecticides anyway, even with a late planting.

Some of the factors that influence the timing of boll weevil emergence are known. However, until recently, there was no way to predict within even two or three weeks when boll weevil emergence would end. In some years, boll weevils emerge in significant numbers only until early June. In other years, emergence can continue into August. In 1989, the first emergence model for boll weevil that seemed able to predict both the timing and magnitude of the boll weevil emergence was developed (Stone et al., 1990). Here we explore the use of that simulation model in combination with a rule-based expert system to advise farmers when to plant their cotton.

5.2 Setting Up the Problem: Classical versus Knowledge-Based Approach

A *classical approach* to the problem of selecting a planting date might involve the following scenario. Develop a simulation model of the cotton plant and link it to a simulation of the boll weevil, including a spring emergence component. Next, add to the model the capability to spray insecticides based on a threshold of weevils per plant or per acre. Since one of the major costs of spraying insecticide for boll weevil is the loss of the natural enemies of other cotton insect pests, this model should also include beneficial insects and other key pests so that the full consequences of insecticide sprays can be accounted for. Finally, run the model under a wide range of weather scenarios, boll weevil emergence rates, and planting dates to generate a surface of revenue less cost of control as a function of initial boll weevil infestations and planting date (Fig. 5.1).

Optimization techniques for problems such as this are called *static*

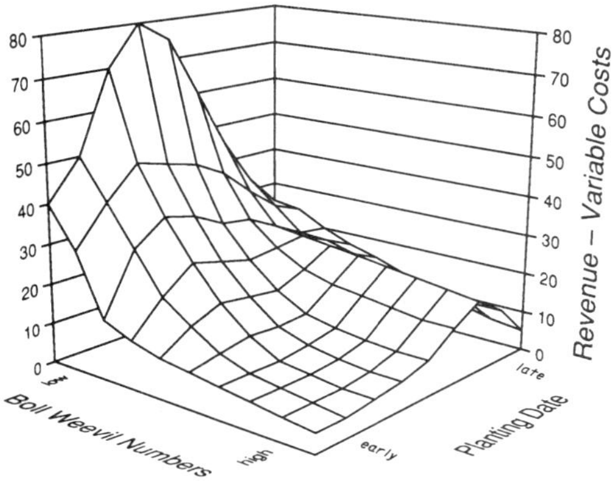

Figure 5.1 Hypothetical simulation results showing profit (in arbitrary units) as a surface function of cotton planting date and boll weevil numbers emerging in the spring.

because they do not involve dynamics, i.e., differential equations. The goal is to find the "highest" point on a surface such as that shown in Fig. 5.1 or on a hypersurface for a problem in higher dimension. Static optimization techniques start with an initial, suboptimal guess and through a series of iterations work their way toward the highest point. The problem of doing this on a computer is like the problem faced by a mountain climber who must climb to the top of a mountain in a fog. As a mountain climber in a fog can only distinguish his or her current elevation and the directions of the slope in the immediate vicinity, so the computer only evaluates the function and its derivatives near the current guess in the iteration. There is a wide variety of static optimization methods available (e.g., Avriel, 1976). The simplest and most intuitive is called the *method of steepest ascent.* In this method the computer evaluates the *gradient* of the function, which gives the direction of most rapid change, at the current iteration and moves a fixed distance in that direction. In our weevil example, static optimization could be used to determine the optimal planting window based on the expected weather pattern and weevil emergence rates.

The problems with applying classical optimization in this case are familiar to modelers of agricultural problems. The system is complicated by limited quantitative understanding, a large number of interdependent processes, and great uncertainty. Just building the models needed to perform the analysis would take considerable time. Even modifying an existing model to fit local conditions would take at least two full seasons. Also, the optimization analysis is dependent on full-season predictions of the growth of the cotton plant, its interaction with the weevil, and the timing and effect of insecticide sprays on the weevil and on other insects. Simulations of agricultural systems have not performed well over such extended periods of time. It is unlikely that the quantitative results of such model simulation runs could be used with confidence.

A more appropriate objective would be to look for trends in the simulation output that could be interpreted qualitatively. For example, the analysis may suggest that either early or late plantings are optimal, but plantings in between are not optimal for some range of boll weevil numbers (e.g., Fig. 5.1). Such a result might be very valuable; however, the reliability of the numbers in specific cases is likely to be quite low.

A *knowledge-based approach* to the problem would focus immediately on the decision that has to be made by the farmer: When do I plant cotton? This decision, we have already seen, is based on many pieces of information: the phenology of the cotton plant, the expected weather pattern for the area, the local weather near the time of planting, the yield potential of cotton planted at different times, the probable damage to cotton by boll weevil, the anticipated boll weevil population pressure and the timing of their emergence, etc. Some of this information is available from Extension Service publications. Some is predicted by a simulation model. Some has to be obtained from expert knowledge and experience.

The design of our system takes advantage of some of the structured techniques described in the last chapter. In particular, we use an information-decision flow model or decision network to show the key decisions and subdecisions involved in solving this problem, and the key pieces of information needed to make those decisions.

5.3 A Decision Network for the Planting Date Problem

In developing the decision network, we simplify the problem somewhat to create a more instructive, although somewhat inaccurate, hypothetical example. The decision to be made is when to plant cotton. This decision needs to be made in the early spring, before the earliest possible planting date. Ostensibly, cotton can be planted on any day of the year. However, elementary constraints limit reasonable planting dates to a few months in the spring. Farmers need to plant cotton into warm, moist soil, so local weather can significantly affect planting decisions. However, local weather in May cannot be predictable before April, so we must ignore those constraints and try to pick the best planting date possible assuming good planting conditions. Extension Service recommendations for the rolling plains area of Texas recommend a uniform late planting after the middle of May. We assume the optimal planting date in the absence of boll weevil would be late April or early May in most years, i.e., planting dates as early as May 1 increase the potential yield, but planting earlier than that results in diminishing potential yield. We assume, then, that the possible choices for planting date begin May 1 and that intervals of five days separate the remaining choices. Finer subdivisions might be more precise, but planting dates are really meant as targets. It normally takes several days to plant, and intervals of fewer than five days are, therefore, not necessary.

The last possible planting time we consider here is mid-June. Cotton must normally grow well into the fall in this area because it is a relatively cool climate. Delaying planting any later than early June would make it very difficult for the cotton to mature before cold fall temperatures stop growth. Furthermore, a planting date of June 1 would result in the first flower buds susceptible to the weevil occurring somewhere in mid-July, safely out of danger from most of the overwintering boll weevils in most years.

At this point, our decision network looks like Fig. 5.2. The goal of the system is a selection of a uniform planting date from a series of dates determined by the rules discussed above. Next, we need to determine what information is needed to select the best planting date. We are not yet trying to identify what knowledge is needed, rather what information a person who has expertise in making this decision would need. Knowledge about how to interpret and manipulate information to make it useful will be added later. At this stage, we are just concerned with information.

We can struggle with this question or we can simply consult an expert. This initial consultation or knowledge engineering session focuses in a very general way on the problem. Suppose we talk with an expert and elicit the following verbal description of the decision process.

> In most years, peak boll weevil emergence occurs about late May or early June, and the emerging adult weevils will live for about two to three

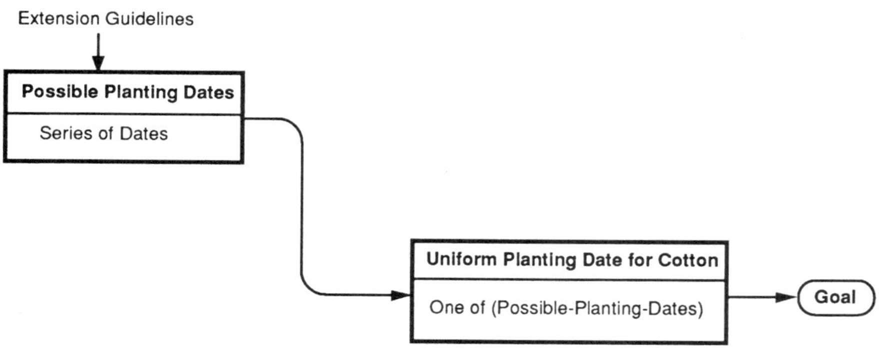

Figure 5.2 The final decision of the cotton planting date problem, the result of which is the goal of our knowledge-based system.

weeks. The first squares susceptible to weevil attack occur about 50 days or more after planting, so by planting in late May, susceptible squares appear in early July. Therefore, most of the emerging boll weevils will not be able to find a host for oviposition and will die without contributing to the next generation. This is called, "suicidal emergence," and it's what the delayed planting program is all about. Ideally, we want about 90 to 95 percent of the weevils to emerge and die before the appearance of one-third-grown square.

The trouble is, in some years weevils do not emerge until later, and our delayed planting does not work. Those years in which there are still significant numbers of weevils emerging from overwintering habitat in late June or early July are the ones that hurt us. In those years, it might be better to plant earlier to take advantage of the longer growing season, since we will have to use insecticide sprays anyway to control the weevil. The main thing is the number of weevils coming into cotton at that key point in the fruiting cycle. In years of heavy pressure, even with the delayed planting we may spray once or twice for the weevil.

If we had a model that could predict the timing of boll weevil emergence for the upcoming year in time, we could recommend an early planting for those years with delayed emergence or very heavy boll weevil pressure. And, if we could determine when 90 percent of the weevils will emerge, we could time our normal plantings more accurately to take advantage of years when the weevils were done emerging early.

Based on the information from our expert, we see that the key variable affecting the success or failure of the delayed uniform planting date strategy is the number of weevils emerging from overwintering habitat at the time that the cotton starts producing squares susceptible to weevil attack. In most years, the natural emergence pattern of the weevils and the delayed planting date ensures that few adult weevils find suitable cotton fields. However, in years of either high emerging weevil numbers or a delayed emergence pattern, the strategy fails to protect the cotton, and insecticides must be applied.

Given knowledge of the longevity of boll weevil adults and the developmental time required for cotton to reach the susceptible stage, our system should be able to make a qualitative judgment about how severe the boll weevil attack will be on cotton at the early squaring stage. We need to know the cotton planting date, the expected pattern of boll weevil emergence, and the expected magnitude of the emerging weevil population. Coincidence of high numbers of emerging weevils and the squaring period of cotton should be avoided if possible, but if it cannot, then the planting decision should be set to maximize potential yield.

The boll weevil emergence simulation model provides much of the information we need. Assuming that the weather data needed to run the model are available, the model predicts the spring emergence pattern, i.e., the percent of weevils that have emerged versus the day of year. It also gives an estimate of the survivorship of overwintering weevils. However, it does not give an estimate of weevil numbers emerging, which is calculated based on sampling data collected in the fall and winter. Those samples are used to categorize the overwintering population of weevils as from very low to very high. Figure 5.3 shows the function model of the problem thus far. Of the information sources, only *Historical Data* has not been mentioned. Some historical information is needed to classify the fall sampling data for a particular year as low, moderate, or high. The same is true for the estimated number of weevils at large at the time of squaring.

Notice in Fig. 5.3 that there are nine decisions or deductions to be made. The simulation model provides information needed in making two of those deductions, estimates of the weevils' survivorship and emergence pattern. The selection of a planting date is made by examining a list of possible planting dates and the expectation of the severity of boll weevil attack on the early fruiting stage for each of the planting dates. The expected severity of attack is based, in turn, on an estimate of how many weevils are seeking cotton (weevils emerged and still alive) in comparison with the timing of the fruiting patterns of cotton crops planted at different dates. This determination of the number of weevils seeking squaring cotton is based mainly on output from the simulation model.

At this point, the overall logical structure of the problem is complete. Now the task becomes translating the boxes into rule bases and providing the required information in suitable structures for the knowledge-based system to use. We begin near the top of the problem, determining the emergence pattern of boll weevils in the spring using a simulation model.

5.4 Piecing Together the System

5.4.1 Boll weevil spring emergence pattern

A simulation model. The model we use was developed by Stone and collaborators from Texas A&M (Stone et al., 1990). It was based on an extensive seven-year study, in which large cohorts of weevils were placed into cages in overwintering habitat during the fall diapause period from September through early November and then monitored during the subsequent year for emergence. These data were unique in that the age of every weevil found emerging in the spring was known, as was the residence time spent in the soil. Previous studies had attempted to use degree-day methods to model the emergence of boll weevil based on trap catch data from the spring, but these models

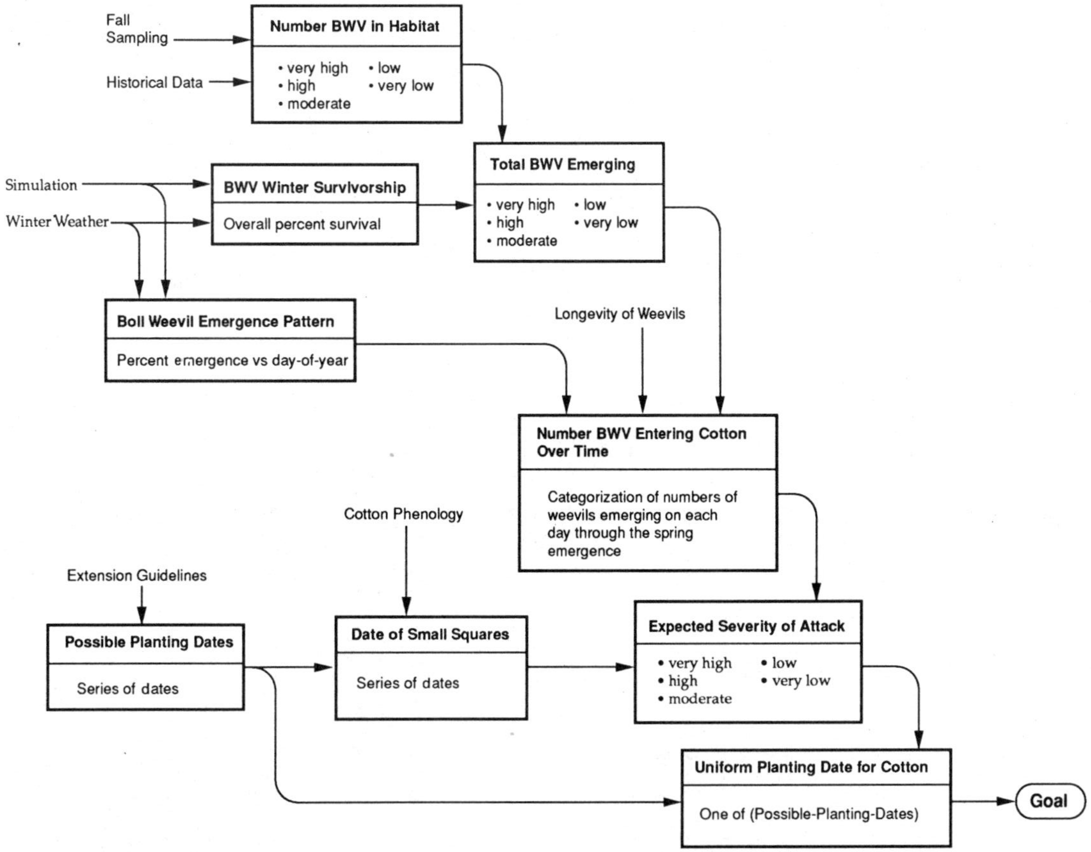

Figure 5.3 A complete decision model of the planting date problem with the links to the simulation model shown in the upper left.

have not been successful. The approach used by Stone et al. assumed a linkage between the severe winter weather that caused weevil mortality and the subsequent timing of the emergence profile.

The model requires as input the maximum and minimum temperature and rainfall data for 100 days (centered on January 1). From these data, the model calculates two indices of winter severity, which are then used to estimate the timing of the 50 and 80 percent points on the cumulative emergence curve and the expected survivorship. The first index uses the concept of negative degree-days to describe how cold it became over an interval of several days. As shown in Fig. 5.4, negative degree-days can be thought of as the area between a threshold temperature and the trace of the ambient temperature over a 4-day period, just the opposite of the classical concept of degree-days (see the appendix and Chap. 6). Negative degree-days therefore measure both how cold it became and how long it remained cold. In the model, the negative degree-days in the longest continuous run of days with temperatures below the threshold was one severity index.

The second severity index deals with the interaction between rainfall and temperature. Boll weevils spend the winter in the leaf litter of wooded habitats. They can be found in depressions near rivers, in shelterbelts, even in the undisturbed leaf litter under trees and shrubs near old abandoned farms and homesteads. The best sites have a deep leaf litter above a rich humus and loose soil. Weevils need the litter for insulation from the cold and the wind, but they also need to avoid desiccation. It seems that very dry weather over extended periods leads to higher mortality, especially when combined with cold temperatures. In the model, the second index of severity is calculated from the product of positive degree-days, a measure of warmth, and the amount of moisture in the leaf litter. The moisture content is estimated based on rainfall, with water content decreasing over time. The product of moisture and positive degree-days is highest in warm, wet

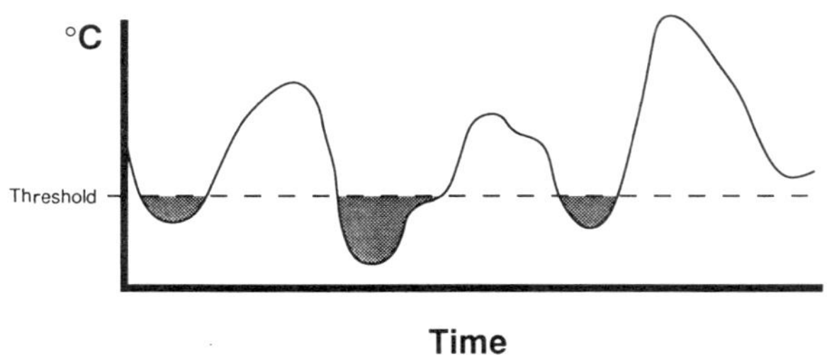

Figure 5.4 The concept of negative degree-days is illustrated as the shaded area between an upper temperature threshold and the trace of the ambient temperature over a 4-day period.

weather and lowest in cold, dry weather. The average value of this product for the overwintering period is calculated and the inverse is taken, giving a number that is highest for cold and dry conditions and that is more sensitive to very low amounts of moisture than to high amounts.

Bivariate functions based on the two winter severity indices provide estimates of the 50 percent point on the cumulative emergence curve and the expected overwintering survivorship for the weevils (Fig. 5.5).

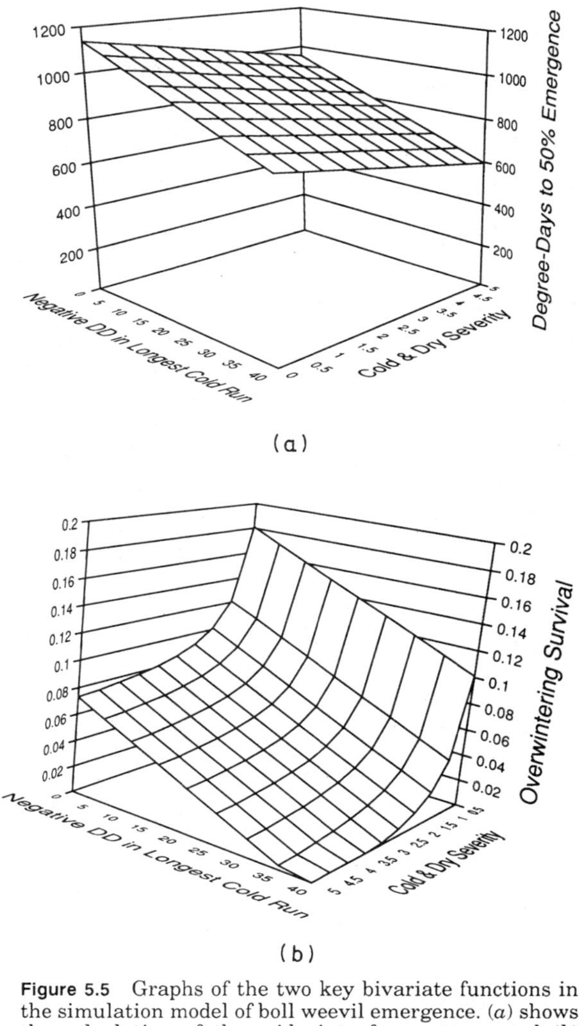

Figure 5.5 Graphs of the two key bivariate functions in the simulation model of boll weevil emergence. (*a*) shows the calculation of the midpoint of emergence and (*b*) shows the estimation of survivorship as a linear function of negative degree-days and the inverse of severity.

The 80 percent point of emergence is then calculated from the 50 percent point and the winter severity. Finally, a sigmoid emergence pattern is calculated by fitting portions of the sine curve through the 0, 50, and 80 percent points. The model uses a degree-day time scale in its calculations, but translates its predictions to calendar days based on actual or historical weather data during the emergence period. Figure 5.6 shows the model's output for the years 1985 through 1987. The output is saved in tabular form to an ASCII file with summary statistics like the expected survivorship recorded first, then a table of day-of-year and the associated predicted proportional emergence.

For our purposes in determining a planting date for cotton, this model is ideal. The data required to generate a prediction can be acquired by the end of February. The model can then be run and the emergence profile for the weevil can be translated back to calendar dates based on historical weather data. This leaves plenty of time between the end of February and the time of planting in May for field preparation and scheduling.

Linking to the knowledge-based system. The required weather data are the daily temperature maxima and minima, and rainfall. These data are read by the model from an ASCII-formatted file. We assume that the process of acquiring the weather data and putting it into ASCII format have already been accomplished. In addition, we assume that the model is written to run automatically from an input file named WINTER.WTH, which contains the weather data. All that is required from our expert system is that it check to see whether a weather file has been created and issue a command to start the model. Most expert system shells allow at least this level of integration between the expert system and other executable programs in the computer. Many shells provide a means to call a program directly as a function. This

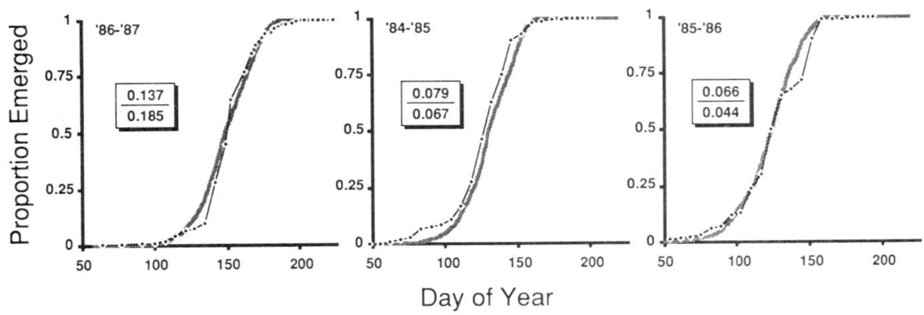

Figure 5.6 Output from the boll weevil emergence model of Stone et al. (1990) showing observed (dot-dashed line) and predicted (thick line) cumulative emergence and the predicted-to-observed overwintering survival for the years 1985–1987.

would be preferable, since the model could pass its relevant output back directly to the expert system, but it is not required.

Transferring the model's output back to our knowledge-based system is accomplished here through a temporary file. The model must be modified to open a file called PREDICT.BWV and to write the relevant information needed by the expert system to that file. The contents of the file should be statements understandable by the inference engine of the knowledge-based system. That is, they should be in the form of assertions or object-attribute-value triples, for example, EMERGENCE MIDPOINT 150, where 150 is the day-of-year, or BWV SURVIVORSHIP 6.8. The inference engine must, then, be capable of opening a file and reading in assertions to the fact base or blackboard.

5.4.2 Boll weevil survivorship and total emerging

Having come up with a curve describing the timing of emergence, we now turn to predicting the number of boll weevils that will emerge. Two factors are needed: the number of weevils that went into diapause the previous fall and the predicted or observed overwintering survivorship. In this system, the former datum must be obtained from the user of the system, the latter is provided by the mathematical model. However, the model is much less reliable when predicting emergence than when predicting timing. Therefore, we recommend using the prediction of overwintering survivorship only in a qualitative way.

On average, the overwintering survivorship of boll weevil in the rolling plains has averaged 6 to 7 percent over the last 10 years, with a low of nearly 0 percent and a high of about 30 percent. For our purposes, it is enough to compare the model's prediction for the current year with the average and range observed over the last decade. We classify the expected survivorship in terms of standard deviations from the mean of the data. However, since the data are percentages, the variance, and hence the standard deviation, is a function of the mean. So, instead of the observed percentages, we use an arcsin transformation which makes the variances independent of the mean (Sokal and Rohlf, 1969). Five categories will be defined: very low, low, moderate, high, and very high, and will correspond to survivorship of 0 to 3.4, 3.41 to 5.8, 5.81 to 12.5, 12.51 to 16.7, and 16.71 to 100 percent, respectively.

The relative number of weevils that entered diapause in the fall must be estimated from fall sampling in comparison to historical data. In addition, the answers to two further questions are needed. How many boll weevils entered overwintering habitat in the fall, and what was the availability and quality of cotton bolls and squares during the time in the fall when the weevils were feeding before going into diapause? The former is needed to estimate the numbers of weevils that will emerge. The latter is important because there is evidence that the heartiness of diapausing weevils and, hence, their ability to survive the winter depends on the quality of their diet prior to entering overwintering habitat. Several conditions, e.g., a dry summer or an early freeze in the fall, can lead to an inadequate food supply for these weevils hoping to fatten up for the winter. These questions are put to the user at run-time since the simulation model has no means with which to guess the answers.

The user is asked to categorize the number of boll weevils in overwintering habitat into one of three categories: low, medium, or high. If sampling data are available, this is fairly straightforward. In addition, the user is asked to describe the general availability and quality of cotton squares and small bolls in the fall. The cotton's quality as a host for the weevil is rated as low, medium, or high.

Finally, these three pieces of information about boll weevil populations in diapause are used to determine a single value for expected boll weevil numbers emerging in the spring, as shown in Fig. 5.3. The elicitation of rules can be accomplished in many ways. For example, we could ask our expert to fill in a table of examples, Table 5.1, from which rules could be derived automatically, or results could be read directly from such a table using a look-up function. Both these strategies, called *rule-induction* and *case-based reasoning,* respectively, have been used in developing expert systems and will be discussed in Chap. 8. Here we continue to focus on more classical knowledge engineering, deriving our rules by asking the expert questions and translating the responses into rules directly.

TABLE 5.1 Number of Emerging Boll Weevils as a Function of Predicted Overwinter Survivorship, Number of Weevils in Diapausing Habitat, and Quality of the Previous Fall's Cotton

Predicted survivorship	Weevils in habitat	Available food	Total weevils emerging
Very high	≥Medium	≥Medium	Very high
Very high	Low	—	High
Very high	≥Medium	Low	High
High	High	≥Medium	Very high
High	High	Low	High
High	Medium	≥Medium	High
High	Medium	Low	Medium
Very low	High	High	Medium
Very low	High	≤Medium	Low
Very low	Medium	High	Low
Very low	≤Medium	≤Medium	Very low

One way to do this is by straightforward use of assertions, as discussed in Chap. 2. An example is

Quotation from expert		Rules generated
Even moderate numbers of weevils going into diapause can cause tremendous problems in the spring if the winter is very mild.	IF	predicted-survivorship is very-high AND weevils-in-habitat is medium
	THEN	total-weevils-emerging is very-high

However, this enumeration of cases with rules can get very tedious and can produce a lengthy and unwieldy rule base. Instead, we try to use rules to describe more general relationships. An example is

Unless you find unusually high or low numbers in habitat, I would use the predicted survivorship.	IF	predicted-survivorship is ?X AND weevils-in-habitat is medium
	THEN	total-weevils-emerging is ?X

This rule uses a variable, ?X, which appears twice in the rule. This means that ?X can take on any value, but it must be the same within both the IF- and THEN-side of the rule. When the inference engine processes this rule, the variable is *bound* to the previously determined value of PREDICTED-SURVIVORSHIP, in forward-chaining systems, or TOTAL-WEEVILS-EMERGING, in backward-chaining systems. The acceptable values for both objects are the same set of values, so the rule indicates that the expected number of boll weevils emerging has the same qualitative value as the expected survivorship when average numbers of weevils are in diapause.

In the next case, two variables are used in combination with a demon to represent a more complex relationship that reduces the value of a variable as follows:

If the cotton was stressed late, not much food will be available for the weevils in the fall, and many of the overwintering weevils will not survive. In such cases, I expect considerably lower survival.	IF	available-food is low AND diapausing-weevils is ?X AND (next-lower (list 'low 'medium 'high) ?X) is ?Y
	THEN	weevils-in-habitat is ?Y
	IF	NOT (available-food is low) AND diapausing-weevils is ?X
	THEN	weevils-in-habitat is ?X

There are two problems here. First, in some cases the value of a variable that has already been determined needs to be modified. Second, the value of a variable that is not numeric needs to be reduced. The inference engine has no way of knowing that low is less than medium, so we need to write a demon that orders a list of possible values for a variable.

The first problem is overcome by introducing the intermediate variable, DIAPAUSING-WEEVILS. This variable and WEEVILS-IN-HABITAT both describe the number of weevils trying to overwinter. We use DIAPAUSING-WEEVILS to hold the user's estimate of the number of weevils entering habitat. Both rules above then set WEEVILS-IN-HABITAT to the same value if the available food in the fall is either medium or high, or to the next lower value if the amount of available food is low.

We can see why the intermediate variable was needed by looking at how the rule would look without it:

```
IF      cotton-quality is low AND
        weevils-in-habitat is ?X AND
        (next-lower (value-list 'total-weevils-emerging) ?X) is ?Y
THEN    weevils-in-habitat is ?Y
```

In other words, if we have previously determined that the level of weevils in diapausing habitat was high, we now reduce that to the next lower acceptable value. Unfortunately, this rule is never evaluated by most inference engines. Once the value of WEEVILS-IN-HABITAT is known, the inference engine does not search for any more values.* If the inference engine needs the value of the variable, it finds it in its memory and does not check for any more rules. By making the value of WEEVILS-IN-HABITAT dependent on an intermediate variable, we ensure that the value can be changed if indicated.

The solution to the second problem requires a demon, the function `next-lower`. We want the effect of cotton stress to be a qualitative reduction in the number of weevils in diapausing habitat. Our demon takes two arguments, a list of the possible values for a variable in ascending order and a specific value from the list. The demon returns

---

*Many backward-chaining inference engines allow variables to be declared *multivalued* or *plural*. For all such variables, the inference engine continues to search even after a value has been determined. In many problem domains several solutions to a problem may be acceptable. In such cases, the expert system should not stop trying to come up with solutions after finding just one. Here, making the variable WEEVILS-IN-HABITAT multivalued would allow the inference engine to evaluate this rule, but the result would be two values for WEEVILS-IN-HABITAT, e.g., high and medium *or* medium and low. We will take advantage of this feature in upcoming sections.

Also, some forward-chaining inference engines allow facts in the global memory to be retracted. Here we would take advantage of this feature by adding the consequent RETRACT WEEVILS-IN-HABITAT IS ?X to the right-hand side of the rule.

Finally, there is an entire class of knowledge-based systems called truth maintenance systems in which the inference engine keeps track of the justification for each assertion in memory. In such systems, the truth value of any assertion is subject to change during the inference process.

the next lower value from the list. If the inference engine is written in LISP, for instance, the next-lower function might look like this:

```
(defun next-lower (value-list selection)
     (let ((lowest (first value-list))
          (second-lowest (second value-list)))
       (cond ((eq selection lowest) lowest)
          ((eq selection second-lowest) lowest)
          (t (next-lower (rest value-list) selection)))))
```

Note that this is a recursive program as described in Chap. 2; it calls itself in each iteration with a new list shortened by the removal of the first element (Fig. 5.7).

In the IF-side of the rule above that uses `next-lower`, ?X is bound to the value, low, medium, or high, given to the object DIAPAUSING-WEEVILS. Then WEEVILS-IN-HABITAT is given the value that is one lower than ?X. If the inference engine allows the acceptable values for a particular object to be specified in a list, then the rule above could be made even more general. Instead of the expression `(next-lower (list 'low 'medium 'high) ?X)` we could use `(next-lower (value-list 'weevils-in-habitat) ?X)`. So, assuming the acceptable values for the object WEEVILS-IN-HABITAT is an ordered list from lowest to highest, (low medium high), the function `next-lower` with `selection` set to high would return medium, to medium would return low, and to low would also return low.

We could now continue in this fashion to develop rules so that we end up with a qualitative estimate, from very low to very high, of the number of boll weevils that will be emerging over the entire spring emergence period.

5.4.3 Number of boll weevils entering cotton over time

Our expert has told us that boll weevil adults live about two to three weeks in the spring without food. For the sake of simplicity, we will ignore the fact that boll weevil aging is temperature-dependent and will stick with a calendar day estimate of 20 days for the lifespan of an emerged adult that cannot find suitable cotton on which to feed. This fact is inserted into our knowledge base directly as an object-value-attribute triple (unfed-weevil lifespan 20) or in the near-English syntax assertion, UNFED-WEEVIL-LIFESPAN = 20. The task now is to determine a characterization for the number of weevils available to enter cotton for any particular day throughout the early season.

The simulation model must be able to produce an output file with pairs of numbers showing the day of year and the percent or proportion of the cumulative emergence achieved by that day. What is really needed is an estimate of the number of weevils emerging each day.

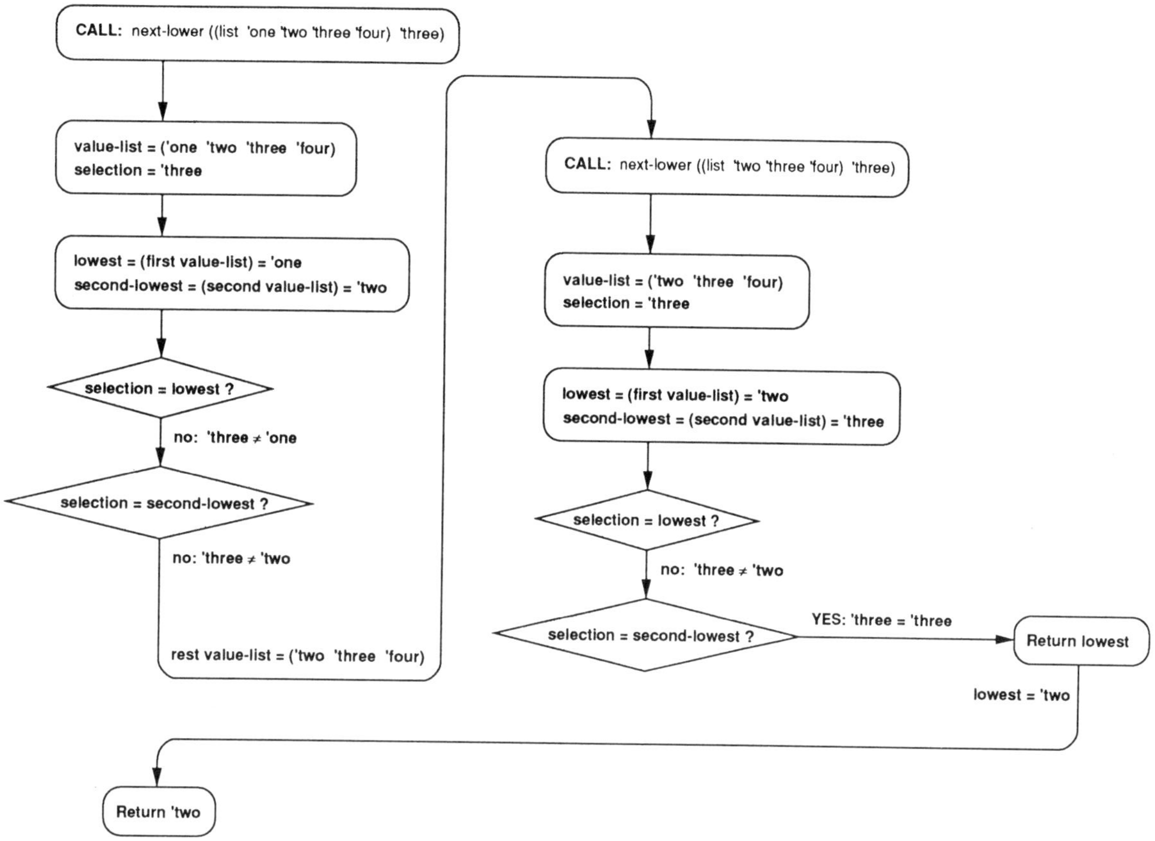

Figure 5.7 A flow chart of the `next-lower` function called with a value-list of `('one 'two 'three 'four)` and a selection of `'three`. The function should return `'two`, the item in the value list one step lower than the selection. The value list must be sorted in ascending order.

These numbers must then be summed over a running 20-day period and the total returned to the expert system. To accomplish this transformation, we need to modify our simulation model to produce output in a readable form, in this case to be used by another demon. For a LISP-based system, the simulation would create an ASCII file consisting of statements like

```
(setf number-of-days 300)
```

Create a variable called number-of-days and set it equal to 300.

```
(setf proportion-emerged
  (make-array 300 :initial-element 0))
(setf (aref proportion-emerged 0) 0.00)
(setf (aref proportion-emerged 1) 0.00)
(setf (aref proportion-emerged 2) 0.00)
(setf (aref proportion-emerged 3) 0.00)
   ⋮
(setf (aref proportion-emerged 165) 0.57)
(setf (aref proportion-emerged 166) 0.59)
(setf (aref proportion-emerged 167) 0.63)
   ⋮
(setf (aref proportion-emerged 297) 1.00)
(setf (aref proportion-emerged 298) 1.00)
(setf (aref proportion-emerged 299) 1.00)
```

Then, set up an array of 300 elements named, proportion-emerged. Each array element is accessed by the 'aref' statement, for 'array reference.'

The last two numbers in each line give the array subscript and the value.

So, in more familiar syntax, proportion-emerged(165) = 0.57.

When these statements are read by a LISP interpreter using the `load` command, the interpreter will create an array of 300 elements called proportion-emerged; each element represents the cumulative proportional emergence on the day of year referenced by the array index. Now, if the simulation output file is called OUTPUT.LISP, a new demon, `calculate-weevils-emerged` can be written. This demon reads the output file from the model, calculates the daily rate of emergence and the running sum of live weevils, and finally inserts the daily running sum into an array called PROPORTION-WEEVILS-ALIVE(*day-of-year*) in the knowledge base. We will assume that the function `assert` asserts its argument into the knowledge base. The new demon will take one argument, the unfed adult weevil lifespan. A flowchart describing this demon is shown in Fig. 5.8, and the implementation in LISP is as follows:

```
(defun calculate-weevils-emerged (age)
  (load "output.lisp")
  (setf proportion-alive
       (make-array number-of-days :initial-element 0))
  (dotimes (i number-of-days nil)
    (setf (aref proportion-alive i)
     (- (aref proportion-emerged i)
       (cond ((> i age) (aref proportion-emerged (- i age)))
             (t 0))
  (assert "proportion-weevils-alive(~a) = ~a"
    i (aref proportion-alive i)))
```

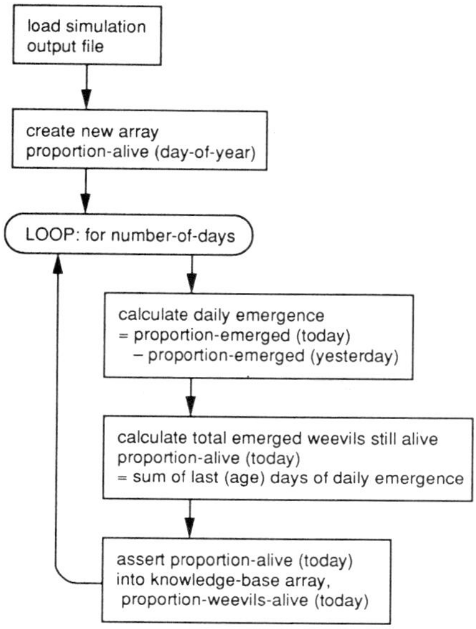

Figure 5.8 Flowchart of the `calculate-weevils-emerged` demon. A LISP implementation is given in the text.

At this point, the relative proportion of total weevils that are alive and ready to invade cotton on any given day of the early production season is known. However, this knowledge about relative numbers still needs to be combined with the previously determined estimate of the total number of boll weevil adults that successfully overwinter. Here both quantitative and qualitative information must be combined to produce what we want. The final result is a qualitative array WEEVILS-ENTERING-COTTON(*day-of-year*) with values from very low to very high.

As we did before, we convert numerical values to classes based on our average expectation. Figure 5.9 shows the average timing of boll weevil emergence during the 1980s and the resulting proportion of weevils alive and seeking cotton over the spring emergence period. The severity of boll weevil attack is estimated from the point at which the solid curve intersects the shaded region, indicating the relative number of overwintering weevils still alive at the time cotton becomes susceptible. Notice that if cotton were uniformly planted at the last feasible planting date (June 15 or day 166) in an average year, virtually all the overwintering weevils would die of starvation.

We make a few assumptions. First, in an average year, planting on May 26 results in low numbers of weevils entering cotton. From Fig. 5.9, this corresponds to a proportion alive of about 0.05 at the time of squaring. Also in an average year, we assume that planting early (on May 1) results in high numbers of weevils entering cotton. Again from

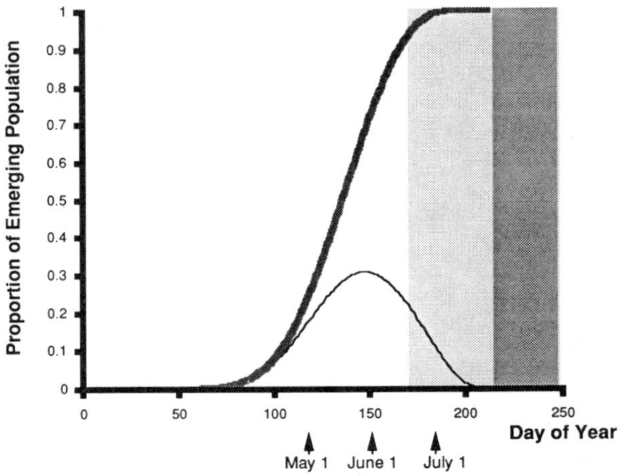

Figure 5.9 Comparing the average cumulative emergence curve (thick line) for boll weevil to the average number of emergent weevils alive (thin line), assuming no cotton is available for feeding. Shaded regions on the right indicate susceptibility of the earliest (light shade) and latest planted cotton crops to boll weevil attack.

Fig. 5.9, this corresponds to a proportion alive of about 0.22. Third, we assume that in a year with very low survival, planting early on May 1 still results in low to moderate numbers of weevils entering cotton, the equivalent of about 0.6 to 0.9 from Fig. 5.9.

Our strategy is to adjust the proportion of live weevils up or down depending on the estimate of total weevils emerging. For years in which the numbers emerging are moderate, no adjustment is made. Depending on the expected number of weevils emerging, the proportion alive values are multiplied by 0.4, 0.65, 1.0, 1.45, and 2.0 for the values very low, low, moderate, high, and very high, respectively. For example, in a year with very low numbers expected but with average timing of emergence, the proportion alive value expected for an early May 1 planting is multiplied by 0.4 to yield 0.22 × 0.4 = 0.088, or in the range required by our assumption. Rules like the following are used for this purpose.

```
IF      proportion-weevils-alive(?X) is ?Y AND
        total-weevils-emerging is very low AND
        ?Y * 0.4 = ?Z
THEN    adjusted-proportion-alive(?X) is ?Z
```

Finally, we need to convert from values of ADJUSTED-PROPORTION-ALIVE into qualitative estimates. Here, we simply divide the expected range into five categories, based on our assumptions. Values around 0.1 are considered moderate; values around 0.05 are low; and values near 0.22 are high. Our ranges are divided as follows: (very low, $0 \leq x < 0.035$; low, $0.035 \leq x < 0.075$; moderate, $0.075 \leq x < 0.14$; high, $0.14 \leq x < 0.25$; and very high, $x \geq 0.25$. This conversion is accomplished by a set of five rules that compute the value of an object EXPECTED-WEEVILS-ALIVE. The rules are of the form

```
IF      adjusted-proportion-alive(?X) is ?Y AND
        ?Y ≥ 0.14 AND
        ?Y < 0.25
THEN    expected-weevils-alive(?X) is high
```

5.4.4 Possible planting dates

We discussed above what planting dates the system will consider. Some rules of thumb were used; for example, that it is never a good idea to plant earlier than the optimal planting date in the absence of boll weevil. We also made some arbitrary choices. We decided that planting after June 1 would never be wise because the grower would sacrifice too much yield potential and that there would be an interval of 5 days between planting dates considered in the system. This makes sense because planting dates are really no more than target dates for farmers. The actual planting date for a given field is usually close to the target date, but generally varies because of local weather, soil conditions, and labor and machinery availability.

While these rules of thumb and numerical planting intervals could be combined into rules, it is simpler just to enumerate the possible planting dates in a static list. This is acceptable here because the list does not change from year to year; it is not dependent on the variables in the system. Furthermore, because this problem is primarily a numerical problem, dividing the acceptable range into intervals of about 5 days, rules are not very well suited to its solution in any case. Writing a demon to return a list of dates, given the first and last days of the range, is better than trying to perform this function with a rule base. So, we proceed with a list of dates starting May 1 with intervals of 5 days until June 15. The list is stored in the system as an object whose value is a list POTENTIAL-PLANTING-DATE = (121 126 131 136 141 146 151 156 161 166). When an object has more than one value it usually is declared as multivalued in the knowledge base using a statement like: MULTIVALUED POTENTIAL-PLANTING-DATE. Normally, an inference engine searches for an object's value only until an acceptable value is found. The MULTIVALUED declaration tells the inference engine to go on searching even after finding an acceptable value until no more acceptable values can be found.

Given the list of possible planting dates in day-of-year format, the dates on which each cotton crop, as represented by a different planting date, reaches the susceptible stage for boll weevil attack must be calculated. Here again, we can use a rule to translate. Cotton growth and development is a complex process, but we assume here the simplest of models and add 50 days to the planting date as the estimate of the date of boll weevil susceptibility. This transformation could be carried out by a demon. In fact, if we were to use a more realistic model of cotton phenology, it would have to be. However, because POTENTIAL-PLANTING-DATE is multivalued, we can create another object SMALL-SQUARE-DATE(*planting-date*) to hold the transformed values. It is an array indexed by the corresponding planting date and is of the form

```
IF      potential-planting-date is ?X AND
        ?X + 50 = ?Y
THEN    small-square-date(potential-planting-date) is ?Y
```

This rule uses addition to bind ?Y to the *day-of-year* that is 50 days after ?X. Since POTENTIAL-PLANTING-DATE is multivalued, this rule will fire repeatedly until the complete array of small square dates has been calculated.

5.4.5 Expected severity of attack

The problem addressed by this portion of the system is now fairly simple. For each potential planting date, we want to evaluate the number of weevils we expect to be entering cotton. All that is missing is the connection between planting date and the estimates of weevil pressure at the corresponding cotton squaring date. The following rule should do the trick.

```
IF      potential-planting-date is ?X AND
        small-square-date(?X) is ?Y AND
        expected-weevils-alive(?Y) is ?Z
THEN    severity-of-attack(?X) is ?Z
```

This single rule will bind to any and all potential planting dates in our static list, with ?X taking on the day of year. For instance, if the potential planting date is 136, May 16, then the variable ?Y is bound to the value of SMALL-SQUARE-DATE(136), or 186, July 5. Then ?Z is bound to the qualitative estimate of the number of weevils emerged and still alive on July 5, which ranges from very low to very high, and that value is assigned to the expected severity of attack for the planting date of May 16. It seems that this particular deduction, which at first appeared to be so difficult, has become trivial. The reason that deciding how severe the boll weevil infestations would be for different planting dates was so daunting before is because the decision had not been broken down into its components. The rule "divide and conquer" is as true in knowledge-based systems as it is in the military.

5.4.6 Uniform planting date for cotton

Deciding which planting date to choose is not simply a matter of picking the planting date with the lowest severity of attack. We have to deal with the trade-offs in the system including yield reductions and risk of complications at harvest because of late planting. This involves judgment and should be founded on expert opinion. Here we choose a strategy that seems to make sense given the new information available from our system.

In most years, the severity of attack expected for the standard recommended planting date of May 21 is borderline between low and moderate (Fig. 5.9). This average value is due to several years with low weevil numbers and occasional years with high numbers. We assume that a severity of low is acceptable but that moderate is not. We also assume that farmers are sacrificing some yield by planting after May 1. Our strategy is to find the earliest planting date for which the expected severity of weevil attack is low. However, planting dates of May 1 through May 16 are not recommended unless the expected severity is very low.

If no planting dates show an expected severity of low, a planting date of May 1 is considered for recommendation. When no planting dates show an attack severity below high, the May 1 planting date is recommended. If some late plantings show a moderate attack severity, our recommendation is to plant early if planting conditions are very good, good moisture in the soil and warm temperatures around May 1, otherwise delay until late spring.

To implement this strategy, we use rules like the following with all new variables being multivalued.

```
IF      severity-of-attack(?X) is very-low AND
        ?X ≤ 161
THEN    good-planting-date = ?X

IF      severity-of-attack(?X) is low AND
        ?X > 161
THEN    good-planting-date = ?X

IF      severity-of-attack(?X) is moderate
THEN    fair-planting-date = ?X

IF      good-planting-date = ?X AND
        good-planting-date = ?Y AND
        ?X ≤ ?Y
```

THEN eliminated-date = ?Y

IF fair-planting-date = ?X AND
fair-planting-date = ?Y AND
?X ≤ ?Y
THEN eliminated-date = ?Y

IF good-planting-date = ?X AND
NOT(eliminated-date = ?X)
THEN recommended-planting-date = ?X

IF good-planting-date is empty
THEN delayed-planting-fails is true

IF if delayed-planting-fails is true AND
fair-planting-date = ?X AND
NOT(eliminated-date = ?X)
THEN recommended-planting-date = ?X AND
recommended-planting-date = 161

IF if delayed-planting-fails is true AND
fair-planting-date is empty
THEN recommended-planting-date = 161

5.5 The Whole System and Conclusions

The integration of a quantitative model with our knowledge base is now complete. Some aspects are mainly quantitative, others mainly heuristic. This hybrid system can now recommend a planting date based on the output from the simulation model, fall sampling, historical averages, and expert opinion. To be sure, in this chapter we have oversimplified such things as cotton phenology and boll weevil adult longevity, but making these calculations more accurate demands only more realistic demons.

We use a structured knowledge base development technique to divide and conquer the complex problem. This technique focuses our attention on key decisions and deductions needed to solve the problem. In dividing the system up according to our decision and information flow model, we keep each subproblem well defined and easy to address. In some cases, the subproblems become trivial once previous decisions and deductions are made. Furthermore, this approach allows us to keep our rules very simple, which makes adding explanation facilities easier.

Finally, we use a simulation model, not as a way to help us understand the system, but as an information source. The model works like any other demon in the system, providing meaningful output, usable by the inference engine. Until we write models using forms and representations that are accessible to a knowledge-based system, our attention remains only on their output and how that output helps us make decisions.

Appendix: Degree-Days

The process of growth for many crops and cold-blooded animals such as insects is often measured using the time unit *degree-days*. The degree-day concept is based on the observation that as a function of temperature the plant's metabolic rate follows a curve having the shape shown in Fig. 5.10.

For computational purposes this curve can be approximated by a straight line segment as shown in the figure. Moreover, the higher temperature at which plant growth ceases, denoted T_u in Fig. 5.10 and called the upper threshold, is frequently so high that it can be safely ignored. The lower temperature at which plant growth ceases is called the lower threshold, denoted by T_l in the figure. For example, the value of T_l for cotton is about 12°C, or about 53°F. The number of elapsed degree-days that have occurred since the start of the growing season is obtained by integrating the growth rate over time, i.e., by integrating temperature above the lower threshold as shown in Fig. 5.11. Mathematically, the formula for degree-days (DD) is computed as

$$\mathrm{DD}(t) = \int_0^t [T(\tau) - T_l]d\tau$$

where the term in the square brackets is computed as the value of the difference if it is positive and as zero if the difference is negative, and τ is a variable of integration.

There are several approximate formulas for computing degree-days; a discussion of these is given by Wilson and Barnett (1983). The term degree-days is sometimes used to represent other computed quanti-

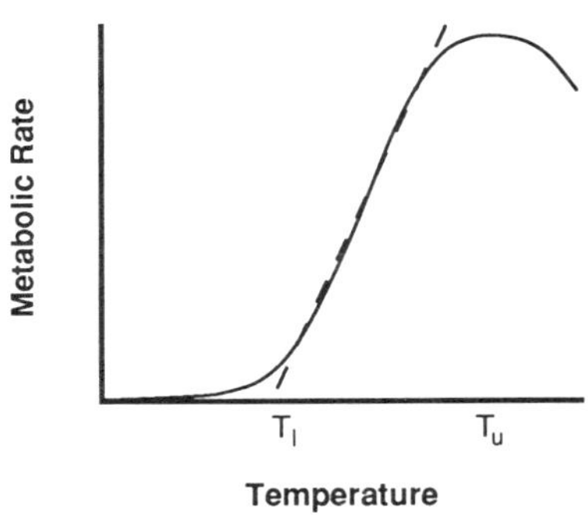

Figure 5.10 Relationship between poikilotherm developmental rate and temperature. Insects and many plants share this temperature-dependent development. The dashed line indicates the typical linear portion of the curve.

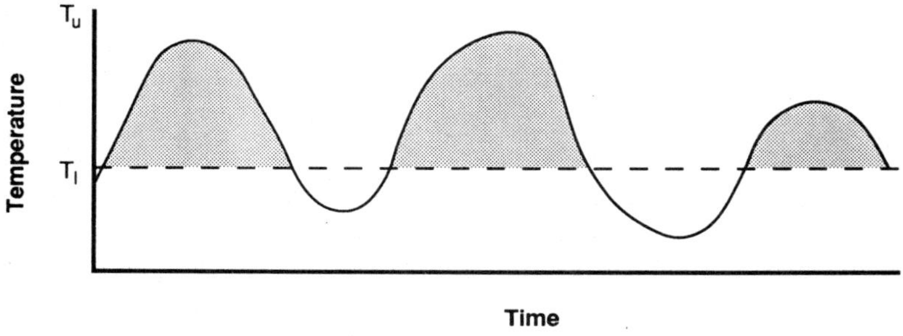

Figure 5.11 Degree-days represented as the area between the time trace of temperature and a lower thermal threshold. Compare to Fig. 5.4.

ties, usually the difference between the mean temperature and the lower threshold. There is some controversy surrounding the use of degree-days as a measure of crop progress (Podolsky 1984), but experience has shown it to be a good indicator variable in the case of cotton and cotton insect pests and we use it here to represent elapsed time.

Chapter

6

Integrating an Expert System with a Regression Model: Tactical Management

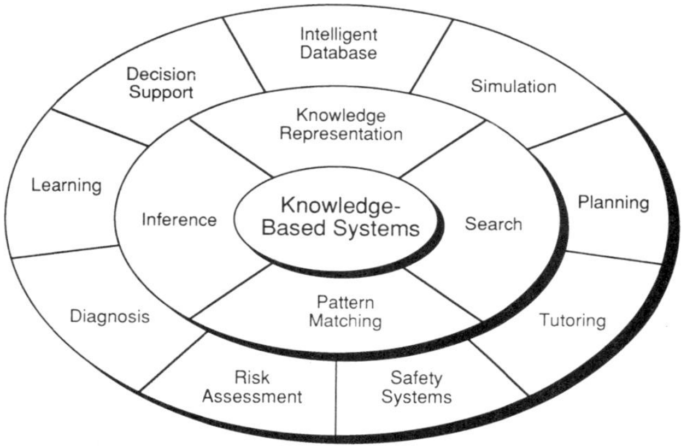

This chapter continues the discussion begun in Chap. 5 on linking expert systems with other computer programs and algorithms. Here we describe the integration of a rule-based expert system with a linear regression program. The regression program fits a curve to a series of points representing values of a parameter varying with time. The regression is then extrapolated to provide a forecast of the parameter value at some time in the future. The expert system interprets the forecast and also determines whether or not the data are suitable for use in a regression.

6.1 Introduction

One of the most promising applications of expert systems is tactical, i.e., seasonal, crop management decision support. A tactical decision support system uses field data obtained during the course of the season to develop guidelines for crop management. Guideline development often depends on the ability to forecast the future behavior of the crop. Such forecasts permit the program to compare alternative management scenarios. Forecasts can be generated in a number of ways. One way to generate a forecast is through the use of a simulation model as described in Chap. 5. This method is discussed in some detail in Chap. 11, but for now we simply note that simulation models have not proven particularly successful as tactical decision support tools. Models often require massive amounts of data and are, thus, impractical in applications other than research. In addition simulation models often provide a good estimate of the behavior of a typical field but, because of extrinsic factors not accounted for in the model, may provide a very poor estimate of the behavior of a particular field. Simple regression models are often more useful devices for seasonal forecasting. When a regression model is integrated into a rule-based expert system for purposes of management decision support, the regression model and the expert system can actually interact to support each other. The regression model can be used to provide information to the expert system for decision making, and the expert system can be used to make the regression model more robust by dealing with cases in which the regression encounters problems or breaks down. For example, the expert system can aid in locating outliers and other "bad" data points and can help the manager to identify special circumstances associated with these points.

This chapter discusses the linking of a regression model with an expert system for tactical decision support. The discussion is presented in the context of a specific example, irrigation scheduling. However, many other such applications are possible. For example, in addition to irrigation scheduling the CALEX/Cotton system (Plant et al., 1988; Goodell et al., 1990) uses regression for pest management and crop termination. However, irrigation scheduling is as good an example as any, and it has the advantage that it will be further discussed in later chapters. The discussion in this chapter begins with a description of the use of extrapolation of leaf water potential data as a means of scheduling future irrigations. A review of the theory of linear regression is then given. The integration of linear regression functions into the knowledge base is then discussed, and an example of the use of these functions to schedule irrigations is presented. This example is based on the irrigation scheduling module in CALEX/Cotton.

6.2 Forecasting the Irrigation Date

Irrigation scheduling involves determining the time at which to apply water to a crop and the amount of water to apply. A useful analogy for understanding the theory of irrigation scheduling is to compare it to the maintenance of a checking account in a bank. Each day the plant withdraws some of the available water from the supply in the ground (the bank). The supply may be augmented naturally by rain. (If only checking accounts operated on the same principle!) If the rain is insufficient to make up for the withdrawals, then irrigation water is applied to replenish the supply when the balance reaches a specified minimum value. The problem associated with irrigation scheduling is, therefore, to determine what the minimum permissible level should be, when the actual water supply will reach that level, and how much water to apply to replenish the store.

The analogy given above focuses on the supply of water in the soil. One can also focus on the plant and its internal water supply. The plant makes use of its water as a reagent in metabolic processes, a means of transport of solutes, and a means of structural support (Meidner and Sheriff, 1976). The banking analogy is less appropriate in this context; a better analogy is that of a sponge. If a sponge is initially full and each day a certain amount of water is squeezed from it, then eventually the sponge will lose enough water that it will need to be replenished. In a similar way, the plant loses moisture through transpiration while the soil is losing water through evaporation. The combined loss is called *evapotranspiration*. One can, in principle, schedule irrigations by measuring the water content of the plant and determining when it is sufficiently low, presumably because of exhaustion of the water in the soil. This method of irrigation scheduling is not feasible for all crops, but it is possible in some cases. For cotton grown in the irrigated Central Valley of California it is not only possible but is generally considered the preferred method (Grimes and El-Zik, 1982).

The water content of the plant is generally expressed in terms of the *water potential*. The water potential describes the energy status of water within the plant (Ray, 1972). Water moves from a region of higher to lower potential, so that the water potential difference between two locations is a measure of the driving force of water movement between these locations. Within the plant, water potential is determined primarily by the concentration of dissolved solutes, the hydrostatic pressure, and matric effects. The latter are due to hydrational forces on water molecules within the cell matrix. Water potential is usually denoted Ψ, and is measured relative to a zero state. By convention this zero state is the water potential of pure water at atmospheric pressure. In the context of the three primary effects on water potential in plant cells the water potential is the sum of the hydrostatic (or turgor) pressure P, the solute concentration effect (or osmotic potential) s, and the matric potential τ. In other words,

$$\Psi = P + s + \tau$$

Water potential may be expressed in units of pressure. The usual unit in applications is the *bar,* which equals approximately 1 atmosphere or approximately 14.7 lb/in^2.

The magnitude of the matric potential is generally much less than that of the other two components. The hydrostatic pressure P is always positive. Since the osmotic potential is measured relative to that of pure water, which is zero, the osmotic potential is always negative. Pure water maintained at a hydrostatic pressure greater than atmospheric has a positive water potential. In a living plant, however, the only source of water is from the soil. Water in the soil is at atmospheric pressure, and therefore must have a negative water potential. Since water moves from regions of higher to regions of lower potential, the water potential of plants must always be negative. The water potential measures the amount of energy necessary to extract water from the plant, and the more negative the water potential, the more desiccated the plant.

This may seem somewhat confusing, but it can be clarified by returning to the analogy of the sponge. The water in a sponge at atmospheric pressure has a negative water potential. To extract water from the sponge one must apply energy by squeezing on it. The more water in the sponge, the less energy necessary to extract a given quantity of water. Conversely, as the water is extracted from the sponge, more energy must be expended to extract the remaining water. This greater amount of energy expended is reflected in the water potential of the sponge becoming more negative.

In the field the water status of the plant is generally estimated by using a device called a *pressure bomb* to measure the leaf water potential (Meidner and Sheriff, 1976). A pressure bomb is a small, airtight container that contains a screw top with a rubber gasket. A leaf is inserted into the gasket, and the screw cap is tightened to form an airtight seal. Nitrogen is then pumped into the pressure bomb until xylem sap just begins to exude from the leaf stem. The pressure in the bomb at which this happens is taken to be the water potential of the leaf. There are a number of complicating factors that can cause errors in this estimation, but it remains the best available method of field estimation of plant water status. If future irrigation is to be scheduled based on leaf water potential, then this potential must be measured at successive times. Field observation indicates that if time is measured appropriately, as described below, then plots of estimated leaf water potential as a function of time form an approximately straight line. Thus, the future irrigation may be scheduled by establishing a threshold leaf water potential and scheduling an irrigation by fitting a line to the leaf water potential data and extrapolating this line to the threshold value (Fig. 6.1).

The basic problem of scheduling an irrigation using leaf water po-

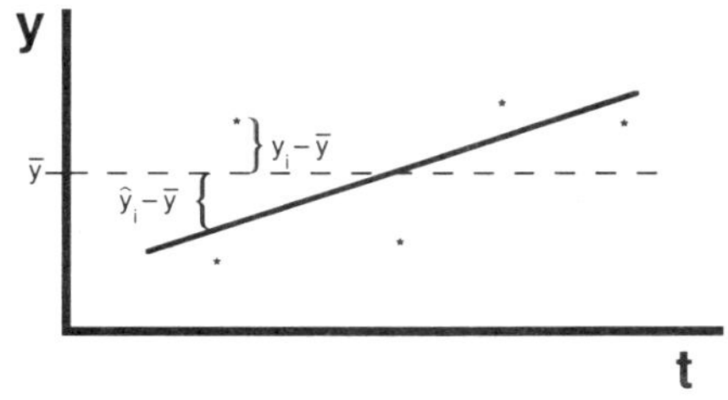

Figure 6.1 A sketch of the various parameters defined in linear least squares regression analysis.

tential data may therefore be expressed as follows. The expert system must assemble the leaf water potential data, which consists of a series of pressure bomb values plotted against elapsed degree-days (see Chap. 5), and attempt to draw a regression line through it. It must determine whether or not such a line can be drawn with sufficient confidence. If so, the system must compute the line and extrapolate it until it intersects with a threshold leaf water potential value. The degree-day value of this intersection must then be converted using forecast weather data to a date that is the next irrigation date. There are a number of complicating factors, some involving the use of the pressure bomb and some involving the practicalities of irrigation scheduling. Regarding the pressure bomb data, temperature variations can have a substantial effect on pressure bomb readings (Grimes et al., 1987). Regarding the practicalities of irrigation scheduling, the procedure just outlined ignores a number of constraints faced by the farmer. Irrigation water may be available only on a fixed rotation that does not coincide with the projected date. Problems such as crop diseases may affect irrigation timing. These and other problems will be considered in later chapters but ignored for now. Instead, this chapter will focus entirely on the integration of the regression model into the expert system.

6.3 Brief Review of Regression Theory

6.3.1 Linear regression by the least squares method

The objective in linear regression is to compute the equation of a straight line that best approximates a set of pairs of data, each of which is a point in the plane. Suppose there are n data points, denoted $(t_1,y_1),\ldots,(t_n,y_n)$. The symbol t is used for the abscissa since, in the application discussed in this chapter, it represents elapsed time in degree-days. Let the straight line approximation to this set of points be denoted by $y = a + bt$. The parameters a and b are called the *regression coefficients*. Let $\hat{y}_i = a + bt_i$, where $\hat{y}_i$ is the value of the dependent variable at t_i if the regression equation were exact. Let the *residual* e_i be defined as $e_i = y_i - \hat{y}_i$. The objective in least squares regression is to choose a and b to minimize the following sum, S:

$$S = \sum_{i=1}^{n} e_i^2$$

The formulas for a and b that achieve this minimization are obtained by computing the partial derivatives $\partial S/\partial a$ and $\partial S/\partial b$ and setting these equal to zero. The resulting formulas are (Draper and Smith, 1981)

$$b = \frac{n\sum_{i=1}^{n} t_i y_i - \left(\sum_{i=1}^{n} t_i\right)\left(\sum_{i=1}^{n} y_i\right)}{n\left(\sum_{i=1}^{n} t_i^2\right) - \left(\sum_{i=1}^{n} t_i\right)^2}$$

$$a = \frac{\sum_{i=1}^{n} y_i - b\sum_{i=1}^{n} t_i}{n}$$

The most common measure of the accuracy of the regression line is the R^2 value, which is called the *coefficient of determination*. This is obtained as follows. Let $\bar{y}$ be the mean value of the y_i, i.e., $\bar{y} = \Sigma y_i/n$ (cf., Fig. 6.1). Then the sum of the squares of the deviations of y_i from $\bar{y}$ may be separated into two components as follows (Draper and Smith, 1981):

$$\sum_{i=1}^{n} (y_i - \bar{y})^2 = \sum_{i=1}^{n} (y_i - \hat{y}_i)^2 + \sum_{i=1}^{n} (\hat{y}_i - \bar{y})^2$$

This is often written as

$$\text{SSM} = \text{SSE} + \text{SSR}$$

where SSM represents the sum of the squares of the deviation of the data from the mean, SSE is the sum of the squares of the residual errors in regression, and SSR is the sum of the squares of the deviations of the regression line from the mean. If we let

$$R^2 = \frac{\text{SSR}}{\text{SSM}} = 1 - \frac{\text{SSE}}{\text{SSM}}$$

then R^2 measures the portion of the deviation from the mean that can be accounted for the linear dependence of y on t.

6.3.2 Inverse regression

A problem such as estimating the irrigation date based on extrapolation of leaf water potential data involves the process of *inverse regression*. Here the regression line is used not to estimate the value of y that corresponds to a particular t, but rather to estimate the value of t corresponding to a particular y. Suppose that n data points have been collected. In the leaf water potential irrigation problem, each data point represents a recorded leaf water potential value y_i, or, more preferably, the average of several such values, taken a day at which t_i degree-days had elapsed. The data set will be denoted (t_1,y_1), $(t_2,y_2),\ldots,$ (t_n,y_n). The objective is to fit a line $y(t) = a + bt$ to this data and determine that value of t^* at which $y(t^*)$ equals y^*, where y^* is a particular value, in our case the threshold leaf water potential value. The basic procedure is to first fit a regression line to the data, and then to determine the value of t^* from this regression line by solving the equation $a + bt = y^*$, obtaining $t^* = (y^* - a)/b$. This algorithm works well provided that the regression on which it is based works well. What is needed first is some means of determining whether the extrapolation can be presumed valid, i.e., whether the data are sufficiently well approximated by a straight line that the extrapolated value of y may be judged accurate.

Two measures of the precision of the regression line are available: the coefficient of determination described above and the confidence interval (Draper and Smith, 1981). The derivation of the confidence interval makes use of the *linear statistical model* (Larsen and Marx, 1981), which has an assumption about the form of the data points (t_i,y_i). The assumption is that the data is capable of being expressed, if there was no experimental error, as the function $y(t)$. Unfortunately, the rigorous use of confidence intervals to do inverse regression requires that the dependent variable as well as the independent variable be random. In the present case the dependent variable t is fixed, i.e., nonrandom, and therefore these techniques do not apply. An alternative method called *time series analysis* is available (Box and Jenkins, 1976), but this method, to provide reasonable predictions, requires considerably more data than is usually available in seasonal agricultural field management situations. Therefore, the coefficient of determination is the only reasonable choice.

The coefficient of determination provides information as to whether the computed value of t^* can be judged reliable. If it is judged unreliable, then some means may exist to "fix" the regression, i.e., to determine why the estimate is not reliable and whether any adjustments can be made in the data or the regression computation to make the estimate better. We can fix the regression by examining in some detail what specifically can cause problems with the regression. There are a number of things that can go wrong. Three of the most common are outliers, trends in the data, and total failure of the regression.

Outliers. Outliers, or, more generally, influence points, are data points that exert a disproportionate influence on the values of regression coefficients. There are two principle reasons why a data point may have this disproportionate influence; these are shown schematically in Fig. 6.2*a* and Fig. 6.2*b*. Figure 6.2*a* shows a true outlier, i.e., a point whose *y* value is very different from the *y* values of data points with similar *t* values or, expressed in another way, the *y* value is very different from that predicted by the regression line. Figure 6.2*b* shows a leverage point, i.e., a point whose *t* value is very different from those of the other points. Since the concern in this chapter is with regression performed on temporal data, in which *t* values are under the control of the sampler and may be spaced at reasonably regular intervals, we will not consider leverage points of the type shown in Fig. 6.2*b* and will instead focus on outliers.

In the case of leaf water potential data, there are a number of possible reasons why a data point might be an outlier. One possibility is that it might correspond to a physically different data set than the rest of the points. For example, leaf water potential readings are strongly affected by the ambient temperature during the reading. If one data point is measured at a temperature that is very different from the others, then its *y* value might be very different from what

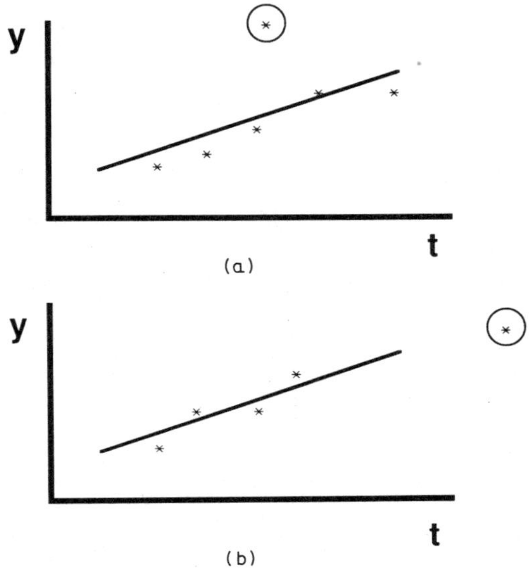

Figure 6.2 Two examples of influence points. (*a*) An outlier, i.e., a point whose *y* value is very different from the *y* values of other points with similar *t* values. (*b*) A leverage point, i.e., a point whose *t* value is very different from the *t* values of other points with similar *y* values.

one would expect. Alternatively, the leaf water potential may have been incorrectly measured, e.g., because of a poor seal in the pressure bomb, or incorrectly recorded, e.g., because of an incorrectly read pressure gauge. A third possibility is that the outlier may in fact be an accurate record of a highly improbable measurement. "The one in a million event happens with roughly that frequency, however surprised we might be that it happens to us." (Fisher, 1935)

Dealing with outliers is a two-step process; the first step is to detect the outlier and the second is to decide what to do with it. The possibility exists, of course, that there are more than one outlier. We will ignore this possibility. This is reasonable since in our application the data points are added, for the most part, one at a time and the regression is recomputed after each point is added. Therefore, any previous outliers will have been dealt with by the time the next one comes along. Detecting the outlier is an algorithmic process and will be discussed here; deciding what to do with it is a job for the expert system and will be postponed for now.

Recall from the discussion above that the residual e_i is defined as $e_i = y_i - \hat{y}_i$. An outlier is obviously characterized by having an exceptionally large residual. A common means of determining if a point is an outlier (e.g., Hawkins, 1980) is by examining its *studentized* residual ε_i, defined as

$$\varepsilon_i = \frac{e_i}{\mathrm{se}_i}$$

where the estimated variance of the residual se_i is defined as

$$\mathrm{se}_i = \sum_{k=1}^{n} \frac{e_k^2}{n-2} \left(\frac{n-1}{n} - \frac{(y_i - \bar{y})^2}{\sum_{k=1}^{n} (y_k - \bar{y})^2} \right)^{1/2}$$

The ε_i is called a studentized residual because it comes from a student distribution. The reason for using studentized residuals to determine whether or not a point is an outlier is that the variance of the ordinary residuals increases at the highest and lowest values of t. Using studentized residuals normalizes this "ballooning" effect. To use a studentized residual, we basically form a statistical hypothesis test, with a null hypothesis that the suspect point is from the same distribution as the rest of the points. Details of this procedure are given by Hawkins (1980) and by Myers (1986).

Trends (nonlinearities) in the data. Trends in the data, also called nonlinearities, occur when the data points have a good fit to some function other than a straight line (Fig. 6.3). A trend in the data is generally interpreted to mean that the regression model, i.e., the assumed linear form $y = a + bt$ of the equation, is inappropriate. This may not, however, be completely true. For example, in the case of leaf water potential versus time data there is ample evidence (e.g., Grimes et al., 1987) that under constant conditions the leaf water potential changes is a linear function of elapsed degree-days. A trend in the data in such a case might represent a change in the field conditions. This might be caused, e.g., by a long spell of unusual weather, the crop's response to an insect infestation, or the application of some chemical.

At this point it is worth setting aside a short paragraph to emphasize an important point. The detection of trends, and outliers as well, in temporal data such as leaf water potential, insect counts, or plant growth data has an importance potentially transcending the use of these data in forecasting, i.e., the outlier or trend can serve as an early warning device for possible problems with the crop. For example, in the case of leaf water potential a trend may be, as was just pointed out, an indication of a detrimental change in field conditions caused, for example, by an insect infestation. If this trend is detected rapidly it may provide an indication of the problem before other symptoms manifest themselves, and this early detection may make it possible, or less costly, to take corrective measures before it is too late. The expert system should, therefore, at least raise a warning flag to the user when such a situation is suspected.

Unlike outliers which involve individual points, trends by their very nature must involve sets of more than one data point. Therefore, they are somewhat more difficult to detect by a computer program. The standard method of detecting a trend is a form of the statistical test called the F-test (e.g., Draper and Smith, 1981, p. 31). This test, however, requires repeated independent plots of the data, something that is impossible in the current context. Alternative methods involve either smoothing the data and examining the curvature of the smoothed curve or comparing the linear model with alternatives such

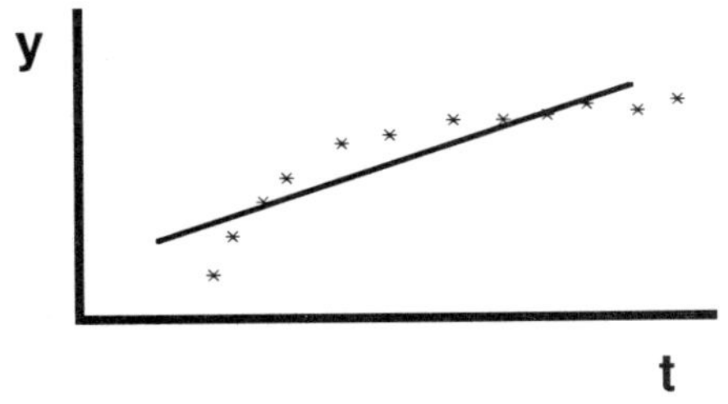

Figure 6.3 A trend, or nonlinearity, is shown. The data can clearly be better fit by some function other than a straight line.

as the quadratic regression model $y = a + bt + ct^2$. These tests are described, for example, by Himmelblau (1970).

Total failure of the regression analysis. As it is used here, the term *failure of the regression analysis* means that the regression line is inadequate to provide a reliable estimate of the value of the time t^* at which the measured data will reach the predetermined value y^*. One possible way that this can happen is that there are too few data points, either zero, one, or two, to fit a regression line. Another possibility is that there is an adequate number of points, but the scatter in the data is so great that the coefficient of determination R^2 is unacceptably small. In the case of outliers or trends, the possibility exists that the data set or the regression can be "fixed" so that the data can still be used to provide an acceptable estimate of the value of t^*. Such is not the case, however, with failure of the regression method. No forecast is available, at least not using extrapolation of a regression line. A decision must still be made, though; for example, the grower still needs to schedule an irrigation whether or not the data can be extrapolated. Therefore, the expert system must be prepared, usually with an alternative method, for this eventuality.

The time has clearly come to leave the domain of statistics and return to that of expert systems. However, before describing the linkage of an expert system with a regression model, another way, the expert system interface, in which expert systems are used in conjunction with regression analysis is described. Considerable research has been done on the use of expert systems as user friendly interfaces to regression packages and to other common software such as linear programming systems, database systems, and so forth. A description of the way an expert system is used as an interface provides a useful contrast with the actual merging of a regression package into an expert system.

6.4 Linking an Expert System to a Regression Program

6.4.1 An expert system interface

One application of expert systems in regression analysis is the construction of a user interface for statistical analysis regression programs such as SAS or BMDP. There is a need for such an interface because powerful statistical analysis programs like these cannot be effectively used by someone completely ignorant about statistics. The goal of an expert system interface to a statistical package is to make the system accessible to a naive user. The expert system interface can in this context serve as a substitute for a statistical consultant who can provide an interpretation of the results of a statistical analysis, an explanation of any problems that might arise, and suggestions as to how these problems might be solved. One expert system interface of this kind is REX (regression expert), an expert system interface to a regression analysis program (Gale, 1986).

Many modern statistical programs have user friendly interfaces, and it is not hard to imagine that tutorial systems for such programs could be constructed. These would be based on the user's manual and might be similar in appearance to the tutorials and on-line help systems that accompany word processors and the like. According to Gale (1986), the distinction between this type of tutorial system and an expert system interface is that the former is passive while the latter is active, pointing out problems and suggesting solutions. Gale demonstrates this with a data set in which the linear regression plot of atmospheric pressure versus the boiling point of water is studied. The expert system first detects a trend in the data and suggests a logarithmic transformation. The system then detects an outlier in the regression of the transformed data and suggests deleting it.

REX serves as an actual interface to a separate statistical analysis package. It maintains an input/output screen with the user and sends commands to the statistical package based on its interpretation of the user's input. The user does not provide any input directly to the statistics program. REX works by following a determined strategy in conducting its regression analysis. This strategy is encoded in a series of *frames* (see Chap. 8) and works in a manner quite similar to the CALEX system described in Chap. 9. Both REX and CALEX are motivated by the CENTAUR system (Aikins, 1983).

6.4.2 Integrating a regression model into a rule base

The REX expert system provides a useful interface between a computer program performing a function such as regression analysis. However, this use does not integrate the expert system with the functional program. The expert system interface is used as a substitute for a trained statistician to aid in interpreting the results of the analysis. The user has some reason for conducting the analysis, but this reason is irrelevant to the function of REX. Moreover, the user must take the results of the regression analysis and apply them, using knowledge of the problem, to a solution. In the application motivating this chapter, irrigation scheduling by extrapolation of a regression line, this would correspond to the user personally using the data to compute the regression line and determining the intersection of this regression line with the threshold leaf water potential. The goal of this chapter is to develop an expert system that will do all this by itself. The regression model is to be integrated directly into the expert system for use in performing the specified task. The user of the expert system may not even be aware that the program is doing a regression analysis. It may only be necessary to present the user with the end result, making an explanation available as an option. The expert system makes all the decisions about interpreting the regression information and altering or deleting points, and it applies the results by itself.

In describing the development of such a system, we will explicitly make use of the fact that the independent variable in the regression analysis is time. Temporal data is often entered sequentially as new data points are collected. When a data point is entered and a set of regression coefficients is computed, the program can save these coefficients together with other information such as the number of data points used in the regression analysis and the coefficient of determination of the regression. These results can then be compared with the corresponding results of the regression analysis performed when the next data value is entered at a later time. This comparison can provide a useful means of detecting potential outliers or nonlinearities in the data. Once a suspected outlier is detected, the expert system can determine whether there is an explanation for it and what should be done in response. We assume that the expert system is always run after each data point is entered, so that there is never more than one new data point. Such might not always be the case in practice, and a real system would have to be prepared for exceptions. However, by ignoring these exceptions here, we can greatly simplify the discussion without sacrificing any essential information. A brief version of this discussion is given by Plant et al. (1990).

6.4.3 The basic algorithm

The first objective is to establish a basic algorithm for computing a regression line through a set of at least three data points. Since this line is to be used for extrapolation of the data set, a minimum acceptable R^2 value is first established. The regression line will not be accepted if its R^2 is not at least as large as this minimum value. Since the data has been entered as a time-ordered sequence, it is reasonable to expect that the most recently entered points are the most representative of the current state of the data set. Older points may reflect conditions that have since changed. For example, in the case of a sequence of leaf water potential values, there may have been an irrigation or rainfall since the collection of the older data. This would invalidate any data taken before the irrigation or rainfall. The algorithm, therefore, works with the most recent data, starting with the three most recent data points.

If a regression line drawn through the three most recently collected data points does not meet the acceptable value of R^2 then a failure of the regression method is indicated. If an acceptable regression line can be drawn through the three most recent data points, then older data points are added to the regression line until the R^2 falls below the minimum acceptable value. The rationale behind this strategy is that the fall in the value of R^2 below its minimum acceptable value is an indication that the validity of the regression method is breaking down, possibly because field conditions have changed since the data were collected. This algorithm ensures that the most recent, and presumably most valid, data are used to construct the regression line.

The detection of outliers and nonlinearities can be accomplished through an extension of the basic algorithm. Consider the situation shown in Fig. 6.4. The initial regression line has been constructed through five data points as shown by the solid line in Fig. 6.4. When another data value is added at time t_{n+1}, it is considerably above the value predicted for this time. Thus, the point may be an outlier or the beginning of a nonlinearity. The basic algorithm attempts, at this point, to construct a regression through the updated data. A newly computed regression, shown by the dashed line in Fig. 6.4*b*, varies considerably from the solid line because of the influence of the most

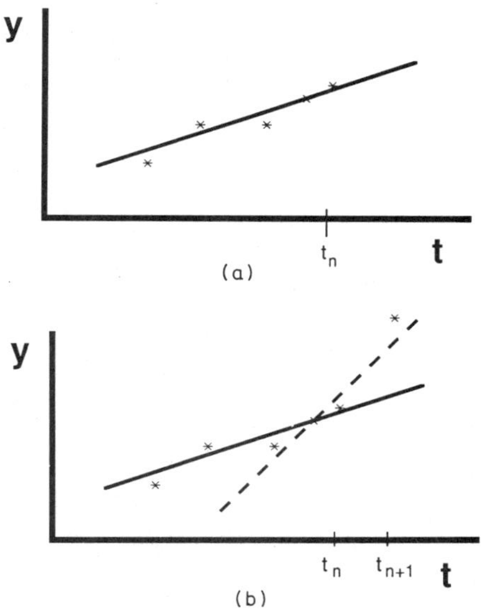

Figure 6.4 Sketch of an algorithm for fitting a regression line to a series of points. In (*a*) the line is fit to five points. In (*b*) a sixth point is added. In order to obtain an acceptable R^2 in a regression line fit to a data set including this new point, it is necessary to exclude one or more of the data points in the old data set. This is an indication that the new point is substantially different than expected.

recently added point. Therefore, the algorithm will not, if the influence of the new data point is strong enough, use all of the data points used in constructing the previous regression line. In Fig. 6.4*b*, the new regression line is based on four data points instead of six. The reason for this is that the value of R^2 fell below the acceptable value for a regression line that included the new data point and, at the same time, the other two data points. Therefore, if the new data point is close to the regression line computed previously, then the number of data points used to construct the new regression line will be one more than the number used to construct the previous regression line. If this is not the case, then the newest point may be identified as a potential outlier.

This algorithm is quite simple and has a "rough and ready" nature. It can be made considerably more elegant and theoretically sound, but experience has shown it to provide sufficiently good performance for the sort of conditions typical in field crop management. The algorithm does not need to be implemented in the expert system's rule base; indeed it would be wastefully inefficient to do so. Instead, it can be implemented as a demon in the manner described in the previous chapter. The expert system must be used to interpret the results of a successful regression analysis and to respond to situations in which the regression analysis is unsuccessful.

If the regression analysis is successful, then the calculated line can be extrapolated to intersect with the threshold value. In the case of irrigation scheduling, this would provide the date on which irrigation should be scheduled based on water potential. There are a number of potential complications, as discussed earlier. Examples are the unavailability of water on the day of the scheduled irrigation and a conflict with some other farm operation such as running a machine through the field. The expert system must search for conflicts and complications such as this and modify the irrigation date accordingly. For example, if water is not available on the date scheduled, then the system should schedule an irrigation as close as possible, but prior to, this date.

The expert system is presented with a more difficult problem if the regression does not succeed. As discussed earlier, the expert system limits its consideration to three possible reasons for lack of success in drawing a regression line that includes a newly entered data point. These are: (1) The point is an outlier, (2) the point is a valid data value but the regression model fails because of a trend in the data, and (3) the regression method fails completely. The expert system must, based on the character of the data and on any other possible information it can obtain, determine which of these three alternatives is the correct one and must make an appropriate recommendation to the user. Let us consider the third alternative first since it is, in a sense, the simplest.

In the context of our one-step-at-a-time algorithm for constructing a regression line through a data set, a total failure of the regression corresponds to the inability of the algorithm to draw a regression line through the most recent three points. There are two possible reasons why this might occur. The first is that there are fewer than three data points altogether; this will be the case early in the season. The second reason is that the scatter in the data points is so great that an acceptable R^2 value cannot be generated. The response to these problems must of course depend on the problem domain. Since the regression line is being used to forecast the date of occurrence of an event in the field, it may be possible to obtain this forecast in some other way. Alternatively, it may be possible to approach the problem in some completely different way. In the case of irrigation scheduling one possibility is to use the water budget method (Grimes and El-Zik, 1982). This approach will be described in Chap. 9. Whatever the approach taken, the problem is moved out of the domain of regression analysis and into the knowledge-based domain described in the preceding chapters.

If the point is judged to be an outlier or the start of a trend, it might be possible to retain the regression approach. Such a judgment can be made if a successful regression line had been constructed through the data set that included all but the most recent point and if it is only by the addition of this point that the algorithm either fails to construct a regression line or constructs one using fewer data points than the previous regression line. The expert system must be prepared, through the inclusion of appropriate rules in the rule base, to determine, if possible, the cause of failure of the regression line construction and the appropriate response. Once again this is a domain-dependent problem, although there are certain aspects that all such problems have in common.

The first step is to determine whether some event has occurred in the field since the recording of the previous data point that would influence the most recent data value. In the case where the data are leaf water potential values, two natural things to look for are precipitation and irrigation. If one of these events is determined to have occurred, based either on data entered by the user or on a direct prompt to the user, then either the regression analysis must be started over again or some adjustment must be made to the data. If no rain or irrigation has occurred recently, the knowledge base might contain rules causing it to look for unusually high or low temperatures on the recording date, in which case the data value might be correctable by an adjustment of its value.

Each of the possibilities discussed in the previous paragraph might be classified as field events that do not warrant further investigation, but rather a simple fixing of the data or a starting over of the regression analysis. A second category of events of this nature is the occurrence of outliers that are due to faulty data recording or entry. For example, leaf water potential values can be incorrectly recorded because of a misreading of the pressure gauge or a failure of the pressure bomb apparatus. Data may be improperly entered because of a simple mistyping. Events such as these can often be caught by displaying the value of the data point in question to the user and asking for verification.

A third, and perhaps most important, category of events are those that indicate the emergence of an undetected problem in the crop. For example, the rate of decline in water potential may change due to stress caused by an insect infestation or by a problem with the crop's root system. The detection of a point that may be an outlier or the beginning of a trend, and the automatic display of the data value associated with this point, can be a simple and effective means of presenting an early warning of potential problems to the user. However, the implementation of such a system *must* be accompanied by the automatic detection and elimination of events of the first category discussed above, those that represent natural field occurrences. For example, the expert system should be "smart" enough to catch the fact that a leaf water potential value is less negative than that predicated by the regression determination if there has been an irrigation between the most recently entered data point and the previous data points. Presenting this information to the user and asking for its cause would be confusing and annoying and would result in a loss of credibility for the expert system. However, if these naturally occurring events can be detected and eliminated, then the regression algorithm can provide a simple and powerful early warning device for the user in a crop management expert system.

6.5 Discussion

The rule-enhanced regression model described in this chapter may be appropriate for a wide variety of seasonal crop management decisions. In the CALEX/Cotton expert system, this model is currently used for decisions involving three activities: irrigation, crop termination, and spider mite management. In the case of crop termination, the data variable is the number of main stem nodes on the cotton plant above the node with a white bloom (flower) on it. Plant et al. (1990) provide a description of this use. In the case of spider mite management, the indicator variable is the portion of leaves sampled that have spider mites on them. This has been established as a suitable damage function by Wilson et al. (1986). They show that a series of data values in this variable can be fitted with a logistic curve, which can then be projected to the economic threshold.

There are a number of features that each of these data types have in common and that seem to be necessary for use in regression-based management decision making. First, each data type consists of data items that are collected with relative ease by the farmer during the course of the growing season, and it is natural to expect the farmer to collect these data on about a weekly basis. Second, each data type, when it reaches a well-defined threshold value, indicates that the time is appropriate for some management action. Third, since each data type can be collected with sufficient reliability, it is reasonable to expect a regression line drawn through a sequence of data points to provide a reasonably good fit to these points.

Regression models are not the only type of model whose function can be enhanced by the linkage, either as an interface or in direct integration, with an expert system. An expert system has been successfully linked as an interface to a complex simulation model for cotton growth in the GOSSYM/COMAX system (Lemmon, 1986). GOSSYM (Baker et al., 1983) is a large scale simulation model for a cotton plant, and COMAX is an expert system interface that was developed to manage the GOSSYM model in precisely the same way that REX manages a regression program. COMAX determines various weather scenarios under which to run the GOSSYM simulation to accomplish such tasks as determining nitrogen requirements, irrigation times, and crop termination dates. The COMAX expert system integrates current weather data and other inputs to run the GOSSYM program in an on-farm situation in the same manner as would a human researcher in an effort to provide management decision support for the farmer. Lemmon (1986) reported substantial yield increases in test applications of GOSSYM/COMAX.

Chapter

7

Knowledge-Based Systems Embedded in Simulation Models: A Strategic Management Example

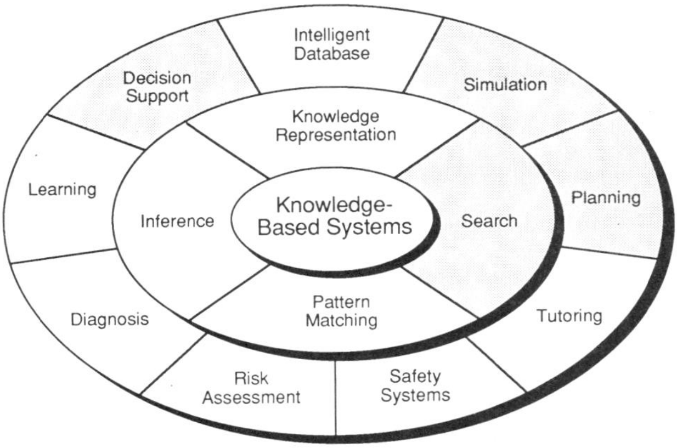

This chapter discusses the general problem of search and the linking of a knowledge-based system with a simulation model. The problem of making proper pest control decisions for *Helicoverpa* spp. in cotton is examined. We compare a classical optimization approach to a strategy that combines algorithmic and heuristic search methods. The search problem is explored with an elementary example; then the more realistic problem, the strategic management of *Helicoverpa* spp., is addressed.

How to make insect pest control decisions has been the topic of much debate, particularly since the introduction of the concepts of integrated control (Stern et al., 1959), later called integrated pest management or IPM, and of computer simulation modeling of agricultural systems (Watt, 1961) in the late 1950s and early 1960s. Since that time, it has been widely accepted that a proper insect control decision should be made using a systems approach. All aspects of the decision, from the cost per ounce of the formulated compound to the potential for secondary pest outbreaks due to disruption of the natural balance of the agroecosystem and for promoting the resistance of the target pest to pesticides, need to be considered (Pedigo et al., 1986).

It has been just as clear to most people that it is impossible to take all these elements into consideration whenever a pest control decision is made. This problem has set the stage for a variety of suggestions for improving the decision process by which chemicals are applied to crops for the control of pests. In this chapter, we look at a hybrid knowledge-based and simulation approach to the problem.

7.1 Strategic Management of *Helicoverpa* spp. in Cotton

Most of the time we think of pest control problems as tactical problems, i.e., there may be insects in the cotton field, and they are doing some damage to the crop. The relevant tactical question is, what does the farmer do about it? The answer usually involves implementing a strategy. Pest management strategies are usually, for example, a set of general rules that indicate the appropriateness of spraying with an insecticide. In the previous two chapters, we used knowledge-based systems instead of general guidelines to solve tactical problems. Now we concentrate on how those strategies are developed and compared. Our system is not designed to solve a tactical problem in the field; it is designed to tell us what set of decision rules seems to be the best strategy.

We are once again using a simulation model in combination with a knowledge-based system. In Chap. 5 we discussed one way to apply a simulation model to the tactical problem of when to plant cotton. In that case, the model was used as a predictive tool and the problem for the knowledge-based system was to achieve a seamless integration between the expert system and the simulation. In this chapter, we discuss a different type of integration between a knowledge-based system and a simulation model. We are not concerned with giving advice to a farmer about a specific cotton field; rather, we want to give farmers a set of decision rules that permit them to make the proper decisions in a variety of tactical situations.

Neither the knowledge-based system nor the simulation model, in this case, needs to be delivered to the farmer. Instead, a knowledge-based system is embedded within a simulation model as an analytical tool to determine which strategies are better overall. The model does not forecast the value of some important variable. Instead, the simulation model takes the place of the natural system in a series of experiments. Ultimately, rule-based pest control strategies are elicited from the analysis, and it is these rule-based systems that are used by a farmer.

The goal of this analysis is to find some nearly optimal strategy represented by a set of rules. An optimal strategy leads to optimal sequences of decisions during the season, so that at the end, profit is maximized. To produce such a strategy, we must first be able to identify optimal sequences of decisions within a given season. In AI parlance, this is a classical problem in *search*. Problems in search are concerned with how to progress efficiently through a sequence of decisions toward an acceptable or optimal solution. Many analytical search methods exist for cases in which the system and strategies can be represented by mathematical equations. These techniques, as we discussed briefly in Chap. 1, form the basis for the science of operations research (OR), and we examine the relationship between AI search and OR search. Finally, as a consequence of solving the search problem, we produce a novel way to evaluate a rule base for its impact on the bottom line, not just its performance in controlled case studies.

7.1.1 Bollworms as pests of cotton

Helicoverpa virescens (F.) and *H. zea* (Boddie) together are commonly called bollworms and as a group are recognized as the principal insect pest complex of cotton in the United States (Head, 1985). Their biology, ecology, and pest management have been extensively reviewed in the literature, most recently by Sterling et al. (1989). Unlike the more famous boll weevil, bollworms are secondary pests. They are normally held in check by a variety of natural enemies, including a rich assortment of predatory insects commonly found in cotton. Bollworm outbreaks occur when the natural control process is disrupted, usually by the application of insecticides to control other pests.

Bollworms are noctuids whose larvae feed on a wide variety of hosts. They attack cotton plants from the time of flowering until nearly the end of the growing season. Adult female moths lay eggs on the foliage near the cotton plants' terminals. The eggs hatch in about 3 days, and the bollworm larvae then feed on cotton fruiting structures, flower buds (called squares), and fruit (bolls), molting four or five times before they pupate in the soil. When natural enemies are present, bollworm eggs and small larvae are subjected to severe mortality. However, once the larvae reach the third or fourth instar, few predators are capable of subduing them and few chemicals are potent enough to kill them.

Because of their secondary pest status, one could guess that bollworm outbreaks would be difficult to predict even given good information on cotton phenology, local weather, and sampling data. In addition, bollworm adults are strong fliers, frequently traveling many miles in search of attractive oviposition sites (Hartstack et al., 1982; Farrow and Daly, 1987; Schneider et al., 1989). This migratory nature adds to the problem of predicting bollworm infestations. Farmers cannot tell when the moths will fly into a field, and once they are there, the farmer does not know whether there are sufficient natural enemies to contain them.

7.1.2 Management practices

Because of their unpredictability, bollworms are managed well only through extensive sampling. Fields must be checked almost daily to determine whether a flight of adults has arrived and laid fresh eggs on the foliage. Recommended sampling programs vary from state to state. Some suggest counting eggs; others suggest counting small and large larvae or "worms," and some suggest using pheromone traps to capture adult males. In addition, natural enemy populations must be monitored to assess the likelihood that any bollworm eggs found will be destroyed naturally, without the use of insecticides. Once insecticides are applied against the bollworms, natural enemy populations will be diminished or eliminated, and unless the natural enemy population can reestablish itself, all subsequent bollworm infestations will have to be controlled by poisons.

Thus, the decision to spray insecticide for bollworm must involve several interacting factors: the extent of the infestation, the age-structure of the bollworms found (if they are already third instars, there is no point in spraying, and if they are all eggs, perhaps the natural enemies will be sufficient), the population of natural enemies in the field, the fruiting stage and condition of the cotton plants (if the plants have suffered previous damage, the farmer may not be able to afford to wait to let the predators act), the moisture content in the soil, the expected weather, and the likelihood that the field will require future control actions because of the considered action. The complexity is mind boggling, and the result is overintensive control.

The problem of overkill. The general economic questions, what are appropriate amounts of insecticide to use to control an insect infestation and what are the threshold densities of insect infestations that would justify an insecticide application, have been the topic of considerable debate in the literature since Stern et al. (1959) defined the concept of *economic damage* to a crop as that amount of injury which would justify the cost of a control measure. Interested readers can refer to Norgaard (1976), McCarl (1981), and Mumford and Norton (1984) for insightful reviews. In general, farmers use more insecticide than society might consider optimal, and at least three factors contribute to this. The first factor is the desire to increase yields. Although it may seem obvious that $10 should not be spent to prevent $9 worth of yield loss, farmers do not always think in terms of marginal profit. Instead, many farmers try to maximize their yield, a practice that is often economically and environmentally detrimental. There are, perhaps, psychological reasons for farmers wanting more yield; in addition, government programs like price supports and crop insurance encourage farmers to bring in large yields by basing program payments on yield not on farming efficiency. The higher the yields, the higher the subsidies.

Avoiding complexity is another incentive to overcontrol pests. It is a time-consuming and confusing task to try to maintain the proper balance in a crop ecosystem by considering the interactions of plants, soil, water, fertilizer, weeds, insects, and natural enemies. It is much simpler to spray the fields, sterilize them of all insects and other pests, and manage only the plant's growth—a difficult enough task. Effectively, the farmer pays a premium to achieve some level of simplicity in the system.

The third factor is uncertainty. Farmers spray as a kind of insurance. When there is uncertainty about whether or not an insecticide ought to be sprayed or about how much insecticide ought to be used, the cost to a farmer of spraying once too many times or of using too high a rate of insecticide can seem like a small price to pay to avoid the potentially devastating losses from letting a pest problem get out of control. Farmers spray more as a way of ensuring that they will not make a mistake by spraying too little or too seldom, and they spray more because their expectation of the effectiveness of insecticides is reduced by uncertainty (Feder, 1979; Plant, 1986).

Traditional approaches to *Helicoverpa* spp. management. There have been two traditional approaches to a pest management problem as complex as that for bollworm in cotton. One is to use a purely heuristic approach in which some consensus is reached on what set of rules ought to be used to control the pest. The second is to use simulation models to account for the complexity in the system. In the first case, the rules generally evolve over time, changing through a slow process of trial and error as well. These heuristics are then suggested to farmers as general guidelines for deciding how to control the pest. The farmers decide to what extent they adopt the rules. When the suggested rules of thumb seem unlikely to be adopted, the rules are often accompanied by some educational program or demonstration.

This is probably the most natural and practical way to approach a problem for which there appears to be no good solution. Left to themselves, farmers develop their own decision rules based on their particular experience, judgment, and understanding. These, as mentioned above, are generally apt to overcontrol pests. By suggesting rules of thumb of their own, researchers and extension service agents can improve the heuristics of the farmers. Also, by using education and demonstration, researchers and extension service agents can directly augment the experience of individual farmers, making them more likely to accept new heuristics.

This approach illustrates what we call *heuristic pest management.* In highly complex systems, farmers and researchers alike avoid trying to understand all the causal mechanisms and interrelationships in the system and focus instead on a shallower view of the problem. Farmers rely on rules of thumb and published recommendations in nearly rule-based format, and researchers focus their experimentation on shallow correlational phenomena, e.g., the relation between insect damage at a key time in the season and final yield reduction. Ideally, heuristic pest management teaches farmers to behave, i.e., make decisions, as if they have perfect knowledge and understanding of the biological and economic systems without actually having the knowledge. Heuristic pest management is perfectly acceptable if farmers follow decision rules that in most cases lead them to make proper decisions. In such cases it is not necessary for them to comprehend the enormity of the whole problem.

There are two basic problems with heuristic pest management. First, farmers and Cooperative Extension Service recommendations are forced to limit the number of rules to a manageable set, a number that people can carry around in their head and that can be read easily in a short publication. The result is often a set of relatively vague recommendations as we saw in Chap. 3 for *Helicoverpa* spp. This is one argument for delivering pest management advice in the form of rule-based systems, which can maintain a simple user interface while incorporating hundreds of rules. Second, shallow rule-based systems cannot anticipate unexpected situations, yet agriculture is so variable that new situations are almost certain to appear with some frequency.

The economic threshold, discussed in Chap. 3, is the commonly accepted basis for virtually all heuristic recommendations for *Helicoverpa* spp. (Benedict et al., 1989). For example, the published recommendation of the Texas Agricultural Extension Service for bollworms in cotton quoted in Chap. 3 describes several economic threshold densities for bollworms, depending on factors like crop development. That recommendation, which has clearly extended the principle of a static economic threshold, is still an oversimplification of the problem. It is sometimes vague about specific thresholds and is clearly meant to leave room for interpretations. In fact, that published recommendation has been translated into a rule base of just 27 rules (Stone and Schaub, 1990), a very small number compared to other pest management rule bases that can contain several hundred rules.

Furthermore, while one might like to believe that economic thresholds are determined scientifically from objective experimentation and unambiguous results, this simply is not the case with *Helicoverpa* spp. There are too many complex interactions in the system to allow anyone to calculate an economic threshold. Such a threshold would certainly have to vary with the phenological stage of the crop as seen in the Texas recommendation. It would also have to change with the extent of previous damage, vigor of the plants, available soil moisture, expected future infestation levels of bollworm, age structure of the bollworms found in the field, number and species composition of beneficial insects, current and recent weather, price expectation for cotton, etc.

In the face of all this complexity, farmers make control decisions for bollworm by relying on a combination of personal decision rules and those published by the extension service, or those recommended by pest management specialists. Most decision makers use the concept of a pest threshold as their basis, but this threshold is a floating target. The actual threshold used varies among farmers, reflecting their personal experiences, their attitudes about risk, and the type of advice they are getting from experts. In general, farmers' rules for modifying their thresholds will be similar. For example, if there are abundant natural predators in their fields, they are more likely not to spray. If the cotton is growing well and there is a heavy fruit load on the plants, they will tolerate more larvae before spraying.

Using fundamental models. These sorts of complexities have led to the second general approach to the problem of bollworm management: building a model to account for the complexity and using the model to make decisions. Although modeling may be the only way we have of ever being able to deal with the complexity of the problem, using modeling to recommend pest management decisions has suffered from various difficulties as well.

One of the biggest problems facing the modeling approach is validity of the model. From our list of factors that influence an economic threshold, we can see that a model should include plant growth in relation to weather and cultural practices, bollworm immigration and in-field development, age-specific feeding rates and mortality rates, beneficial insect population dynamics, plant compensation, fruit shed rates, and lint production functions. There must be a component that converts the amount of cotton lint produced to harvested yield and another that estimates cotton price to calculate revenue. Finally, the model must accurately describe pesticide applications, including the effects on all age-classes of bollworm and beneficial insects. Very few models even approach this level of complexity. Most leave out one or several features. For example, some cotton models do not include a water component, many do not include the influence of fertilizer, and most do not vary all these various parameters with cotton cultivar.

Those models that do include the key factors needed to analyze the problem of bollworm pest management face further problems. First, such models are very complex and require considerable data to parameterize them for a particular growing region, soil type, and cotton variety. In order to help a farmer make a tactical decision in a specific field, the model must be further parameterized to respond like that particular field. This is extremely difficult and beyond the capabilities of virtually all plant models. A particular field has characteristics like slope, drainage, salinity, soil compaction, water gradients from irrigation, and neighboring field and edge effects that are absolutely specific to that field. Modifying a crop model to fit the growth of one specific field would take as much effort as making the model fit a whole growing region. It simply is not feasible. As a result, virtually all models of *Helicoverpa* spp. populations do not attempt to predict on a field-to-field basis for more than a few days. Instead, models have focused on predicting general or average populations over time based on estimates of areawide population densities (Stinner et al., 1974; Hartstack et al., 1976; Brown et al., 1983; Ives et al., 1984) or have limited their predictions to a short time span (Hopper and Stark, 1987; Legaspi et al., 1989).

The problem of parameterization is ameliorated if we discard the idea of using the simulation to model specific fields. In looking at a strategic problem like how to discover nearly optimal control rules for *Helicoverpa* spp. management, we are interested not in specific fields, but in typical fields. A simulation model parameterized to mimic the dominant variety of cotton in the growing region of interest will suit the purposes of strategic analysis. So, let us assume that we have a model that responds just like a typical field on a farm. We are now faced with the problem of decision making. Simply being able to predict the crop-pest dynamics does not automatically solve the problem by any means. We still need a decision-making component to run the model, collect the appropriate output, and make a decision.

A simplistic approach to the related tactical problem of controlling specific bollworm infestations illustrates our new problem. Suppose we find some bollworms in a cotton field and we want to use our model to see what to do. We run the model up to our current situation using weather data we have been collecting. We tell the model to apply an insecticide to the simulated field, and we then run the model to the end of the season and write down the model's estimate of our revenue less the cost of the spray. We then do the same thing again, only this time we schedule no insecticides. Now we compare our final profits. Suppose, as is likely, the simulation run in which we sprayed with insecticide shows a higher profit, have we solved our problem?

The answer is no. Our analysis has one major and one minor flaw. The minor flaw is that we ran the model only once for each decision, but predicting what will happen in nature is a tricky business. What we should have done was simulate a whole range of likely future scenarios and gathered a whole series of paired profits for comparison. It might be that, on average, spraying produces a higher profit but that the variance is very large, whereas not spraying has a lower average but the variance is much smaller. If that is the case, a farmer might prefer the lower yielding decision because it has a lower probability of very low profit.

The major flaw, however, is that we ignored all the future decisions that our farmer will be called upon to make under our two scenarios. We have mentioned that spraying an insecticide makes it much more likely that in the future a farmer will have to spray again because the beneficial insects as well as the pests have been killed. Also, just because a farmer chooses not to spray today for bollworm does not mean that the farmer should not spray at some later date.

The situation is illustrated in Fig. 7.1 for a hypothetical season in which only four decisions are ever made. Our simple example compared the two sequences of decisions shown in gray. However, to be fair, we should compare the best profit we could make given that we spray today with the best profit we could make given that we do not spray today (Fig. 7.2). That is, we need to compare the 16 possible re-

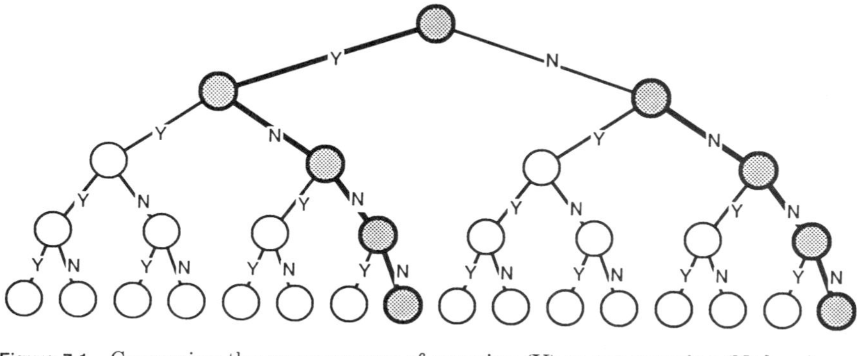

Figure 7.1 Comparing the consequences of spraying (Y) or not spraying (N) by simulating both options but not allowing for further sprayings. Several plausible options are ignored even when there are only four decision points in the entire season.

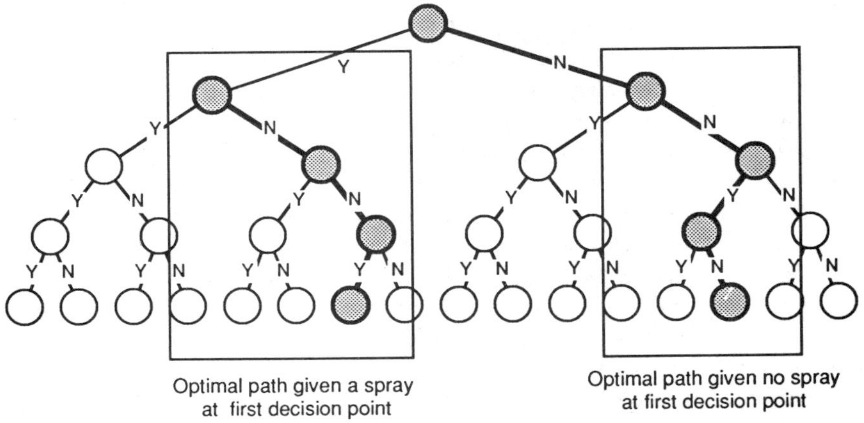

Figure 7.2 The proper comparison, assuming that the choices indicated by the two boxes represent the best we could do given the first decision to spray (Y) or not to spray (N).

sults that represent all possible combinations of the current decision and future decisions. In this simple case, we would run our model 16 times to determine the best choice for any particular expectation of future weather patterns and bollworm immigration. If we use a stochastic model to simulate multiple future scenarios, we would run the model 16 times for each scenario and compare the means and variances of the profits for each of the 16 different sequences of decisions.

Assuming that 100 scenarios could capture the range of expected future events and that each simulation run takes 30 seconds of computer time, the 1600 required runs would take more than 12 hours. A more realistic scenario would involve more than four decision points in a season and would take much longer. Clearly, this approach cannot work for a tactical problem in which a decision is required almost immediately.

The time required for the analysis could be reduced greatly if we had some idea of the best way to proceed in the future. For instance, if we could simulate each of the tactical options today, but use a nearly optimal rule base to make all subsequent decisions, we would be able to compare just two simulation runs for each scenario of future weather and bollworm infestations. However, this is a catch-22. In order to make the appropriate comparison to make a recommendation today, we must know how to identify the best decisions in the future. But if we knew how to choose the best decisions in the future, we could simply use the same method to choose the best decision today.

We are facing what AI researchers would call a problem in *search*. We are searching for optimal sequences of decisions through the course of a season. We would like to generate either (1) a methodology that allows us to apply a model to solve today's problem of whether or not to spray or (2) a set of rules that allows us to identify the best choices at all decision points. Solving the second problem, the strategic problem, effectively eliminates the need to solve the first, and for the rest of this chapter, we will concentrate on the second problem.

7.2 Search

The problem of developing a rule base to provide farmers with a near-optimal decision process for *Helicoverpa* spp. control in cotton consists of two different problems in search. The first is how to choose the proper sequence of decisions within a single production season so that the end result can be considered nearly optimal. The second involves the refinement of the set of decision rules being developed: How should changes be made in such a rule base to achieve near-optimal performance as efficiently as possible?

7.2.1 The general search problem

The classical search problem involves finding a path through a network, from one node to another. In our example, the nodes represent decision points in the growing season, but for now we will use a much more elementary example: finding a path from a starting node to a goal node. Consider the problem of driving a tractor from field A to field B on an unfamiliar farm. The fields are connected by a variety of paved and unpaved roads, and we want to go as quickly as possible, i.e., travel the shortest possible distance. The network in Fig. 7.3 shows the roads and intersections, with field A being the starting point and field B, the goal. The problem is now to find the path from A to B so that the total distance traveled is acceptably low.

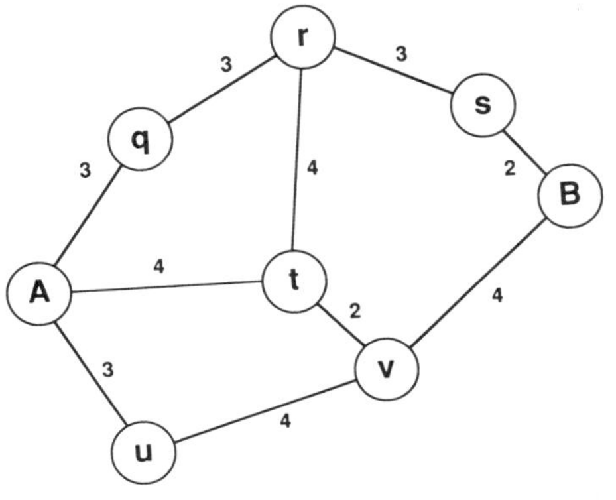

Figure 7.3 Problem of moving from A to B shown as a network. Paths are indicated by lines connecting the nodes. Each node is an intersection where a decision must be made. The numbers on the paths represent the distance between the indicated nodes.

There are now several questions we can ask. What is a possible path from A to B? What is a good path? What is the best path? Note that the process of finding a path takes time and energy. It would not be worth spending an extra 10 minutes to find the best path if the average path takes only 5 minutes less than the best one. At times, it would be sufficient just to find any path from A to B.

It is much easier to talk about different paths if we have a way of representing them discretely. If we start at A in Fig. 7.3 and show all possible paths from A except those that repeat or circle back, we will produce the tree shown in Fig. 7.4. In the tree representation, the starting node is called the *root,* and the nodes immediately below and branching from it are its *children.* Nodes with no children are called *terminal nodes* or *leaves.* We are justified in eliminating the circular paths because they are clearly inefficient. The general search problem can now be thought of in terms of how to traverse the tree, starting at the root A and moving along the branches until reaching an acceptable terminal node B. If it were possible in a reasonable amount of time to enumerate all the paths and sum up their costs of traversal,

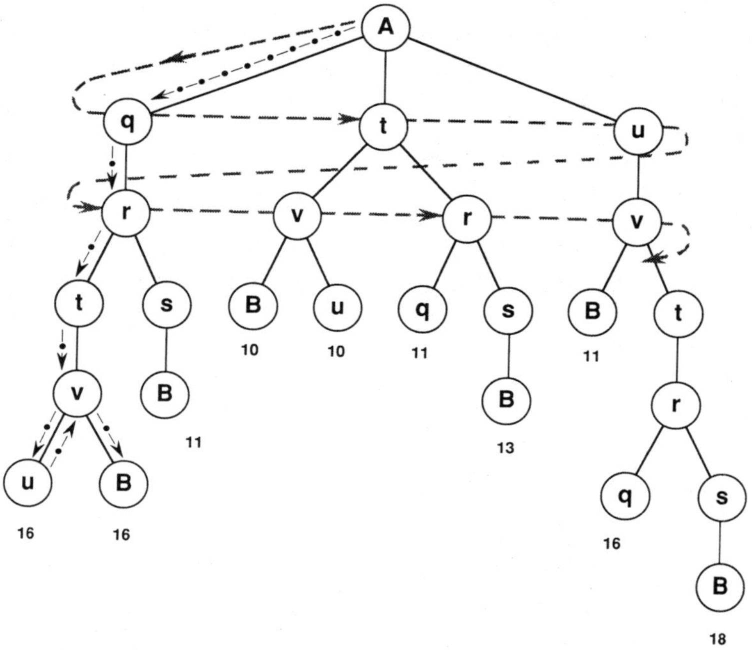

Figure 7.4 The search network from Fig. 7.3 is shown rearranged as a search tree. The root node A is the starting point, and all paths are listed as branches with all circular paths eliminated. The total cost for each path is shown at the terminal nodes. The dot-dashed line indicates the sequence of the depth-first search, and the dashed line shows the breadth-first search.

the solution would be trivial. One could just pick the path with the lowest total cost. This strategy has been called by various derogatory names including the *brute-force method* and the *British Museum procedure* (Winston, 1984). The trouble is that even in fairly small problems, the tree can get very large and enumerating all the possible paths gets to be very inefficient.

Sometimes, we only need to find an acceptable path, e.g., when differences in cost are negligible or when finding even one solution is difficult. There are several common strategies for finding an acceptable path on a search tree, including *depth-first, breadth-first, hill-climbing, best-first,* and *beam search* strategies. Strategies for finding the optimal path are more complicated regardless of whether one uses an AI approach or an OR approach like dynamic programming. Finally, there are strategies that are useful for search in games, i.e., problems in which players take turns searching the same tree but for opposite goals. While the problem of pest management may be thought of as a game in which pests and humans take turns countering each other's tactics, in fact, the pests and humans are pursuing different goals using different tactics. They do not play by the same rules. Therefore, while these strategies have typically been used in AI programs that play games like checkers and chess, they are not applicable to our problem of pest management.

7.2.2 Finding acceptable paths

One simple method of finding a path from the root to the goal is depth-first search. The idea is to follow a single path down the tree until you reach a terminal node. If that node is a goal, you have successfully solved the problem. If the node is not a goal, you retrace your steps to the last decision point and take another path at that point.

An algorithm adapted from Winston (1984) can be used to implement depth-first search. It uses the concept of a *last-in, first-out* (LIFO) stack that we first encountered in Chap. 2 while discussing backward-chaining inference.*

1. Push the root of the tree onto the LIFO stack.
2. Proceed until there are no more nodes on the stack or the goal is reached.
   a. Check whether the first element on the stack is the goal; if it is, the search is complete and skip to step 3.

---

*The classical model of goal-oriented search or backward chaining is a type of depth-first search. A particular branch of the dependency network is followed as far as possible until the problem is solved or there are no more rules to process.

   b. Pop the first node off the stack.
   c. Push that node's *children,* nodes immediately below the current node, onto the stack.
3. Report a failure if the goal has not been reached; otherwise, report a success.

Figure 7.4 shows the path taken by a depth-first search for our tractor example. We start with just A on the stack, shown as {A}. Since this is not the goal, we pop A off and put A's children on so that the stack now looks like {q, t, u}. Since q is not the goal, we pop it off, and replace it with r, and the stack now looks like {r, t, u}. We then do the same with r, leaving {t, s, t, u}. Continuing gives us {v, s, t, u} and {u, B, s, t, u}. Since u is not the goal, it is popped off the stack, but since it has no children, i.e., it is a terminal node, nothing replaces it. The stack then changes to {B, s, t, u}, and the problem is solved at the next step.

Depth-first search is an efficient method only if most paths terminate in an acceptable solution. One danger of depth-first search is that if the branches of the tree are uneven, the search can get bogged down in a deep but unproductive branch. In our example, an acceptable solution was found quickly, but it is by no means the optimal solution.

The opposite approach to depth-first search is breadth-first search, which is useful if the search tree is very uneven or if several long branches are expected to yield no solution. In breadth-first search, one steps down the levels of the tree one at a time, moving from branch to branch at each level in search of the goal node. The algorithm for breadth-first search is very similar to that for depth-first search except that instead of a LIFO stack we use a first-in, first-out (FIFO) queue, similar to a line at a bank teller. New items go in at the back of the queue and items are removed at the front. The following is also adapted from Winston (1984).

1. Place the root node of the tree at the front of an empty FIFO queue.
2. Proceed until there are no more nodes in the queue or the goal is reached.
   a. Check whether the first element in the queue is the goal; if it is, the search is complete and skip to step 3.
   b. Remove the first node from the queue.
   c. Put that node's *children,* nodes immediately below the current node, at the end of the queue.
3. Report a failure if the goal has not been reached; otherwise, report a success.

Figure 7.4 also shows the path taken by a breadth-first search for our tractor example. We start again with just A in the queue, shown as {A...}. Since this is not the goal, we remove A and put A's children in so that the queue now looks like {q, t, u...}. Since q is not the goal, we remove it and put r, its child node at the back of the queue, leaving us with {t, u, r...}. We next remove t since it is not the goal, and place v and r at the end of the queue. Now we have {u, r, v, r...}. After removing all these and placing their children at the end each time, we have {v, t, s, B, u, q, s...}, which will yield a solution after three more iterations. Notice that the queue begins to get very long as the search proceeds down a few levels. This strategy will become very inefficient for deep, even trees.

Adding some intelligence. You have probably noticed that these algorithms are not in the least intelligent; they merely plod along. Fortunately, there are ways to apply knowledge about the problem to speed up the search process. One is to guide the ordering of the nodes examined based on some knowledge about how close each node is to a solution. Another is pruning branches of the search tree that do not seem to be promising. We will look at two examples of the first type called *hill-climbing* and *best-first* search. Examples of the second type include *beam* search, which is related to breadth-first search, and *heuristic pruning.*

Hill-climbing search is just like depth-first search except that the child nodes are pushed onto the stack after being sorted according to some heuristic rule. The idea is to try first those nodes that seem closer to the goal. In our example, the notion of closer can be measured in line-of-site distance (Fig. 7.5). In other applications it might be some more abstract or heuristic measure of closeness. The algorithm for hill-climbing is very similar to that for depth-first search.

1. Push the root of the tree onto the LIFO stack.
2. Proceed until there are no more nodes on the stack or the goal is reached.

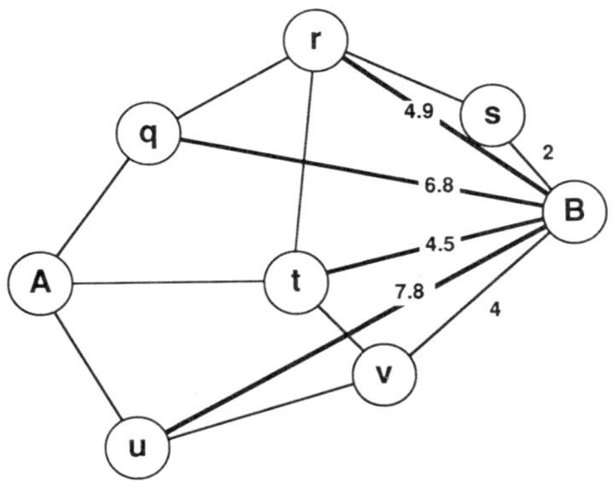

Figure 7.5 The tractor problem showing the visual distances from each node to the destination field.

a. Check whether the first element on the stack is the goal; if it is, the search is complete and skip to step 3.
b. Pop the first node off the stack.
c. Sort the children of that node according to closeness to the goal.
d. Push that node's *children* onto the stack in order from the farthest to the closest node, i.e., the closest node is the last in and first out.

3. Report a failure if the goal has not been reached; otherwise, report a success.

In our example, we start by putting A in the stack and then replace it with q, t, and u. However, we first push u (distance from B equal to 7.8), then q (distance of 6.8), and t (distance of 4.5). The stack is then {t, q, u}, and we then remove t, replacing it with r and v pushed in that order to yield {v, r, q, u}. Then we remove v and push u and B in that order to give us {B, u, r, q, u}, which yields a solution at the next iteration. Notice that this method gives us the optimal solution, a total cost of 10.

Best-first search is like the hill-climbing search except that all the nodes in the stack are sorted at each iteration. You might want to convince yourself that in the tractor example, best-first search will give exactly the same result as the hill-climbing approach. Figure 7.6 contains an example in which a best-first search is better than a hill-climbing search.

Search methods with pruning. The beam search method uses the same measure of proximity to the goal state as hill-climbing and best-first search, but beam search uses the measure to eliminate whole branches of the search tree from consideration. Beam search is otherwise like a breadth-first search strategy. At each level of the search

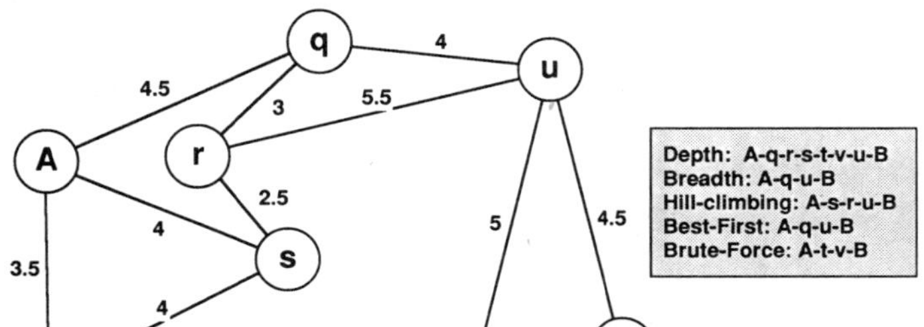

Figure 7.6 Another configuration of roads in which a best-first search is better than a hill-climbing search. Distances are represented proportionally. The paths selected by five different search methods are shown at right.

tree, only a predetermined number of nodes are examined further. The nodes that are examined are those deemed closest to the goal by some heuristic measure. Again, in our example we can use the visual distance as the measure of closeness to see how the search operates.

With a maximum node number of 2, beam search starts with just A in the queue, {A...}. This is removed and replaced by A's children along with estimates of the distance from each to the goal {q(6.8), t(4.5), u(7.8)...}. Only the two best are kept; u is, therefore, removed. Then, since q is not the goal, it is removed and r, its child node, is placed at the back of the queue, leaving us with {t, r...}. Next t is removed and its children placed at the back of the queue, and we find ourselves at the next level with the queue, {r(4.9), v(4.0), r(4.9)...}. The first two of these are kept based on the distance estimates, and then a breadth-first search continues from {r, v}. The first is replaced with t and s, and the second with B and u, which yields {t(4.5), s(2.0), B(0.0), u(7.8)...}. Only s and B are kept, and after one more iteration, B is discovered as the solution.

Even with a very deep and wide search tree, beam search guarantees that the number of paths explored will never be greater than the width of the beam multiplied by the depth of the tree. However, there is no guarantee that the better paths or even all possible solutions are found. The next section on choosing the optimal path addresses this problem and also discusses heuristic pruning, another way to eliminate unlikely paths from further consideration.

7.2.3 Finding the optimal path

Finding the optimal path in a search tree is a problem not usually associated with AI but with OR. As we develop this section, however, we will see that dynamic programming, the OR methodology that deals with problems involving sequences of related decisions, can be thought of as simply another search problem like those discussed above.

As we have seen, one sure way to find the optimal path is to examine all paths and pick the best one. This brute-force method is only helpful for very small trees. For larger trees we will use the A^* search algorithm (for a LISP implementation, see Tanimoto, 1987), a method that combines the optimization techniques with heuristics to find optimal paths through trees. To understand A^*, one must first understand its component parts, *branch-and-bound* search and the *principle of optimality* from dynamic programming (Bellman, 1957).

We begin with the branch-and-bound method. As the name implies, the idea is to move from branch to branch and to do some pruning. Basically, instead of keeping track of nodes in a list, we keep track of partial paths from the root node to some current node within the tree. For each partial path, we also keep track of the total costs incurred. At each step, the search continues from the last node of the partial path with the lowest cost incurred so far. That partial path is extended by adding the last node's children to the path, creating one or more new partial paths. These are then added to the list, the list is sorted by total cost, and the process is repeated until all paths are complete or the goal has been found.

Recalling Fig. 7.4, a branch-and-bound method would begin with the partial paths from the root node to q, t, and u. Our queue of partial paths and their costs is {A-q (3), A-t (4), A-u (3)...}, which is sorted by cost to give {A-q (3), A-u (3), A-t (4)...}. The first path is removed and extended, and the new partial path added to the list after sorting. Now we have {A-u (3), A-t (4), A-q-r (6)...}. The first path is extended again, leaving us with {A-t (4), A-q-r (6), A-u-v (7)...} after sorting. Another iteration gives four paths as the two children of the t node are added: {A-t-v (6), A-q-r (6), A-u-v (7), A-t-r (8)...}. At the next step, the goal node appears in a path in the queue but not at the head of the list: {A-q-r (6), A-u-v (7), A-t-r (8), **A-t-v-B (10),** A-t-v-u (10)...}. We cannot stop yet because it is still possible that one of the three partial paths with costs less than 10 could reach the goal at the next step and have a total cost of less than 10. The search must continue as follows:

{A-u-v (7), A-t-r (8), A-q-r-s (9), **A-t-v-B (10),** A-q-r-t (10), A-t-v-u (10)...}
{A-t-r (8), A-u-v-t (9), A-q-r-s (9), **A-t-v-B (10),** A-q-r-t (10), A-t-v-u (10), A-u-v-B (11)...}
{A-u-v-t (9), A-q-r-s (9), **A-t-v-B (10),** A-q-r-t (10), A-t-v-u (10), A-u-v-B (11), A-t-r-q (11), A-t-r-s (11)...}
{A-q-r-s (9), **A-t-v-B (10),** A-q-r-t (10), A-t-v-u (10), A-u-v-B (11), A-t-r-q (11), A-t-r-s (11), A-u-v-t-r (13)...}
{**A-t-v-B (10),** A-q-r-t (10), A-t-v-u (10), A-u-v-B (11), A-q-r-s-B (11), A-t-r-q (11), A-t-r-s (11), A-u-v-t-r (13)...}

Now, at last, we can stop, declaring A-t-v-B the optimal path, since all other possible paths, even the partial ones, are already at least as costly. Figure 7.7 shows the paths searched and the sequence of searching. Notice that while not all the nodes were evaluated, this search method is not as efficient as it might be. For example, there is really no point in expanding both partial paths A-t-r (8) and A-q-r (6). If the optimal path goes through r on the way to B, then obviously we should expand only the shorter of the two paths to r. The path A-q-r-(best path from r to B)-B will always be two units shorter than A-t-r-(best path from r to B)-B because A-q-r is two units shorter than A-t-r, and there is only one best path from r to B.

We can now write a general method for performing branch-and-

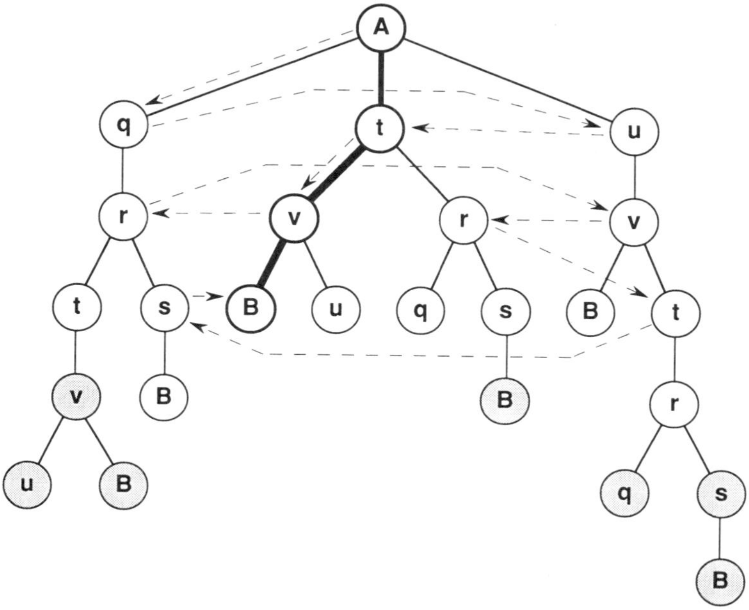

Figure 7.7 Branch-and-bound search sequence (dashed lines with arrows) and solution (heavy path) for the tractor example. The shaded nodes were never searched.

bound search amended with a pruning rule to eliminate suboptimal partial paths (modified from Winston, 1984).

1. Place the root node of the tree at the front of an empty FIFO queue of partial paths.
2. Proceed until there are no more paths in the queue or the goal is reached.
   a. Check whether the first partial path in the queue reaches the goal node. If it does, the search is complete and skip to step 3.
   b. Remove the first partial path from the queue.
   c. Extend that path by adding the *children* of the last node of the partial path to the end of the path.
   d. Add the new path(s) to the queue, and sort the queue by the cost accumulated so far by each path.
   e. If two or more partial paths end in the same node, eliminate all redundant paths except the one with the lowest accumulated cost.
   f. If a partial path ends in the same node as a previously evaluated partial path with a lower cost, eliminate that partial path.
3. Report a failure if the goal has not been reached; otherwise, report a success.

Using this algorithm on the tractor problem speeds up the search considerably.

{A (0)...}
{A-q (3), A-u (3), A-t (4)...}
{A-u (3), A-t (4), A-q-r (6)...}
{A-t (4), A-q-r (6), A-u-v (7)...}
{A-t-v (6), A-q-r (6)...}
—[A-u-v (7) and A-t-r (8) eliminated]
{A-q-r (6), **A-t-v-B (10)**...}
—[A-t-v-u (10) eliminated because of A-u (3)]
{A-q-r-s (9), **A-t-v-B (10)**...}
—[A-q-r-t (10) eliminated because of A-t (4)]
{**A-t-v-B (10)**, A-q-r-s-B (11)...}

Done!

Dynamic programming. The last example illustrated the principle of optimality, namely that the optimal path from A to B through a point X is the optimal path from A to X plus the optimal path from X to B (Bellman, 1957). This might seem like little more than common sense. However, when applied to a search tree, it is a powerful pruning rule. In the tractor problem, it lets us prune the search tree as we proceed from the start to the goal. In dynamic programming it can also be used to prune when searching the tree in the opposite direction.

Dynamic programming is a quantitative approach to finding an optimal sequence of decisions over time. Typically, as in classical optimization analysis (see Chap. 5), there is an objective function that one wants to maximize or minimize. In agricultural examples, the objective function is generally a profit function, and the sequential decisions are crop management decisions (see Chaps. 5 and 9). Shoemaker (1981) has used dynamic programming extensively in agriculture. For example, Shoemaker and Onstad (1983) look at the long-term consequences of decisions that must be made once each year in alfalfa, such as the timing of the first cutting. This single decision has a large impact on one of the key pests of alfalfa, the alfalfa weevil. Each year, the same decision must be made, with the same options. However, the results of the previous year's decision affect that of the current year.

When thought of as a search problem, the multiyear (n years) crop management problem includes three elements. First, the tree has a repeated branching pattern in which the level of the tree is equivalent to an individual year, and the branching points represent the same annual decision with the same options. Second, the profit for a given decision in a given year replaces our notion of the cost of traversing the path between two nodes and is calculated by some set of equations that depend on the previous year's choice and the current year's choice. Third, the goal is to maximize the profit over the complete path from year 0 to year *n*, rather than to minimize the cost.

Because the branching sequence is repeated, it is possible to proceed backward from the final year of the analysis to the first, using the principle of optimality. Figure 7.8 shows a hypothetical problem over four years in which at each year there are three choices. We assume the choice is one level of an input like water or nitrogen, and the choices are *none* (N), *some* (S), and *lots* (L). Consider the final year's problem. There are 81 possible paths, but there are only nine different possible combinations for that year. The previous year's choice was either *none, some,* or *lots,* and this year's choice must be from the same

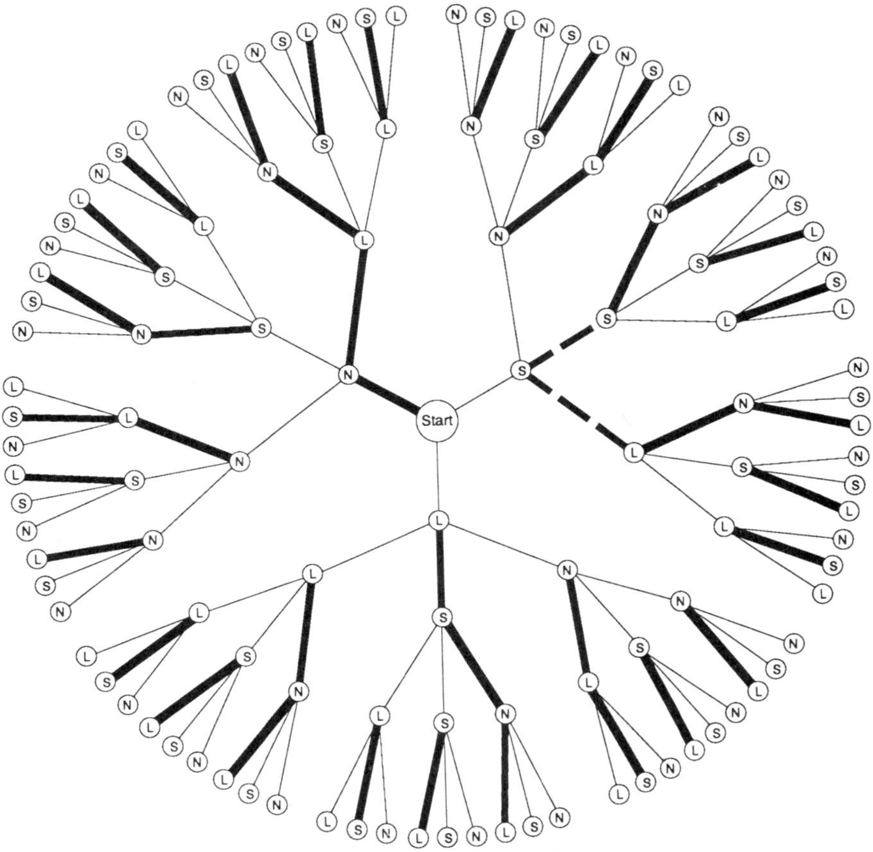

Figure 7.8 A search tree for the dynamic programming problem. The optimal paths are shown in heavy black starting with the outermost subpaths and moving inward. Each path is optimal assuming the previous node has been reached. The dashed lines indicate a tie, where each path is equally as profitable.

list. There are really only nine evaluations to make to calculate the profit expected in the last year.

Therefore, for each of the previous year's choices, there is only one optimal decision for the final year. Let us assume our profit function gives the following values. Notice that using *lots* of the input gives the highest values for profit, but that if *lots* of the input was the selection for the previous season, then profits in the current season are reduced. In other words, using *lots* of the input is good in the short term but bad in the longer term (Table 7.1). The numbers in bold in Table 7.1 are the maximum profits given the previous year's input (indicated in the left column of the table). Thus, in the final year, if the previous year's input was *some* or *none, lots* should be used, and the profit will be 1.3 or 2.2 units, respectively. The principle of optimality states that the optimal path through any node is the optimal path up to that node plus the optimal path from that node to the goal. We have now computed the optimal paths from the 27 nodes at the next-to-last year to the terminal nodes.

In Fig. 7.8, this result is reflected in the subpaths leading directly from the terminal nodes of the tree. Notice in each group of nine terminal nodes the same pattern is repeated. Once we know the last decision, we can more easily compute what to do at the second-to-last decision. We must keep track of the optimal paths and the accumulated profits for the optimal choices as we proceed.

In the next-to-last year, the situation is much the same. We can choose any of nine combinations. However, the profit for each calculated path now is added to the appropriate optimal path that we have just calculated. So if the input for the third year is *none,* we add 2.2, the best we are able to do in the fourth year based on the third year's decision of *none,* to each row of the *none* column. The rest of the table is filled in the same way, and the optimal choices for each category of the third year's input is noted in bold (Table 7.2).

Calculations for the second year proceed in the same way. We already know the best we can do in the third and fourth years, so all we need to do is look at the nine different choices, evaluate their profits,

TABLE 7.1 Expected Profits in the Fourth Year of the Dynamic Programming Example*

	Fourth year		
Previous year	None	Some	Lots
None	0.6	1.4	**2.2**
Some	0.5	1.1	**1.3**
Lots	0.3	**0.9**	0.8

*Profits are based on inputs from the third and fourth years.

TABLE 7.2 Profits for the Third Year of the Dynamic Programming Example*

	Third year		
Previous year	None	Some	Lots
None	0.6 + 2.2	1.4 + 1.3	**3.1** = 2.2 + 0.9
Some	**2.7** = 0.5 + 2.2	1.1 + 1.3	1.3 + 0.9
Lots	**2.5** = 0.3 + 2.2	0.9 + 1.3	0.8 + 0.9

*Profits are based on inputs from the second and third years.

TABLE 7.3 Profits for the Second Year of the Dynamic Programming Example*

	Second year		
Previous year	None	Some	Lots
None	0.6 + 3.1	1.4 + 2.7	**4.7** = 2.2 + 2.5
Some	0.5 + 3.1	**3.8** = 1.1 + 2.7	**3.8** = 1.3 + 2.5
Lots	0.3 + 3.1	**3.6** = 0.9 + 2.7	0.8 + 2.5

*Profits are based on the inputs from the first and second years.

and add them to the previously obtained optimal profits (Table 7.3). Notice that we have a tie. This means that if in the first year we chose *some,* we can either put on *some* or *lots* in the second year. Both choices yield the same maximum profit.

In order to calculate the best choice for the first year, we need to know what was done the year before our first year. We assume that *some* was applied then. We now must make only three more calculations to determine the total profit for the first year's decisions (Table 7.4). The optimal choice for year one is to apply no input, which leads to a maximum profit of 5.2 units by the sequence *none-lots-none-lots.* Just for fun, you might want to convince yourself that if the owner of the land put out none the year before our first year, the best path would be *lots-some-none-lots,* or that in general the best long-term strategy is alternating between *none* and *lots* from year to year.

TABLE 7.4 Profits for the First Year of the Dynamic Programming Example*

	First year		
Previous year	None	Some	Lots
Some	**5.2** = 0.5 + 4.7	1.1 + 3.8	1.3 + 3.6

*It is assumed that in the year before the start of the analysis a moderate amount of input was used.

Adding heuristics to optimal search. It might seem that finding the optimal path and using heuristics to help are not compatible. Heuristics are by nature qualitative and subjective, unlikely to have the kind of precision needed to assure optimality. However, heuristics can be used to speed up search, even when an optimal path is sought.

For example, it may be possible to estimate the distance to the goal from any node in the search tree as we did before for hill-climbing and best-first search. Given such an estimate, we can compute an estimated total path length for each subpath from the distance, profit or cost, already accumulated plus the estimated distance to the goal node. If the estimates were perfect, sorting our queue of partial paths in a branch-and-bound search by this total length estimate would keep us on the optimal path at all times.

Of course, we must expect these heuristic estimates not to be accurate. In that case, overestimating the distance to the goal can be a major problem. An overestimate could result in the optimal path being overlooked for a long time or even throughout the search. However, underestimating the distances to the goal causes no such problem. Bad underestimates can move a suboptimal path ahead of the optimal path in the sorting for a time, but as the suboptimal path is expanded, its estimated length will approach its true total length, and the true optimal path will be restored to its proper position.

We can now write the algorithm for branch-and-bound search using both the principle of optimality and heuristic (under)estimates of remaining path length.

1. Place the root node of the tree at the front of an empty FIFO queue of partial paths.
2. Proceed until there are no more paths in the queue or the goal is reached.
   a. Check if the first partial path in the queue reaches the goal node. If it does, the search is complete and skip to step 3.
   b. Remove the first partial path from the queue.
   c. Extend that path by adding the *children* of the last node of the partial path to the end of the path, and add the new partial paths to the queue.
   d. If two or more partial paths end in the same node, eliminate all redundant paths except the one with the lowest accumulated cost.
   e. If a partial path ends in the same node as a previously evaluated partial path with a lower cost, eliminate that partial path.
   f. Estimate the cost or distance from each partial path to the goal node.

   g. Sort the queue by the cost accumulated by each path so far plus the estimated cost to the goal node.
3. Report a failure if the goal has not been reached; otherwise, report a success.

This is the A^* search algorithm that produces optimal paths if the heuristic estimate of remaining path length or cost is always an underestimate. When applied to the tractor problem, we can use line-of-site distance as our heuristic underestimate since the straight line distance is always less than or equal to the actual distance. The A^* search sequence for the tractor problem is

{A...}
{A-t (4 + 4.5), A-q (3 + 6.8), A-u (3 + 7.8)...}
{A-q (3 + 6.8), A-t-v(6 + 4), A-u (3 + 7.8), A-t-r (8 + 4.9)...}
{A-t-v(6 + 4), A-u (3 + 7.8), A-q-r (6 + 4.9)...}
—[A-t-r (8) eliminated because of A-q-r (6)]
{**A-t-v-B** (10 + 0), A-u (3 + 7.8), A-q-r (6 + 4.9)...}
—[A-t-v-u (10) eliminated because of A-u (3)]

Done!

The paths searched using the A^* search algorithm are shown in Fig. 7.9. Notice the great increase in efficiency in terms of the number of nodes never added to the queue.

7.3 *Helicoverpa* spp. Pest Management as a Search Problem

Recall from Figs. 7.1 and 7.2 that the problem of controlling bollworms in cotton throughout a season can be structured as a search problem by considering each day of the season as a decision point at which a choice is made between spraying or not spraying an insecticide. This tree will look similar to the dynamic programming example above. The ultimate goal is to find the path that maximizes profit, and the decisions at each node are the same so that the tree takes on a repeating pattern. The layers of the tree represent time, and at each layer of the tree each node has two branches. Assuming that bollworms are a problem in cotton for July and August, the search tree will have 2^{62} terminal nodes, far beyond the capacity of brute-force methods for solution.

The *tactical* problem of bollworm management can be thought of as determining the appropriate decision at any particular node of the tree. The *strategic* problem involves developing and refining the search algorithm itself. Ultimately, we want to be able to apply pruning rules so that only a single, nearly optimal path remains. To do this

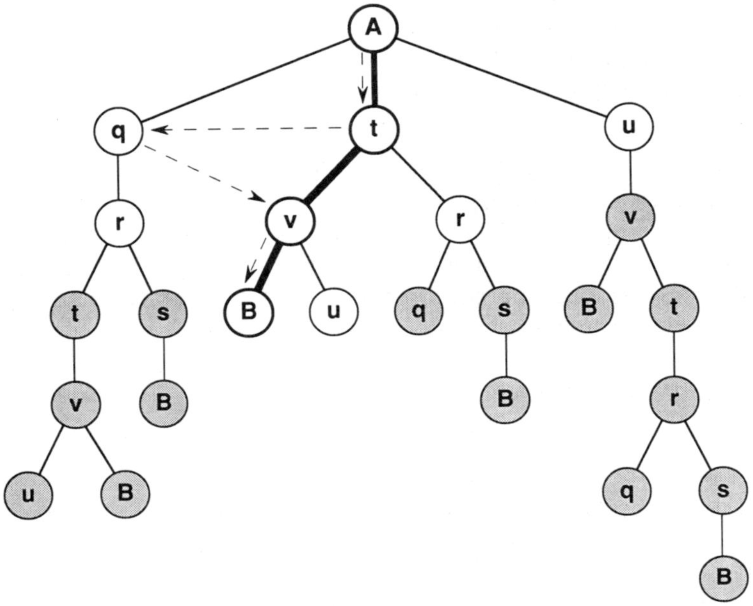

Figure 7.9 Search sequence using A^* search on the tractor problem. The heavy line indicates the optimal path, the dashed line is the searching sequence, and the shaded nodes of the search tree indicate nodes not checked during the search.

we must have at our disposal a fully defined search tree for the problem. In fact, if we want to ensure that our search algorithm works over the expected range of conditions found in the field, we would need several fully defined search trees, representing the results of all the possible decision paths for a large set of seasons. The only possible way to meet this requirement is with a simulation model. Instead of generating all possible paths with the model, we generate paths only as needed in our analysis. We use the model to keep track of the state of the system along all the paths we search and to keep track of the decisions made at each decision point or node in the search tree. The model has to be dynamically linked with a decision-making system so that the effects of decisions are reflected in the system's state. Finally, we need to convert the model's end result from yield to profit, so we need to have fairly detailed information about the number of cotton bolls produced and their size and quality.

A simulation model provides us with the search space we need to implement search strategies. However, finding even the best path through a search space that represents only one season does not solve the strategic problem. Our approach to the strategic problem is to find optimal paths through a set of different simulation scenarios representing a spectrum of natural conditions. The results from these optimization searches are then used as an example set in a rule-induction system to generate decision rules that produce optimal behavior.

7.3.1 A hybrid simulation model and knowledge-based system

The simulation model we need must have several features. The model must be stochastic to be able to represent realistic field conditions, not just mean expectations. It must include the cotton plant, the bollworm pests, and their dynamic interaction so that feeding by the pests results in appropriate physiological compensation by the plant. To simulate pest management, the model has to include a tactical component that mimics the control practices undertaken in the field. The simulation model must be able to generate the type of sampling information that decision makers would normally be able to gather to make control decisions in the field and to calculate the costs of pest control decisions, converting biological yield to revenue. Finally, from a structural standpoint, the model must be able to branch, i.e., stop at any point during a simulation run and record its state variables, so that it can proceed down first one path and then a second path from the same starting point. It must also be able to evaluate a rule base from within its simulation loop, and that rule base must be able to control the generation of branching paths.

An example of this type of model was developed by Schaub and Stone (1989). The biological system was represented by three linked models that were independently developed: (1) a random weather generator, WGEN (Richardson and Wright, 1984), which produces daily maximum and minimum temperatures, rainfall, and solar radiation based on longitude and latitude; (2) the cotton model developed by Gutierrez and his coworkers (Gutierrez et al., 1984) and modified to simulate short-season cotton cultivars under Texas conditions (Sequeira et al., 1989); and (3) a stochastic model of *Helicoverpa* spp. migration and population dynamics described below.

The temperature data shown in Fig. 7.10 compares generated to actual temperature data and illustrates the difference between simulating for prediction and simulating for realism. Notice that the average weather data (Fig. 7.10*b*) follow relatively smooth curves compared to either the field data (Fig. 7.10*a*) or the generated data (Fig. 7.10*c*). The generated weather data show patterns that look much more like actual field data, but on any particular day, the generated weather is much less likely to match the true field conditions than the average data. Even though the average curves provide better estimates of the temperature on any given day, the average season is not a realistic

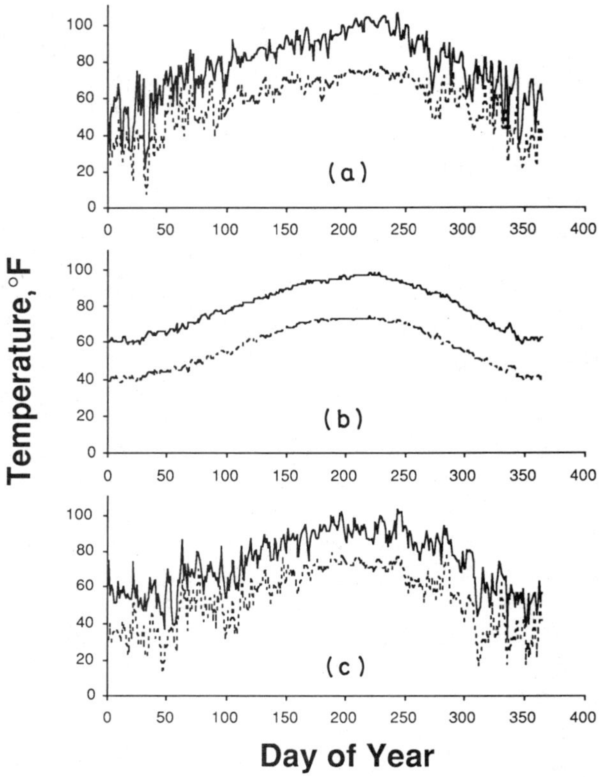

Figure 7.10 Comparison of (*a*) field-recorded temperatures, (*b*) averaged historical weather, and (*c*) computer-generated weather for similar regions in Texas. (*From Stone and Schaub, 1990.*)

approximation of any particular season's weather. There can be no unusual cold spells, no droughts, no storms, etc. In short, there is no such thing as a "typical year."

The cotton model is a deterministic model of cotton growth and development as influenced by weather. Driving variables include temperature, rainfall and irrigation, nitrogen applications, and solar radiation. A cotton crop is modeled based on the predicted growth of a single plant with interplant competition modeled as a function of planting density. All rates are functions of physiological time, measured in degree-days over a lower threshold of 12°C. The plant is modeled as a series of submodels, one each for the leaves, fruit, stems, and roots. Subpopulations are linked through a single metabolic pool and a priority scheme for the allocation of carbohydrates and nitrogen. The model calculates daily numbers and weights for all fruit age-classes,* as well as final yields of lint and seed.

Various insect models have been linked with this cotton model in the past, and the linkage to the *Helicoverpa* spp. model (below) was accomplished based on linkages to Lygus bug (Gutierrez et al., 1979b), boll weevil (Gutierrez et al., 1979a), and pink bollworm (Stone and Gutierrez, 1986). Bollworm damage is reflected in the cotton model as a reduction in the number of fruit in the attacked age-classes. Cotton plant compensation for damage is included through the metabolic pool model. Fruit loss decreases demand for carbohydrates and, therefore, increases the available supply for other fruit and for vegetative growth.

The bollworm model simulates two processes: immigration of bollworm into cotton fields and development of bollworm within cotton fields, subject to weather, natural enemies, and management actions. Development through age-classes is deterministic and is modeled using a rate-summation algorithm and a constant distribution describing the variation in aging rates. Bollworm migration into the system is modeled stochastically, with the number of freshly laid eggs per plant on any given day of the simulation determined from a negative binomial probability density function in which the mean density and aggregation parameter are functions of time. Figure 7.11 shows the probability of finding any given number of freshly laid eggs per 25 plants throughout the cotton growing season. Other stochastic elements of the bollworm model include the original number of beneficial insects in each field and the effect of each insecticide application on the bollworm and beneficial insects. Bollworm damage is modeled after the method of Wilson and Gutierrez (1980). The probabilities of bollworm attack and larval feeding rates are determined as a function of the density and age-structure of both the bollworm population and the cotton fruit.

The three model components are linked together and to a rule-based system as shown in Fig. 7.12. To simulate one field season, the weather generator is run once and the output is stored. The initial number of beneficial insects in the field is chosen randomly, and the bollworm oviposition pattern for the season is generated and stored. The stored weather pattern and immigration (oviposition) are then used as lookup tables during the simulation runs for that year.

The cotton and bollworm models are run in parallel. On each day, the cotton plant growth is updated, taking account of the bollworm

*An age-class is a range of physiological ages. For example, the first age-class for fruit would include all squares up to 100 degree-days old.

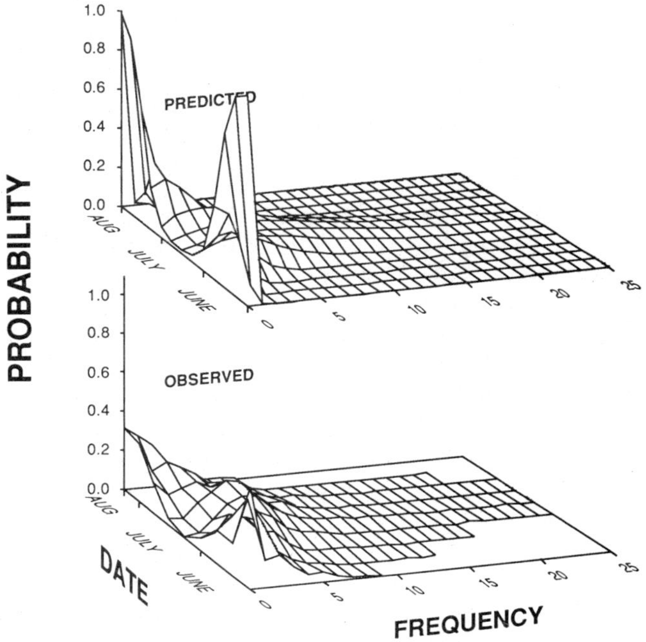

Figure 7.11 Model of the probability of a given number of *Helicoverpa* spp. eggs being laid per 25 plants in a cotton field. (*Modified from Stone and Schaub, 1990.*)

population and age-structure as updated on the previous day. After the cotton fruiting arrays are updated by the cotton model, the bollworm model is run to update the bollworm population and age-structure, as well as the number of fruiting structures damaged. Summary variables representing the state of the system at the end of each day are stored by both the cotton and bollworm submodels and are accessed by a rule-based system written in CLIPS. The rule-based system uses the daily output of the models to make decisions about the future course of the simulation.

Based on its conclusions, the CLIPS component returns information to the simulation. Depending on the instructions returned, the simulation could save its current state variables and schedule the running of another simulation starting with those parameters, and/or schedule a spray for the subsequent day of simulation. This hybrid simulation and rule-based system provides all the functionality for implementing a search algorithm. Heuristics based on the state of the system can be put into the CLIPS component, and branching can likewise be controlled by rules.

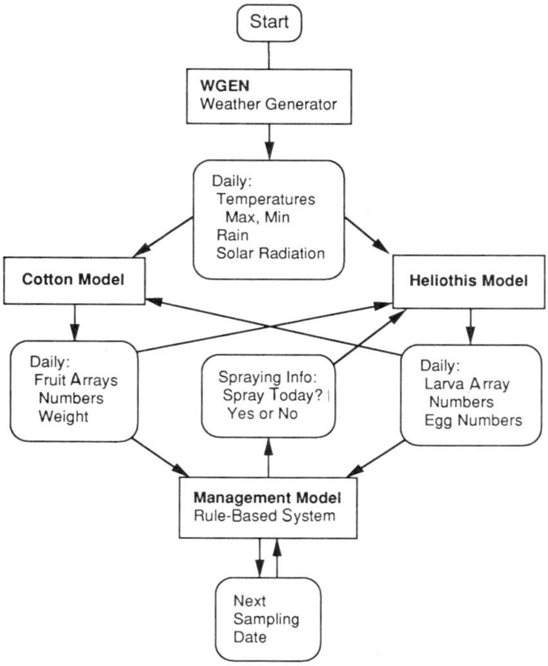

Figure 7.12 Linkage among the various components of the hybrid knowledge-based system and simulation model for *Helicoverpa* spp. and cotton. (*Modified from Stone and Schaub, 1990.*)

7.3.2 Search strategies in *Helicoverpa* spp. management

Now that we have a structure for carrying out our analysis, we can return to the search problem itself. Although the problem is set up much like a dynamic programming problem, the complexity of the system makes dynamic programming intractable. The biggest limitation is that the system's state is different at each decision point and is dependent on the whole series of decisions made in the past. Working backward will not work in this case. Pruning using the principle of optimality in a forward direction is likewise impossible because no two paths or branches will ever lead to the same node. There will always be only one path leading to each node in the search network.

Using AI search in this biological and economic system poses some unique challenges as well. First, there is no specific goal state, there is only a relative goal, i.e., choose the path that gives the best profit. The actual profit that corresponds to the best profit is unknown and changes during the season. For instance, if we are halfway through the damage window with little or no fruit loss and no spraying, our expected net profit should go up. Coming up with a heuristic measure of how close we are to the goal at any particular node is thus impossible. Second, determining the cost of partial paths traversed is very difficult, although it has been attempted for cotton (Stewart and Sterling, 1989). This involves assigning a dollar value to the state of the crop at a specific point in the season. Third, we have to include some way to account for the inherent variance and uncertainty in the system as we evaluate the effects of decisions on the state of the crop and pest.

7.3.3 Heuristic measures for sorting or hill climbing with an expert system

Although we cannot describe the final goal state for our search problem in any specific way before we get there, we can define heuristics that could be used to select the most promising path during the search process. When we developed the hill-climbing search algorithm, we used the straight line distances from all nodes in the search net to the goal to sort the new nodes before pushing them onto the stack. Distance was a convenient choice in the tractor problem, but we could have used any heuristic method for sorting, provided that the most promising of the new nodes ended up at the top of the stack. In the bollworm management problem we can make some general observations about the state of the cotton plant and use them as a sorting criterion, e.g., the more squares per plant the better or the fewer insecticide sprays made the better. Once we have created a heuristic sorting procedure, then we could use any of the search strategies that employ sorting, e.g., hill-climbing, best-first, or beam search strategy.

In order to search for the optimal path without the principle of optimality, we could implement the branch-and-bound search method if we have a good way to estimate the monetary value of a crop only partially grown. As mentioned, some research has been done along these lines. However, without some help, the branch-and-bound procedure can be very slow.

Our approach will be to develop something like the A^* algorithm, replacing the use of the principle of optimality with heuristic pruning. The A^* algorithm sorts all partial paths under consideration by the distance traveled plus an underestimate of the total distance to the goal. That is, the paths are sorted by a heuristic estimate of the total path length, some of which are known with certainty. As we saw, by using an underestimate, the optimal path is never overlooked. In our example, we are not trying to minimize a cost or distance, we are trying to maximize profit. We want to sort our partial paths by an estimate of the total profit each path will achieve when it is finished. This estimate will get better as the season progresses because we will have more information about the state of the crop. Estimating final yield and cost one day before harvest is going to be much more accurate than estimating final yield and cost one day after planting. Instead of requiring underestimates of distance, something we want to avoid, we must use *overestimates* of the profit that will be attained, something we want as much of as possible. That way, since the best path will have the highest real profit, it should always be near the front of the queue of partial paths. Furthermore, as the season nears its end and the overestimates approach the actual values, the best path will move to the front.

Our goal then is to develop a heuristic overestimate of final profit based on the current state of the crop-pest system. This can be achieved by asking the question, "What is the maximum profit possible given the current situation?" That is, assuming optimal conditions from here on, what is the best that one can expect? At times it might be easier to answer the converse, i.e., by at least how much has final profit been reduced up to this point? A simple example occurs after a spray. The spray has cost money, perhaps $20 per hectare. At the very least, the final profit has been reduced $20 from what it could have been at the start of the season. It certainly has been reduced more because of the pests' feeding before the spray, but this amount may be much harder to estimate.

7.3.4 Heuristic pruning

Even with a reasonably good method to estimate the maximum possible profits for partial paths through the search space, the search process could easily get bogged down. For example, there are unlikely to be many appreciable differences in conservative estimates of losses until the season has progressed a couple of weeks into the damage window. The number of paths to be searched in just two weeks would yield 2^{14}, over 16,000, simulations to keep track of, estimate profits for, and sort.

Since we cannot use the principle of optimality to help us, we turn to heuristics again. Experts in *Helicoverpa* spp. pest management may not always make the correct decisions or even the same decisions as other experts in the difficult cases. However, there are many decisions that can be made with virtual certainty and upon which experts will agree. A trivial example of such a decision is that if you have sprayed your field yesterday, you will not spray again for another two days. That rule alone can trim the search tree by half (Fig. 7.13). Another might be that if there is a very high population of small bollworms in

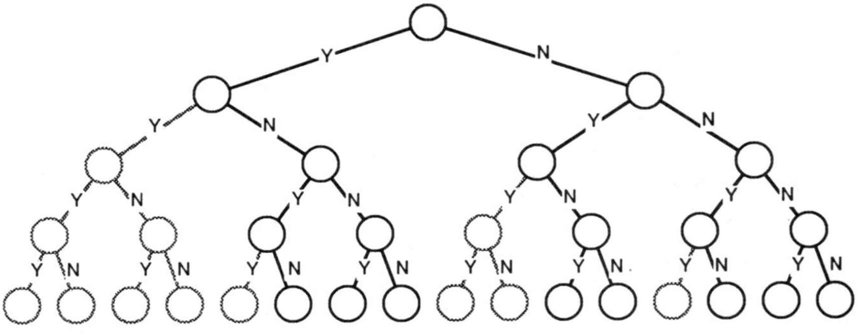

Figure 7.13 Demonstration of heuristic pruning rule: IF you spray today THEN do not spray for the next two days. This rule cuts the number of paths searched from 16 (solid and patterned branches) to 8 (solid branches only). *Y* indicates a spray; *N* denotes no spray.

a field with a large number of small bolls and there are no beneficial insects in the field, then a spray is essential. Even very simple rules can be used to prune unprofitable paths from the analysis.

The final algorithm that will create a report relating the state of the system to the optimal decisions made at each decision point is shown in Figs. 7.14 and 7.15. This algorithm assumes that the simulation model is running in a multitasking environment so that stopping and running another simulation is done by temporarily halting the currently running process, simulation, and forking to a new process to run another branch of the simulation.

7.3.5 Rule induction from example tables

By using the modified A^* search algorithm with heuristic pruning and heuristic overestimation of final profits, we can generate a series of thousands of optimal decisions and the conditions under which they were made. In addition, by delaying the starting date of the damage window, one can artificially trick the program into starting with very bad conditions, so that optimal control decisions can be calculated for situations that should never occur under optimal management.

Our ultimate goal is achieved by using these thousands of examples as input to a rule induction process or as the basis for a case-based reasoning system (see Chap. 8). Both would provide optimal decision rules to farmers in a computer program completely distinct from the hybrid simulation and knowledge-based system that created the examples.

In principle, if a model is capable of recreating conditions in a particular field during the growing season, then the hybrid system could be run to determine the tactical solution to the pest management

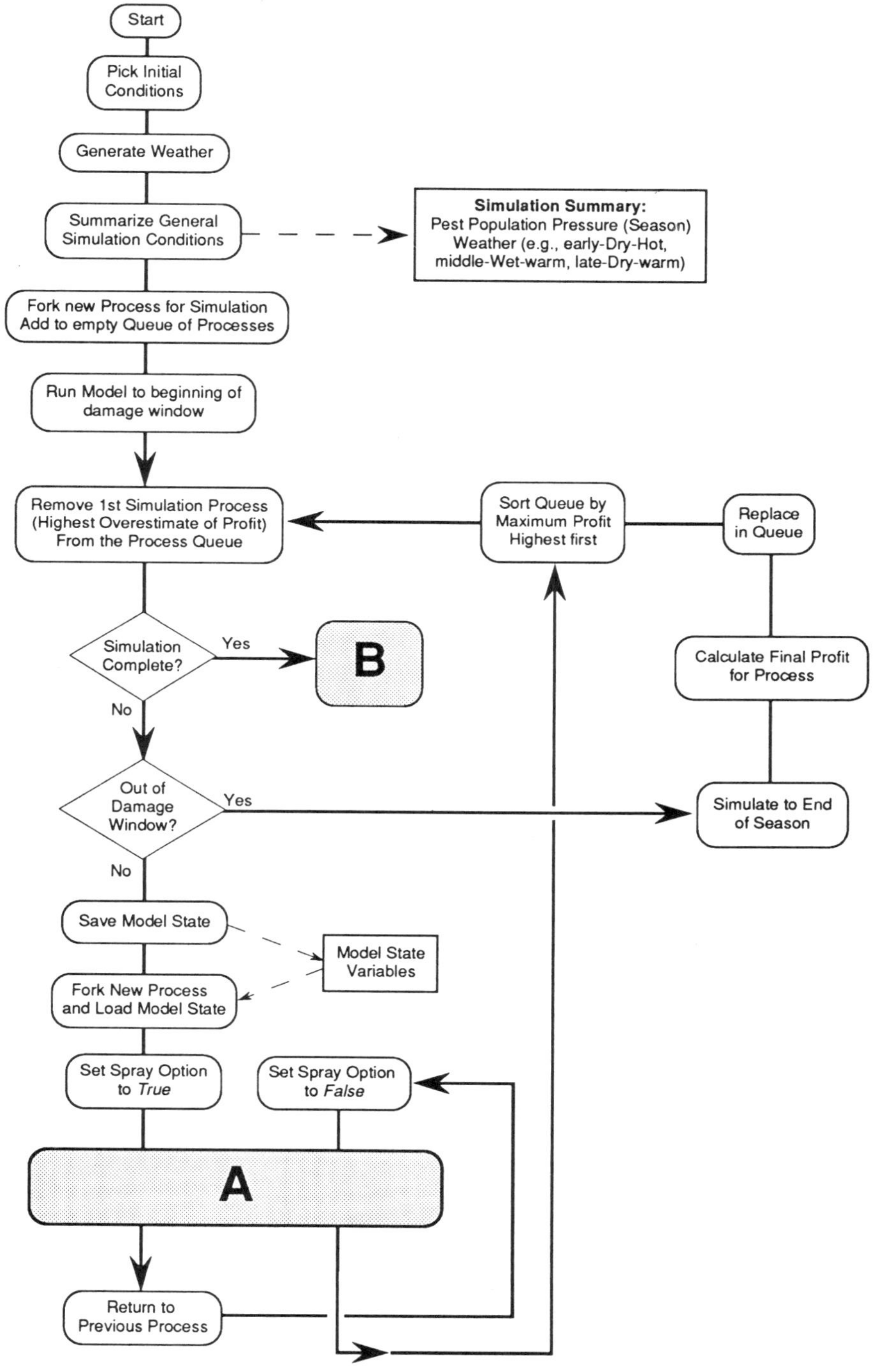

Figure 7.14 Flowchart showing the search algorithm used to evaluate *Helicoverpa* spp. management with a simulation model using a multitasking computer. A and B refer to Fig. 7.15.

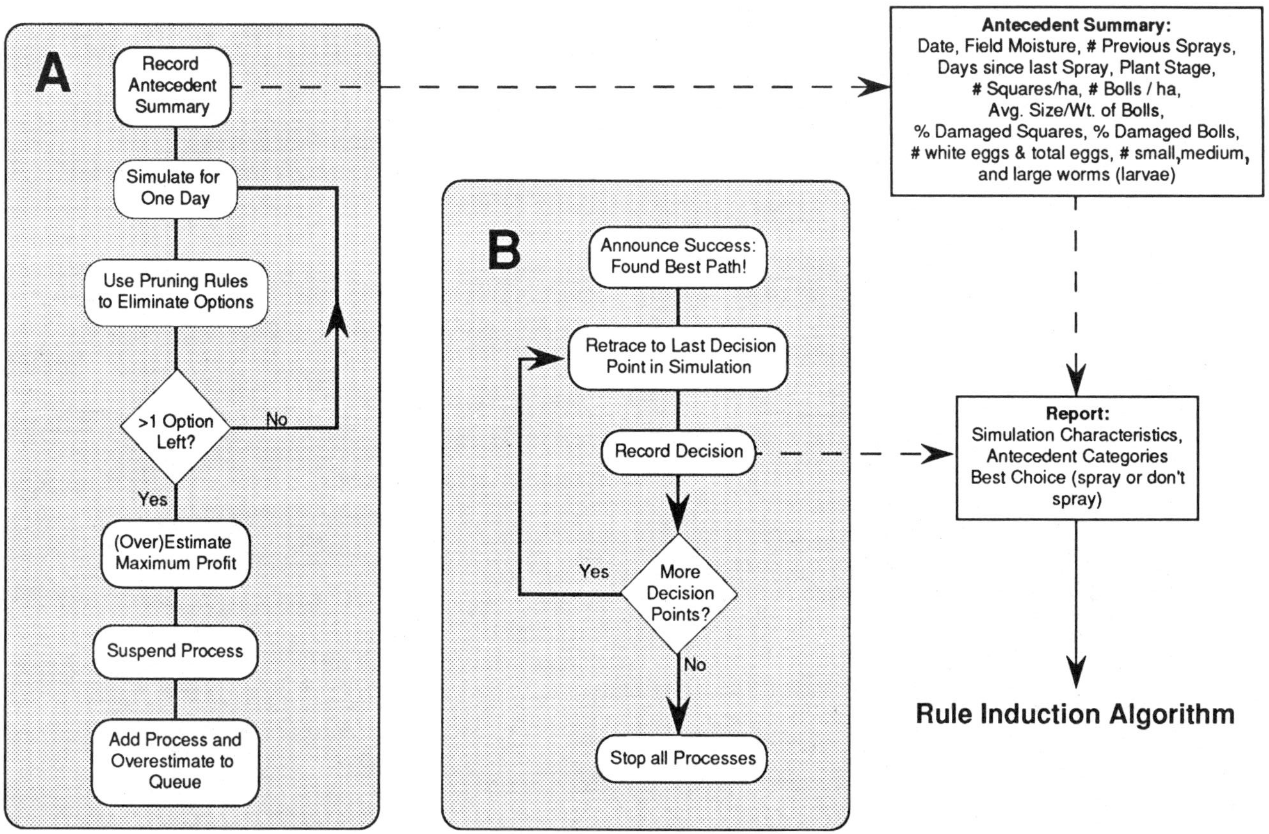

Figure 7.15 The remaining portion of the flowchart shown in Fig. 7.14.

problem. However, this type of analysis takes much too much time to give a farmer an answer in an on-line setting. Furthermore, fitting models to specific field situations has always been more difficult than developing models that behave in general the way natural fields behave. This analysis has the advantage that only the derived decision rules need be delivered to the field.

7.4 Summary and Conclusions

Pest management science is concerned with improving tactical and strategic pest control decisions. We have focused in this chapter on representing the control of a persistent key pest of cotton as a search problem and have seen how a computer simulation model with an embedded capability to control the simulation using heuristics can be used to find the optimal path through a very large search tree. We also saw how the results from several experimental search problems can be used to generate nearly optimal control rules for the pest.

AI search methods provided the model for our heuristic optimization procedure, here called *heuristic programming*. It has the potential to derive deliverable decision rules from a variety of complex simulation models of agricultural systems. However, we must emphasize that the key to the success of this kind of optimization is the knowledge embedded within the system. The simulation must be an adequate representation of the natural system and its variability, or the analysis will be invalid. In addition, the knowledge used for heuristic pruning and heuristic overestimation must be accurate and conservative. The more conservative the heuristics, the higher the probability of finding optimal paths. Unfortunately, conservative heuristics also lead to a higher probability that the system will be overwhelmed with paths to search.

Finally, in the process of building the hybrid model and knowledge-based system to perform this analysis, we have constructed a tool for evaluating knowledge bases. Any set of rules for making control decisions can be used as the heuristic pruning algorithm. A set of decision rules that will solve tactical problems will *always* prune the search tree to a single path. Therefore, this system can be used to compare several different sets of decision rules in a controlled system by systematically using the different rule bases as the pruning heuristics over a set of simulations.

Chapter

8

Knowledge Representation

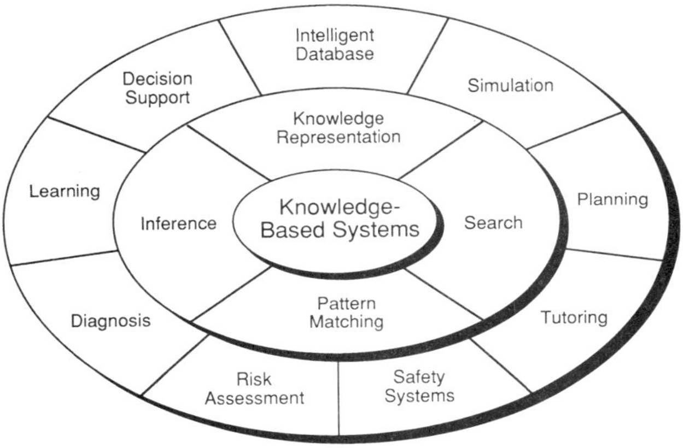

The term *knowledge representation* refers to the study of how to encode human knowledge into a computer program. A good deal of human knowledge takes the form of rules of thumb, e.g., "if it is very cloudy in the morning, then there is a good chance of rain and so I'd better take an umbrella to work," but not all of it can be easily represented this way. For example, such rules are not adequate to represent our knowledge of the laws of physics or of mathematics; there is no natural way to put a statement such as force equals mass times acceleration into rule form. Rules are also inadequate to represent the knowledge of how to perform a complex, step-by-step procedure. For example, while rules can be used to represent many of the decisions we must make in driving a car, e.g., if you come to a stop sign, then you must stop and wait for other cars to pass, they cannot be so easily used to store the knowledge of the many mechanical actions we must make in driving, e.g., start the car by inserting the key in the ignition, pump the accelerator a bit, turn the ignition key clockwise, etc. Moreover, rules are not suitable for storing the large body of *associative* knowledge that all humans possess. For instance, if we see a dog in the street we automatically associate with that dog the fact that it is a mammal, which in turn carries many associations; we might associate it with a dog that we own or once owned, once again carrying many associations of memories connected with the dog, and so forth.

Not all forms of human knowledge need be represented in the construction of knowledge-based systems for use in agriculture. For example, it is unlikely that a detailed representation of the theories of modern physics will ever prove necessary. Nevertheless, the representational inadequacies of simple production rules can quickly become a problem, particularly in the construction of something more than the simplest expert system. For this reason it is useful to look into the theories of knowledge representation and to examine how expert systems have made use of some of these theories. We begin by examining in greater detail the theory associated with the simple production rule, which is called propositional logic. We then discuss other forms of logic such as predicate logic and inductive logic that have been used in expert systems. Next we discuss some of the means of associating and organizing rules, such as frames, scripts, and semantic nets.

8.1 Logic in Rule-Based Systems

8.1.1 Deductive logic: Propositional and predicate

The statement

IF The animal has feathers
THEN The animal is a bird

may be represented abstractly as follows. If we let the propositions P and Q be defined as

P = The animal has feathers
Q = The animal is a bird

then the statement can be written

IF P THEN Q

This is a *sentence of propositional logic.* Propositional and other forms of logic are discussed in books such as Rich (1983), Manna and Waldinger (1985), and Nilsson (1971). The capital letters P and Q in the sentence are called *propositional symbols.* They represent statements, or assertions in the terminology of Chap. 2, that may be either true or false. Sentences in propositional logic are built up from propositions by application of the propositional connectives NOT, AND, OR, IF-THEN, IF-THEN-ELSE, and IFF, i.e, IF AND ONLY IF. The single proposition P is a legitimate sentence. Some more complex examples of propositional logic sentences are (Manna and Waldinger, 1985)

IF (IF P THEN Q) THEN (IF (NOT P) THEN (NOT Q))
IF ((NOT P) OR (NOT Q)) THEN (NOT (P AND Q))
(IF P THEN Q) IF AND ONLY IF ((NOT P) OR Q)

Besides propositions and connectives, the other basic items of propositional logic are the words *true* and *false,* which are called *truth symbols.* Each sentence has a truth symbol attached to it, i.e., each sentence is either true or false. The primary objective of propositional logic is to determine the truth value, i.e., truth or falsehood, of logical sentences from the truth values of other propositions and sentences. Since individual propositions are also sentences they have truth values. Some sentences, for example

P OR (NOT P)

are true regardless of the truth or falsehood of their propositions. Such sentences are called *valid.* Other sentences, such as

P AND (NOT P)

are false, independent of the truth of P. Such sentences are called *contradictory.* For example, the first of the complex sentences above is contradictory and the second two are valid. Note that a sentence can seem to "make sense" in English and still be contradictory in logic. The first of the above sentences might be interpreted in English as "If the statement 'if the animal has feathers, then the animal is a bird' is true, then the statement 'if the animal does not have feathers, then the animal is not a bird' is also true." While this seems reasonable, the equally valid interpretation "If the statement 'if $x < 0$, then $x < 10$' is true, then the statement 'if $x \geq 0$, then $x \geq 10$' is also true" is clearly false.

Some sentences in propositional logic are neither valid nor contradictory. For example, the sentence

IF P THEN Q

is neither valid nor contradictory; it has both true and false instances. However, if the proposition P is known to be true, then the above sentence establishes the truth of Q. The fact that the truth of P together with the sentence IF P THEN Q establishes the truth of Q is called *modus ponens* and is the simplest example of a rule of propositional logic.

Another rule of propositional logic is *chaining,* which says that the two sentences S_1 and S_2 are together equivalent to the single sentence S_1 AND S_2. Suppose we are given the two sentences IF P THEN Q and IF Q THEN R, and suppose we know that P is true. We can then use modus ponens together with chaining to establish the truth of R. That the truth of R follows from these two sentences may seem so obvious as not to require proof but these and similar laws can be pieced together to establish the truth of statements that are far from obvious. This process is called *deduction* since the truth of the statements is deduced from the sentences and the rules of logic. Rule-based expert systems, however, only use the simplest form of deduction in propositional logic, modus ponens coupled with chaining, given above. This is why there was no need to introduce logic theory in Chap. 2; the logic is obvious.

Logical propositions ordinarily represent statements that have some meaning in reality, e.g., "The animal is a bird." In the application of logic to expert systems, propositions represent assertions as they are defined in Chap. 2. However, it is important to remember that as far as the theory of logic is concerned the "meaning" of propositions P and Q is irrelevant. Proposition P might represent the statement "The moon is made of green cheese," and Q might represent "The Russians are coming;" it makes no difference in propositional logic. Real sentences do have meaning, however, and a theory of logic that had some capability of dealing with the meaning of a sentence would clearly be very useful. The theory used to deal with the meaning of sentences is called *predicate logic.*

The use of predicate logic in expert systems allows them to consider the relationships among assertions. For example, the four assertions "The animal has feathers," "The animal has hair," "The animal is red," and "The breast is red" have obvious relationships. Each expresses something about some attribute of the animal, with two describing the skin covering, one describing the color of the whole animal, and one describing the color of the breast. The first two describe different values of an attribute of the same class, while the second two ascribe the same value to different classes. However, pure propositional logic is too crude to express these relationships among the assertions; it can only state the assertions themselves. Predicate logic makes use of *predicate functions* to express relationships such as these.

Predicate functions can be used to ascribe a value, e.g., feathers or red, of some attribute, e.g., skin covering or color, to a class, e.g., breast or animal. We can use this to characterize and organize assertions. The predicate functions p and q are used to ascribe a value to an attribute of the class x with the notation

IF $p(x)$ THEN $q(x)$

For example, one can say

IF feathers(skin,animal) THEN bird(animal)
IF bird(animal) AND red(color,breast) THEN robin(animal)

Here the functions *feathers, bird, red,* and *robin* are predicate functions representing values. *Skin* and *color* represent attributes and *breast* and *animal* represent classes. As mentioned in Chap. 1, this form of representation of the rules was used in the earliest expert systems, partly because earlier work on production systems beginning with that of Post (1943) had been expressed in terms of predicate logic. Many currently used expert systems retain this use of predicate functions as a means of expressing rules. In addition, the computer language Prolog is based on predicate logic.

If the expert system makes use of predicate rather than propositional logic, the detailed functioning of the inference engine is changed slightly from that described in Chap. 2. The inference engine described in that chapter, which uses propositional logic, requires a list of hypotheses such as "The animal is a duck," "The animal is a horse," and so forth. The inference engine accumulates verified assertions in an effort to verify one of these hypotheses. This process is similar to the *parameter-driven* inference process described by Engel et al. (1989).

An inference engine that uses predicate logic specifies a goal class and a list of possible instances of that class. For example, the goal class in the animal identification system is *animal* and the possible instances are *duck, horse,* etc. The rule base does not need to independently specify all the possible hypotheses as is the case with a propositional system; only the goal class needs to be specified. The inference engine can accumulate all the possible instances of this class as it reads the rule base. The inference engine itself functions as in Chap. 2 except that it tries to associate values with attributes and objects. The values are the predicate functions and the arguments of these predicate functions are the objects and attributes. This process is similar to the *pattern-matching* inference process described by Engel et al. (1989). The expert system MYCIN (Buchanan and Shortliffe, 1984, Chap. 5), often regarded as the original rule-based expert system from which all others are descended, uses predicate logic in its inference engine.

In the study of logic, predicate logic provides a significant advance in power over propositional logic. This is because predicate logic provides a means of examining the meaning of statements in a sentence rather than simply their truth value. This is accomplished through the use of the quantifiers *for all,* denoted $\forall$, and *there exists,* denoted $\exists$. Quantifiers can be used, for example, to ascribe a property to something if that thing is known to be a member of a class of things that have the property. There is a famous example of this use that is quoted in all logic texts, and we can do little better than to repeat it: Suppose we want to prove that Socrates was a mortal. We can do this by noting that Socrates was a man and that all men are mortal. In words the argument is as follows:

Socrates was a man.
All men are mortal.
Therefore, Socrates was a mortal.

Using quantifiers and predicate functions the proof would look like this:

man(Socrates)
$\forall x$, IF man(x) THEN mortal(x)
therefore mortal(Socrates)

Much of the increased power of predicate logic comes from this use of quantifiers. Unfortunately these are rarely if ever useful in expert systems, so in expert systems the increase in power provided by predicate logic is not quite so great. Whether or not predicate logic is used, the same two basic logical operations, modus ponens and chaining, are all that is needed to work the inference process. This is such a small and simple subset of the available rules of logic that it prevents, or renders unnecessary, the use in expert systems of the full representational power of predicate logic. In general the fact that it is possible to ascribe different values, e.g., red, green, and brown, to an attribute, e.g., color, of a class, e.g., breast, does not enable the rules to be expressed with any more subtlety than by simply forming different assertions, each with the attribute having one of the values, e.g., breast is red, breast is green, etc. In some specific instances, however, the separation of value from attribute is useful in the expert system inference process.

The most common reason to separate an attribute from its value is in the case of attributes that take on numerical values, particularly those involving measurement, e.g., length, height, or weight. Frequently in the inference process it is necessary to compare these numerical values either with each other or with some reference value in order to ascertain the truth of some assertion. For example, we might wish to include rules saying that if the animal is a bird and the length of the bird is less than 10 cm, then it is not a duck, and if the length of the bird is less than 20 cm, then it is not an eagle. This requires the use of two assertions *length < 10 cm* and *length < 20 cm*. The simplest approach would be to associate a separate prompt with each assertion. The problem with this is that it entails asking the user two separate questions about the length of the bird. Moreover, since control over the order of questioning is not always complete, it may happen that the user is asked whether the length is less than 10 cm after the user has already informed the system that the length is not less than 20 cm. This is a case in which the *meaning* of the attribute, the fact that it represents a length and, as such, must obey certain relationships such as being greater than 10 cm if it is greater than 20 cm, is important. In any such situation the best approach is to separate the value from the attribute within the assertion. In the case of numerical valued attributes, this is possible if we keep a list of such attributes and permit prompts that ask about their value, e.g., "What is the length of the specimen in cm?" One must also permit assertions that include numerical comparisons, such as length < 20. Finally, the inference engine must be capable of determining that an assertion involves a numerical comparison, and after making such a determination, the inference engine must find out the numerical value of the attribute in question and perform the requested comparison.

Predicate functions can be used to represent the current certainty associated with a particular assertion. This is often useful in constructing rules that are *self-referencing.* A rule is called self-referencing if the same assertion appears in the antecedent and in the consequent. For example, suppose that an animal identification system is to be constructed that carries the identification all the way to the species level, as in the case of a typical key. Suppose further that in order to distinguish between two species of very similar appearance an expensive test such as electrophoresis is required. We would not wish to call for this test unless we were fairly sure that the specimen was one or the other of these species. This can be accomplished by incorporating into the expert system a predicate function that measures the certainty with which an assertion is currently known. A rule calling for electrophoresis can then be written in the form

IF certainty of species A > 0.5
AND electrophoresis indicates species A
THEN animal is a member of species A

The rule is called self-referencing because its consequent refers to the same statement as its antecedent. With this form of a rule, the costly electrophoresis test is avoided unless the user is already fairly certain, with certainty greater than 0.5, that the specimen is a member of species A. An analogous rule could then be written for the other, similar species.

There are also some advantages to be gained from separating attributes from values and objects as was done in the rules about feathers and red breasts given above. One is that the process of questioning the user may be speeded up and streamlined. This is possible because a prompt can be written in such a way that the user is asked the value of a certain assertion, e.g., the color of the feathers. That is, given a rule such as "If bird(animal) and red(color,breast) then robin(animal)," one can write a prompt asking the user for the color of the breast, rather than simply asking whether the breast is red. However, such prompts are also possible when a purely propositional system is used. The prompt must simply permit answers, rather than simply answering yes or no, that allow the user to select from a number of options, each of which corresponds to an assertion, e.g., breast is red or breast is green.

One feature that is possible with the separation of values from attributes and is impossible with pure propositional logic is the use of *metarules* (Davis and Lenat, 1982). The concept of metarules is well known to expert system theorists although it is rarely implemented in practice, at least at the microcomputer level, because of the complexity it introduces into the system. Nevertheless, metarules do provide a useful way of organizing and controlling the inference process that is impossible with a pure rule-based system. A metarule is, simply put, a rule about rules. Metarules can be used, for example, to determine which features of an animal to ask about first. The process works like this. Suppose that the animal has been determined to be a bird. Probably the most easily distinguishable feature of a bird is its size. The knowledge base could therefore include a metarule that says

IF the animal is a bird
AND there are any rules about birds that have length in the antecedent
THEN ask about length first

Next to length, color is probably easiest to determine. Thus another metarule could be included that has the form

IF the animal is a bird
AND the length has been checked
AND there are rules about birds that have color in the antecedent
THEN ask about color

At each stage in the inference process, the inference engine checks to see if any metarules apply to the current situation. Once the animal is known to be a bird, the inference engine looks for rules of the form

IF bird(...) AND length(...)
THEN...

After all rules of this type have been tested the inference engine checks for rules that have bird and color as predicate functions. Although this sort of control can be achieved without the use of metarules by a careful ordering of the rules, dependence on such an ordering violates the declarative spirit in which expert systems are constructed and, more importantly, may prove difficult to maintain and modify.

In addition to making possible the use of metarules, the use of predicate logic and the idea of separation of the attribute and value make possible the use of natural language interfaces in knowledge acquisition and knowledge base modification and debugging, as discussed in Chap. 4.

8.1.2 Inductive logic

The most commonly known form of logic, and the form that people associate with the term *logic,* is the deductive logic described above.

However, there are other forms of logic, one of the most important being *inductive* logic. Inductive logic is not used in rule-based expert systems but has been used in other forms of expert systems and, thus, provides a nice bridge to some other forms of knowledge representation. Unfortunately, the terminology in this area is inconsistent and confusing. Some authors (e.g., Pople, 1982) use the terms *abductive logic* and *abduction* to mean basically the same thing as what we call *inductive logic* and *induction*. Our use of the term is consistent with its use in applied statistics (e.g., Fisher, 1935) and with its use by many authors in artificial intelligence.

Induction differs from deduction in the following way: While deduction is used to reason from the general to the particular (e.g., Socrates was a man; all men are mortal; therefore Socrates was mortal), induction is used to reason from the particular to the general. An example of induction is to turn around the proof that Socrates was a mortal and "prove" that Socrates was a man because he was mortal. That is,

Socrates was a mortal.
All men are mortal.
Therefore, Socrates was a man.

Clearly this conclusion is not necessarily true; women are mortal too. The argument merely establishes the *possibility* that Socrates was a man. Unlike deduction, induction is used to establish *conjectures* rather than *facts*. While inductive logic cannot be used to prove the truth of a statement, it can be used to establish hypotheses that may be promising. Most creative scientific thinking is inductive logic.

In the example of animal identification, suppose a person is asked to identify an animal and is told that the animal lives in the water. The person might first consider the possibility that the animal is a fish since all fish live in the water, even though other animals live in the water as well. This is a case of the use of inductive logic to establish a hypothesis. An expert system called INTERNIST developed by Pople (1982), which will be described later, makes use of inductive logic. Before we can discuss these types of systems, we must introduce some structures for representing human knowledge on a computer.

Rules and their associated forms of logic are not the only forms of knowledge representation that have been used in the construction of expert systems. In this section we look at four other methods of knowledge representation that, while not used as often as rules, nevertheless provide many advantages when they are either used alone or together with rules in the construction of expert systems. The primary purpose of each of the knowledge representation structures, at least as they relate to expert systems, is to provide a mechanism for ordering and structuring the knowledge base. In discussing the various concepts in knowledge representation, we have to contend with the fact that definitions in the field are often vague and imprecise. We follow wherever possible Winston's (1984) terminology, which is also in some cases imprecise.

It is interesting to note that each of the four knowledge representation structures originated in the field of natural language understanding. This is the branch of artificial intelligence concerned with translating input to the computer program from a human language such as English into a form intelligible to the computer, thus permitting humans to communicate with the computer in ordinary discourse as they would with another human. An example of a natural language system is KLAUS, discussed in Chap. 4. Natural language is just one of many branches of artificial intelligence, and one might speculate as to why so many of the commonly used methods of knowledge representation have come from this one field. Presumably at least part of the answer is that in communicating with a computer (or another human), a human is transmitting knowledge, and if the computer is to make use of this knowledge, it must first have a structure to represent it. In any case, this uniformity of origin provides a natural means of introducing the structures. We will initially discuss each of them in their original context, understanding a phrase in a natural language. To be specific, suppose a computer program is to understand the simple story "The farmer saw a chicken and bought it."

8.2 Semantic Nets

One of the first things to recognize in understanding how a human processes the information in a phrase, such as the one describing the purchase of the chicken, is that the human brain contains a complex network of associations for words such as "chicken." Upon reading this word, a person calls to mind a bird that pecks and clucks. (Another equally valid interpretation, obviously incorrect from the context, is that a chicken is a cowardly person.) Pursuing the association, the brain would recognize that a bird is a kind of animal that is covered with feathers. An animal, in turn, is a living thing that breathes oxygen. Another form of animal is a cow, except that cows are covered with hair instead of feathers. And so on. *Semantic nets* provide a means of representing the associations that humans form in their minds. They are networks in which relationships of *meaning* are specified between items. Figure 8.1 shows a semantic network organizing the information about birds and cows contained in this paragraph.

As with any network, the individual items, represented by ovals in Fig. 8.1, are called *nodes*. The links emanating from nodes to connect them with other nodes are called *slots*. For example, the bird node in Fig. 8.1 has skin covering and a_kind_of slots (the *a_kind_of* slot is a

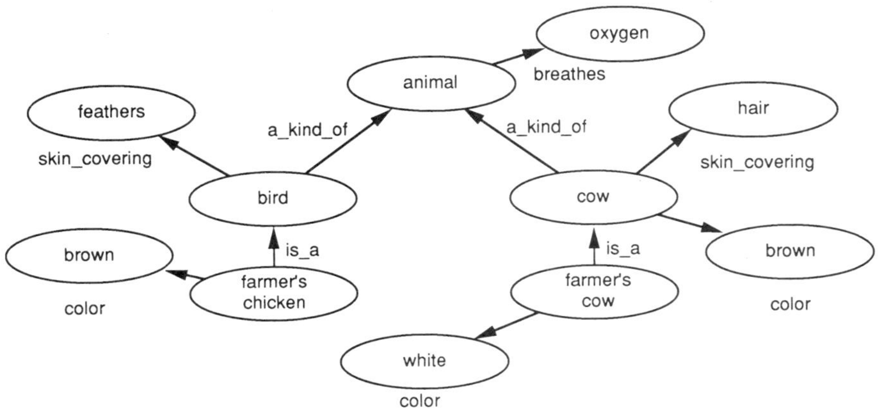

Figure 8.1 Semantic net for farmyard animals.

special sort of slot discussed below). The node at the head end of the arrow is called the *value* of the slot. For example, the value of the skin covering slot of the bird node is *feathers*. The process of understanding a phrase in a natural language is connected with finding values for the various nodes of the network. The human equivalent to this is the process of putting a statement into context. Thus, when presented with the sentence "The farmer saw a chicken," the computer must place the farmer's chicken in the appropriate semantic network, i.e., context, and find the appropriate values for the slots. Semantic nets have methods for determining values if they are not specified. (The sentence "The farmer saw a white chicken" is a case in which a value, color, is specified.) Three of the most common methods are inheritance, if-needed procedures, and defaults.

8.2.1 Properties of semantic nets

Inheritance. Inheritance in semantic nets is a representation of the process by which humans, in understanding language, implicitly associate properties with an object because that object is a member of a class of objects all of which possess the same property. For instance, a human can immediately recognize the incongruousness of the phrase "The farmer saw a chicken. She bought it. Then she filled a bowl with water and put it in." On the other hand, if the word "goldfish" is substituted for "chicken" in the phrase then it makes perfect sense. Everyone knows that chickens have lungs and breathe air while fish have gills and get their oxygen from water. In the context of knowledge representation, chickens *inherit* the property of breathing air by being birds, which in turn inherit this property by being, along with mammals and reptiles, animals with lungs. In a semantic net, this sort of inheritance is ordinarily represented by two special kinds of slots, the *is_a* and *a_kind_of* slots (Winston, 1984). Both of these slots specify that the node at the tail end of the arrow can inherit a property from a node at the head of the arrow.

Is_a slots. If two nodes are connected by an is_a slot then the node at the tail end of the arrow (e.g., *farmer's chicken* in Fig. 8.1) is a *specific instance* of the node at the head of the arrow (e.g., bird in Fig. 8.1). This is analogous to the membership relation of set theory, in which, for example, the farmer's chicken is a specific member of the set of birds. Node *a,* the farmer's chicken, automatically inherits any specified properties of node *b,* bird. For example, it inherits from the bird node the property of being covered with feathers. Nodes may have more than one *is_a* relation. For example, the farmer's chicken might also be connected with an *is_a* slot to a node *domestic fowl.* From this node it might inherit information about the manner in which is it fed, its habitat, and so forth.

A_kind_of slots. The a_kind_of slot is very similar to the is_a slot. In fact, the two are functionally identical; the difference between them is in the kinds of node they connect. While the is_a slot connects a specific instance of a class to that class, the a_kind_of slot connects a class to a larger class. For example, in Fig. 8.1 the *bird* class is connected to the *animal* class by an a_kind_of slot. In words, a bird is a kind of animal. The mode of inheritance for a_kind_of slot is identical to that of the is_a slot, i.e., it flows in the opposite direction to the arrow. In Fig. 8.1, the *bird* node inherits the property *breathes oxygen* from the *animal* node, and the *farmer's chicken* in turn inherits this property from the *bird* node. The distinction between the is_a and a_kind_of slots is identical to the distinction in set theory between an element that is a member of a set and a set that is a subset of another set.

If_needed procedures. The if_needed procedure is a special slot that contains a procedure for doing a calculation (Winston, 1984). This represents the idea that a person presented with some information will often do a quick calculation or deduction to make sense of the information or gain additional knowledge from it. For example, if the farmer bought 50 chicks at $1.50 each, a quick calculation indicates that the total amount paid was $75.00.

Defaults. In the absence of any other information, people will use defaults to assign value to a slot. An example of assigning a value by default is the color slot of the farmer's cow in Fig. 8.1. Not all cows are brown, but a person thinking of a cow might, unless otherwise informed, assign that color to it. If the color of the specific cow is given as white or if some other information such as the breed is given that would allow the person to deduce the cow's color, then this value replaces brown in the color slot. Default values can be indicated in a semantic net by attaching them to classes remote from a particular specific instance node. Thus in Fig. 8.1, although the default color for cows is brown, the farmer's cow is specified to be white.

8.2.2 CASNET, a semantic net–based expert system

CASNET (Weiss et al., 1978), which stands for *c*ausal-*as*sociative *net*work, is a diagnostic expert system that uses semantic networks. It was developed as a decision-making aid for the treatment of glaucoma. As usual, in our description of the CASNET program, we translate the domain into one more appropriate to agriculture; in this case, the identification of a crop disorder. Weiss et al. (1978) refer to CASNET as a model-based system, and the model is the semantic network itself. This network consists of three coupled levels, representing three levels of abstraction in describing the processes of the crop disorders (Fig. 8.2). At the highest level are the disease categories, i.e.,

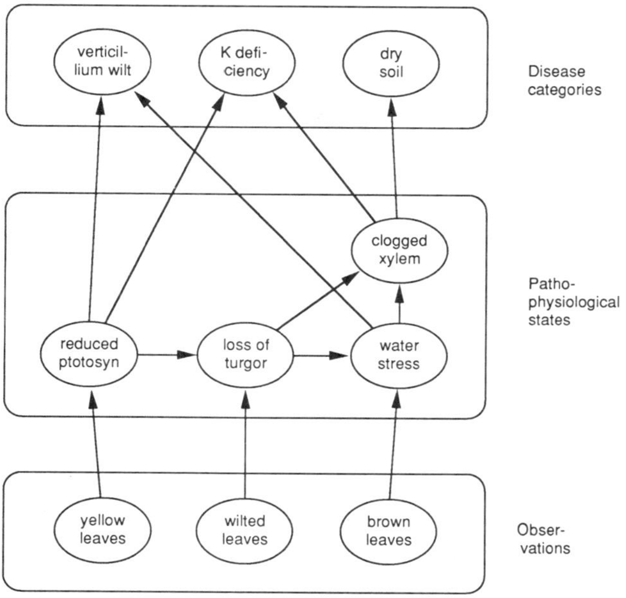

Figure 8.2 Associative network for a CASNET-based expert system for diagnosis of plant disorders.

the various identified causal agents. In our example we will limit these to a (1) plant pathogen, verticillium wilt, (2) nitrogen deficiency, and (3) water stress. At an intermediate level are the pathophysiological states, i.e., the various pathologies such as a blocked flow of xylem, reduced level of chlorophyll, reduced turgor, etc. At the lowest level are the observations, i.e., the symptoms and measurements associated with the disorders. The links in the CASNET network do not represent meaning as in an ordinary semantic network. What they represent depends on the level or levels containing the nodes they connect. Links connecting nodes in the pathophysiological states to nodes in the disease categories are called classification links. Those connecting nodes within the pathophysiological states are called causal links, and those connecting nodes at the observations level to nodes in the pathophysiological states are called associational links.

The object of a CASNET consultation session is to determine, based on which nodes are instantiated at the observational level, the most likely disease category level node. Each link has a strength associated with it, and each node can be instantiated to a given degree based on the strength of the links pointing to it. The degree to which a node is instantiated is computed based on mathematical formulas that are very similar to Bayes' theorem, although they do not represent probabilities. In Fig. 8.2 the strengths of association are represented by the thickness of the connecting arrows. The CASNET consultation system begins with some preliminary observations to establish the most likely disorder category and then attempts to confirm this category by attempting to instantiate observation-level nodes most closely associated with it.

The CASNET architecture, while it represents an interesting departure from the rule-based system, has not been widely copied. An inspection of the network shown in Fig. 8.2 indicates that although the CASNET system appears superficially to be very different from that of a rule-based system, in many respects they are quite similar. Each node in CASNET can be identified with an assertion in a rule-based system, with the disease category nodes corresponding to hypotheses. The consultation strategy corresponds to initial forward chaining, the preliminary observations, followed by backward chaining, the confirmation steps. While the splitting of the nodes into three levels does provide a considerable amount of organization to the rule base, it is highly specific to the diagnostic domain and does not necessarily have any analogy in other domains. Thus, we can legitimately question whether the semantic network formulation of CASNET really provides any significant advantage over the more flexible and intuitive rule-based formulation.

8.3 Frames

Although semantic nets provide a useful way of pictorially representing human mental associations, they can quickly become overwhelmingly complex. Frames are a means of organizing the information in semantic nets by structuring of slots. The frame concept was originally developed by Minsky (1975) as a natural language understanding tool, but the organizational power of the frame soon led to its being used in a wide variety of applications. The frame is a representation of the human ability to conceptualize an abstract entity such as animal and provide this conception with certain attributes that all, or most, instances of that entity possess. A frame is similar to a semantic net except that the nodes are eliminated and the slots themselves become the representational objects. Figure 8.3 illustrates the frame idea. The example is similar to that used to illustrate the semantic net and consists of distinguishing the similarities and differences between the concept of a *bird* and the concept of a *mammal.*

A frame represents an individual conceptual entity, e.g., a bird. The structure of the frame consists of a collection of slots, represented by the rectangles in Fig. 8.3. Each slot represents a fundamental aspect

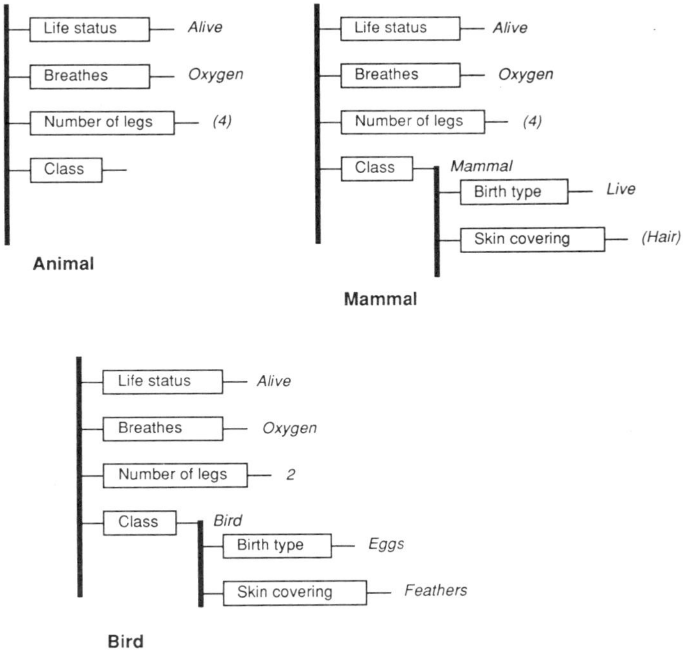

Figure 8.3 Frames for types of animals.

of the frame; the value that the slot takes on in a particular instance is called the *filler*. For example, the filler of the *breathes* slot in the *animal* frame in Fig. 8.3 is *oxygen*. The filler of a slot can be a frame itself. In this case, the filler is said to be a subframe of the parent frame. For example, the filler of the *class* slot of the animal frame is either the *mammal* subframe or the *bird* subframe.

Like the semantic net, the frame supports inheritance and defaults. In Fig. 8.3 default values are shown in parentheses. For example, the animal frame has the default filler *4* for the slot *number of legs*. This represents the fact that a typical person, in thinking of an animal, pictures that animal as having four legs. The default value for the mammal frame is also four legs, but in the case of the bird frame this value is two for all instances of the frame. The concept of inheritance in frames is illustrated by the fact that every instance of an animal frame inherits *oxygen* as a filler for the *breathes* slot.

8.3.1 INTERNIST/CADUCEUS

The expert system CADUCEUS, later renamed INTERNIST, is one of the earliest frame-based expert systems. As the second name implies, it was developed as a tool for diagnosing disorders in internal medicine. The expert system's knowledge base does not contain rules as such; instead it contains two large sets of frames. One set, called *disease entities,* represents models of each disorder; these frames contain slots representing the effects of the disorder. A second set of frames, called *manifestations,* contains representations of various standard symptoms and other manifestations of the disorder. Each manifestation frame has a data structure called EVOKES, which records with each manifestation (symptom) the list of disease entities associated with that manifestation together with an evoking strength. Each disease entity has a MANIFESTS data structure containing all manifestations of the disease entity together with a scaled frequency of occurrence.

The basic idea behind INTERNIST, and a later revision called INTERNIST-II (Pople, 1982), is to make use of inductive logic in the same way that an actual physician presumably does. The physician presumably carries in his or her head models of each of the various disorders of internal medicine, each model being represented conceptually by a frame. In attempting to diagnose a disorder, the physician would then take an initial set of symptoms and use inductive logic to formulate an initial set of possible diseases. Inductive logic would be used to hypothesize a disease from a set of symptoms whenever those symptoms were associated with the disease. Thus, INTERNIST is to some extent a matching and search device. It searches among its disease models for a differential diagnosis in which one hypothesis, i.e., disease, is relatively more likely than all of the others.

INTERNIST initially uses some key symptoms to establish a preliminary set of hypotheses. Once these preliminary hypotheses are established, it uses one of three strategies to search among the disease models for the correct one. If there are more than five preliminary hypotheses, a strategy called RULEOUT seeks to eliminate some of them. If there are two to four preliminary hypotheses, a strategy called DISCRIMINATE focuses on the leading two contenders and looks for information to distinguish between them. If there is only one preliminary hypothesis, a PURSUE strategy seeks confirming evidence for it. The search is conducted by questioning the user as described before for a rule-based expert system except that instead of attempting to instantiate rules, the program is attempting to fill slots in a manifests frame. The prompts to the user are generated by the manifests frame, and once the slots are filled, they are compared with the corresponding slots in the hypothesized disease entities frames. The three strategies are used to determine which slots to attempt to fill next.

Once the search has been followed through to completion, the surviving disease is presented in the form of a diagnosis. If there are more than one surviving hypotheses, then they are all presented. Uncertainty is permitted, and each symptom has an "evoking strength" associated with each disease that represents the degree of certainty that one can conclude a disease given the symptoms. Several advantages over rule-based programs are claimed for this approach. One of the most important is that rule-based systems "fail to aggregate diagnostic possibilities into decision sets; instead, each considered diagnosis is evaluated as though independent of all other alternatives" (Pople, 1982, p. 131). INTERNIST to some extent adopts a more holistic approach. Two of the main disadvantages seem to be that the approach may be hard to program and may be quite domain specific.

8.3.2 CENTAUR

CENTAUR (Aikins, 1983) is a rule-based expert system that uses frames to organize knowledge and place it in context. Context is defined by Aikins as "the set of facts or preconditions which when taken together describe the situation in which the knowledge is applicable." CENTAUR was created to address problems that were encountered in the development of an early expert system called PUFF (Kunz et al., 1978). PUFF was developed to diagnose lung diseases and used an architecture very similar to that of MYCIN, i.e., what is now considered a "standard" expert system architecture. The problems addressed by

CENTAUR were primarily those associated with the lack of structure in the rule base and were described in Chap. 2. One primary problem is the "amorphousness" of the rule base, i.e., the rule base is basically a sequence of rules with no structure or organization imposed on these rules except by their location in the sequence. This lack of structure can cause a number of problems. One principal problem is that the modification of existing rules or addition of new rules may have unexpected side effects. Another problem is that, since all inference is accomplished by rules, it is impossible to distinguish control rules, which control the flow of information in the expert system, from inference rules, which are used to make actual deductions.

CENTAUR makes use of frames to deal with these organizational problems. The fundamental frame is called a *prototype*. Each possible disease is considered to be a hypothesis and is represented by a prototype. In addition, there is a prototype representing a consultation with CENTAUR itself, a prototype for knowledge review and explanation, and a prototype to represent the knowledge common to all hypotheses. Prototype frames in turn have subframes called *components*. Components contain control information, data, and rules. Rule bases are represented as slots in a component frame.

Rules are explicitly divided into the following five categories: triggering rules, inference rules, fact-residual rules, refinement rules, and summary rules. The consultation session with CENTAUR is controlled by the consultation prototype through an agenda that the prototype establishes. The disease hypotheses prototypes themselves add items to the agenda. Triggering rules are used to determine which hypothesis prototypes to "trigger," i.e., consider next. Inference rules are the standard expert system rules, used to infer facts and values. Fact-residual rules are used to account for residual facts, i.e., facts that have been acquired but not explicitly used in the inference process. Refinement rules are used to refine the diagnosis, and summary rules are used to summarize information for display.

When the hypothesis prototype is triggered, it becomes active, and its rules attempt to match data with expected values. This inference process is superficially not unlike that of INTERNIST. As with this latter program, CENTAUR proceeds to accumulate possible hypotheses and attempts to eliminate them, and, when only one hypothesis remains, CENTAUR attempts to increase the certainty of that hypothesis. In sharing some of the superficial features of INTERNIST, CENTAUR also appears to share its advantages and disadvantages.

8.4 Blackboards

A blackboard is a special kind of frame that helps organize other frames. The concept of the blackboard originated with the HEARSAY project (Erman et al., 1980), an early speech understanding effort. Whereas a frame is a representation of the human ability to conceptualize an abstract entity, a blackboard is a representation of the human ability to integrate knowledge from different sources in order to determine the content of something that is perceived. This ability is especially essential in understanding speech. For instance, suppose you hear the sentence "Let's wait till Bob rings us up on the phone." This sentence, adapted from Erman et al. (1980), contains the phrase "till Bob rings." Although this phrase would be obvious in context, without this context it would be very difficult to distinguish this from phrases such as "till Bob brings" or "tell Bob rings." The way this distinction is made is from the context imparted by the surrounding phrases. The sentence "Let's wait till Bob brings the news," for example, would make the phrase "till Bob brings" easy to interpret. The listener interprets a phrase by placing it in context, i.e., by linking its meaning with other information that the listener has at hand. The blackboard structure was originally developed as a means of representing diverse information such as context, word order, and so forth that the listener uses to interpret the meaning of a sentence or phrase.

The blackboard is a useful concept in the application of knowledge-based systems to integrated crop management. Each individual knowledge base can be considered as a human expert, and the sequential nature of digital computers imposes the restriction that only one expert can be "thinking" about the problem at any one time. However, crop management is, or should be, a holistic effort in which the impact of pest management decisions on irrigation, and of irrigation decisions on nutrition, and so forth must be considered. The blackboard allows each knowledge base to communicate with all of the other knowledge bases while maintaining its own individual identity (Fig. 8.4). In the blackboard architecture, however, the blackboard is more than simply a passive slate on which the experts write their information. It is more like an administrator who, in addition to serving as a means of communication among individual technical experts, also assigns tasks to the experts and, because of knowledge of the individuals' expertise, assigns the appropriate expert to each task at the appropriate time.

The CALEX expert decision support system uses the blackboard concept to integrate crop management information in this manner. One of the design goals of the CALEX system was the maintenance of completely independent and self-contained knowledge bases for individual field actions while permitting these knowledge bases to communicate with each other in the generation of guidelines for integrated crop management. At one level, independence of knowledge is maintained through the physical separation of knowledge bases. CALEX uses a blackboard that allows the knowledge bases to communicate their information to each other while maintaining complete

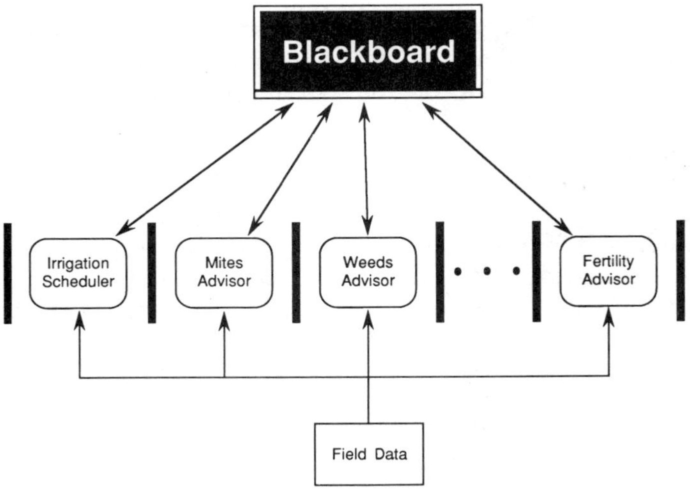

Figure 8.4 An arrangement of a group of individual experts offering guidance in field crop management analogous to the CALEX expert system. The experts cannot communicate directly but instead write information to and read information from a blackboard.

separation of the inference processes. Activities are called upon sequentially in an iterative algorithm to perform their respective inference processes. The *activity critic* is used to implement the iterative algorithm of determining which knowledge base to call next. The iterative process consists of a number of loops. At every iteration, each knowledge base is examined in order, and it is determined whether that knowledge base should be called, i.e., processed. For example, suppose that there are three activities: spider mites, irrigation, and verticillium wilt, each with its own knowledge base. In every iterative loop the critic determines in a fixed order, say, first mites, then irrigation, then verticillium wilt, whether each activity should be called upon to perform its inference process in that iteration. The order is not under the control of the knowledge bases, so the results of the overall process must be independent of the order in which the activities are called in the iterative loops.

Figure 8.5 shows a flowchart of the scheduling algorithm under the control of the activity critic. In the first iteration, each knowledge base is called. Each knowledge base represents a single activity. In addition to attempting to generate one or more management guidelines through its inference process, any information determined during the course of the inference process that might be relevant to other activities is placed on the blackboard. In particular, an activity may impose conditions on the field that influence other activities. For instance, the

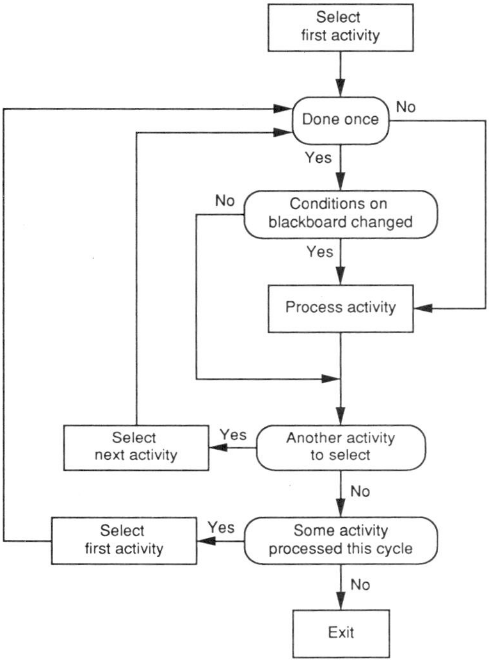

Figure 8.5 Flowchart for the CALEX blackboard iteration algorithm.

irrigation activity causes the field to be wet at certain times. Since scouting for mites cannot be carried out when the field is wet, the irrigation activity influences the mites activity. To convey the fact that at a certain time the field is wet, the irrigation activity places this information on the blackboard. After the initial iteration cycle of activities, the activity critic uses this blackboard information to determine whether or not an activity, i.e., a knowledge base, should be called. The action of repeated iteration of the activities involves at each step the examination of every activity by the critic to determine if that activity should be reprocessed because of some new information on the blackboard that influences decisions about that activity.

Communication between the activity critic and the individual activity frames is maintained through two slots in the activity frame: the *influence list* slot and the *conditions* slot. These are the "read to," influence list slot, and "write from," conditions slot, mechanisms of the activity. The conditions slot contains all of the items that the activity has done that might influence other activities. For example, the conditions slot for the irrigation activity would contain the information that the field will be wet on certain dates. The contents of the conditions slot change each time the activity is called. The influence list contains all of the items that might influence the activity. For example, the fact that the field is wet would be in the influence list slot of the mites activity. The contents of this slot are fixed. The use of these two slots preserves the modularity of knowledge within each activity. When the knowledge bases are constructed, the experts constructing a given knowledge base, e.g., irrigation, need only know what aspects of their activity influence other activities, not the manner of this influence.

After an activity completes its inference process, the contents of its conditions slot are written to the blackboard. At each iteration the activity critic examines the influence list of each activity. If an activity's influence list contains a condition that has been written to the blackboard since the activity was last called, then the activity is called again. For example, if the dates on which the field will be wet have been changed since the mites activity was last called, then the mites activity is called again. The iterative process terminates when a complete loop passes without the critic calling any activity.

There is in this process the potential for an endless loop if two activities influence each other. For example, if some condition affected by the spider mite activity is on the irrigation influence list, an endless loop might occur. To deal with this situation the blackboard permits only four cycles of iteration to occur. Anything that would force another processing of an activity after this is instead marked as a warning of a potential conflict to be passed to the user when the guidelines are displayed. While this method of handling infinite loops may seem like a "cop-out," it is the result of careful consideration and is arguably the best option for dealing with these situations. It provides the user with the maximum possible amount of information and allows the user to make the final decision.

8.5 Scripts

To understand the motivation for scripts, let us return to the story that our computer is supposed to try to understand "The farmer saw a chicken and he bought it." A person reading this story fills in a lot of details to aid in understanding it. The farmer was at an auction, looked in a cage, made a bid that won, paid the auctioneer for the chicken, took the chicken and brought it to the farm, and so forth. Most of this information is contained not in the story but in our own experience, which tells us how a transaction such as the farmer's would normally be conducted. Scripts (Schank and Abelson, 1977) provide a way of describing the way in which this experience is organized. They describe sequences of actions that occur in a predictable way. In essence a script is like a frame in which the slots are encountered in sequence rather than all at once. Scripts are used to represent everyday occurrences that are done so frequently by so many people that a complete description of them is unnecessary. They represent the method people use to understand cryptic "stories," such as the story about the farmer's chicken, that do not explicitly provide enough information to make complete sense. For example, upon seeing the story of the farmer's chicken, we would ordinarily invoke a "going to the auction" script that includes going to the auction, looking for an item, making a bid, and so forth. The fact that the farmer was at an auction is inferred (the farmer could, for example, have been at the farm of someone with a chicken to sell) by the listener to "flesh out" the story and allow the listener to form a mental picture of the action. If the story were changed to "The farmer saw a chicken and then went to the lion and tiger cages," we would in response invoke a "going to the zoo" script. However, there is not a script for the story "The farmer saw a chicken and went into the theater," so this story does not on the surface make any sense.

Each element of a script, e.g., going to the auction and looking for an item, is represented by a frame. Schank and Abelson (1977) developed an ingenious set of basic descriptions of actions in terms of transfers, e.g., going to the auction involves a transfer of oneself to the auction, that describe many of the actions of everyday life in terms of a few basic descriptors. These descriptors make up the slots of the frame representing the element of a script. The script is a sequence of these frames that represents the sequence of actions performed by the protagonist of the story in order to achieve his or her (or its) goals. In the case of the farmer and the chicken, the farmer is the protagonist and, upon hearing the story the listener recognizes that the farmer's goal at that moment in the story was to acquire a chicken and achieved this by being at the auction and purchasing it.

The CALEX system (Plant, 1989) of implementing agricultural management guidelines is essentially script based. The individual script frames are called *guidelines* and represent the action that should be taken in response to a particular state in the field. Guideline frames are integrated with rule bases to construct management recommendations. The guideline frame is used to construct the appropriate management recommendation once a hypothesis in a rule base has been verified. The advantage of using a script-based system such as this one is that it can synthesize recommendations for a wide variety of management decisions independent of the domain-specific nature of the decision itself in the same way that the inference process in the expert system is independent of the domain of the rule base. Every hypothesis in a rule base is associated with one guideline frame.

The generation of management recommendations for spider mites in the case when mites are over the action threshold, the level of infestation at which a pesticide application is recommended, is a good example of how the script-based guidelines work. Figure 8.6 shows a schematic diagram of a guideline frame with its slots filled by information from the hypothesis *over_threshold* in the spider mites activity. Each guideline frame contains the following slots: hypothesis, when, action, amount, info, show data, and plot. Not all of these slots need be filled for each guideline. The hypothesis slot holds all of the hypotheses associated with that guideline. The when slot holds the date on which a given action should be taken. The action slot is used to specify an action to be taken. The amount slot is used to specify a quantity associated with that action, e.g., the amount of water associated with an irrigation. The info slot contains the name of a text file that may be displayed at the option of the user to obtain further, more detailed information. The show data slot contains a data set to be displayed to the user. The four most recently entered data values for this data set are displayed. The plot slot is used to specify data for which a graphical plot is to be made available to the user.

Suppose that the level of spider mite infestation is found to be above the action threshold by a verification of the hypothesis *over threshold.*

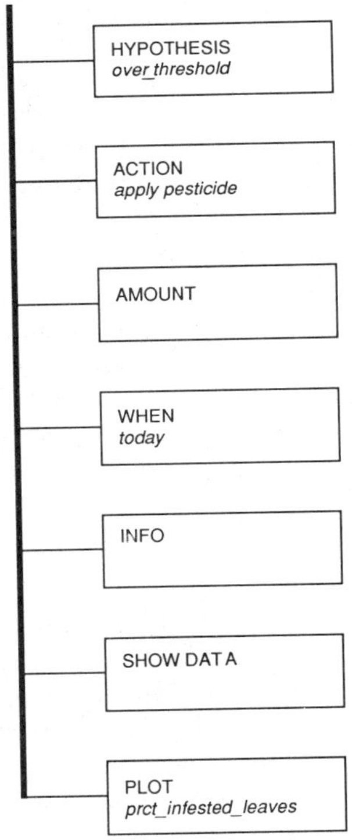

Figure 8.6 Guideline frame in CALEX. Each slot provides information that the script generation system uses to formulate a guideline script.

The appropriate action for the farmer is to apply a pesticide immediately and to scout the field for mites one week after application. CALEX locates the guideline frame shown in Fig. 8.6 and identifies that it is associated with the hypothesis *over threshold.* Therefore, this guideline is put on a stack for eventual display to the user. The next step is to generate the recommendation that the field be scouted in one week. This is an example of the sequential aspect of a script-based system described above. The script controller must determine the next action in the story. In the case of the farmer's chicken this was returning home after buying the chicken; in the case of spider mites it is scouting the field after it is safe to reenter. The manner in which CALEX accomplishes this will be described in more detail in Chap. 9. Essentially it involves looking at the most recently verified hypothesis and searching for rules that apply to a system whose state includes the verity of that hypothesis. In the case of spider mites, the system finds a rule that says, in effect, "If the current state of the system is that spider mites are over the economic threshold, then verify the hypothesis *scout* and set the date to do this to today's date plus 7 days." The script manager then finds the guideline shown in Fig. 8.6 and adds this to the stack for display to the user.

8.6 Statistical Pattern Matching

As discussed in Chap. 2, statistical pattern matching is a commonly used technique in taxonomic identification. Statistical pattern matching differs from the kind of pattern matching described earlier in this chapter in that the former permits matches that are close but not exact. Statistical pattern matching has been used in what are nominally called expert systems as well (Frey, 1986). Frey's expert system is based on a classification scheme called the *Holland classifier* (Holland, 1986). These classification systems are particularly appropriate for problems of identification, such as taxonomic identification or symptom diagnosis. Consider the example of diagnosing a crop disorder based on a set of symptoms. In its simplest form, the Holland classifier is based on a string of *bits,* i.e., numbers that can take on the value 0 or 1. Each bit string is called a *message.* For example, one group of bits might be associated with leaf symptoms, e.g., wilted, chewed, brown spots, and chlorotic, another with bud symptoms, another with root symptoms, and so forth. A message associated with a particular disorder would have 1's in each bit corresponding to a symptom of that disorder and 0's in all other bits. If a disorder has more than one collection of symptoms associated with it, then there can be more than one message associated with that disorder.

It is often the case that some disorders are "diagnostic," i.e., they are present if and only if the disorder is present, while some disorders are strongly or weakly associated with a disorder. Frey accommodates this by subdividing the association of a value of 1 in a particular bit with the presence of a disorder into three categories denoted A, B, and C. These categories correspond to the three degrees of association of a symptom with a disorder.

With a pattern-matching system such as this, the "inference" process consists of determining which disorder most closely matches a particular pattern. Unless all questions are put to the user in advance of the matching process, there must be some strategy for determining which symptoms, i.e., which bits, to determine next. Frey describes the following simple strategy. First, certain basic bits, e.g., the organ that is displaying the symptoms, are evaluated. Second, the system determines which disorder most closely matches the current message and attempts to further validate this match. If the match is invalidated by an answer, then the remaining closest match is pursued and so forth. Frey describes a simple yet very fast algorithm for testing the message against various disorders, a process that must be repeated for each disorder after each question.

The principal advantage of the classifier approach seems to be its simplicity. This simplicity of the knowledge base and of the inference process may mean that other aspects of the expert system, such as knowledge acquisition and machine learning, can be accomplished more readily with a pattern-matching system than with a rule-based system. There seem to be two fundamental disadvantages. One is that fundamentally, all of the knowledge encoded in the knowledge base is associative. In the animal identification problem of Chap. 2, for example, the knowledge in the system would be much more akin to the simple, one-level-deep rule base of Fig. 2.3 than to the deeper rule base of Fig. 2.4. This may be appropriate for some domains but is probably a significant restriction. Features such as explanation, maintenance, and so forth would be more difficult to accomplish with a pattern-matching system than with a rule-based system.

Finally, just like the situation with pattern-matching taxonomic keys in Chap. 2, there is little if any difference between such keys and the "expert systems" described here. Bacteriological systematics have for a long time relied on such systems; the study of these systems is often called *numerical taxonomy* (e.g., Sokal and Sneath, 1963). Evidently the overlap between expert systems and keys in this area is quite substantial.

8.7 Case-Based Reasoning

Case-based reasoning has evolved in recent years from the idea that human beings do not actually use rule-based reasoning when they solve complex problems. Instead, humans seem to rely much more on experience, solving specific problems by recalling similar problems they have solved in the past and adopting the past solution to the new context. There is no doubt that people use cases in problem solving, the question is to what extent. For example, when the green bug, a common pest of wheat, was first noticed building up populations in sorghum in the panhandle of Texas, farmers demanded advice about how to control the pest. Entomologists solved the problem by treating the green bug as if it were just like another aphid pest of sorghum, using the same economic threshold. Similarly, when Virginia farmers began growing cotton again in the late 1980s as a rotation crop for peanuts, they needed advice about pest management. The problem was solved by adopting the advice published by the North Carolina Agricultural Extension Service for cotton growers in that neighboring state.

In these examples, the experts did not apply any fundamental rules about how to recommend pest management strategies, they simply found the most similar examples and adapted their solutions. Even when rules can be elicited from an expert, there is disagreement about whether the expert actually uses those rules when solving problems. For example, a seasonal pest control problem can be described in terms of the state of a field: The crop has reached stage-3, there are 12 insects per sweep net sample, and there are only minimal visual signs of damage. Proponents of rule-based reasoning would say that the expert applies rules to make inferences and solve the problem. Thus, if the crop is in stage-3, then it does not readily show pest injury. If the chances of pest damage are high and the number of insects per sample is greater than 10, then a spray is indicated.

The argument for case-based reasoning would go something like this: An expert has seen several situations much like this one before. It is this experience that makes the individual an expert. Without going through any rule-based reasoning, an expert would simply be reminded of a set of previous similar cases and would recommend the solution that worked then. Only when no case seems to apply would an expert resort to rule-based reasoning, and then only to reconcile the differences between previous experience and the current situation.

The case-based reasoning model is illustrated in Fig. 8.7. All case-based systems have a case library that serves as the system's memory. Two key features of these systems are the process of assigning indexes or features to cases as they are entered into the system, and the adaptation of cases to specific problems. Indexes are assigned to cases to allow them to be recalled easily when an appropriately similar case arises. When a problem is described, it goes through the same indexing procedure, and cases with common indexes are recalled from the case library. Adaptation involves reconciling the differences between

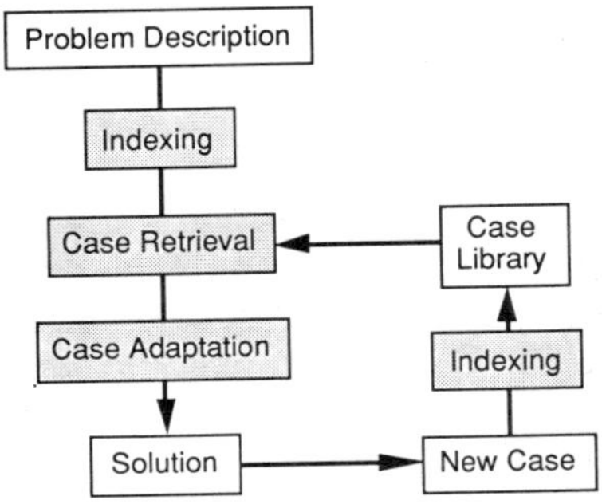

Figure 8.7 General flowchart of a case-based reasoning system. Shaded boxes indicate the need for rules, e.g., indexing rules and rules to adapt a retrieved case to current circumstances.

retrieved cases and the specific problem. Both of these processes use rules, although they should be called metarules since they are rules about manipulating cases in general.

The use of metarules in adaptation can best be illustrated by a simple example. Suppose we have done a χ^2 goodness of fit test with eight degrees of freedom and determined a χ^2 value of 15.05. We now want to determine the probability of getting that number under the null hypothesis, i.e., we want to know the p-value for the test. Unless we have a computer or calculator that does this calculation, we would use a table. The table tells us, for example, that 13.36 is the value associated with a p-value of 0.10 and 15.51 is the value associated with $p = 0.05$. Given these two examples, we would then use a very simple rule of interpolation to estimate the p-value of 15.05. The rules needed for interpolation are very simple and depend for their validity on finding cases reasonably close to the specific problem. Even though rules are used in solving the problem, the solution is still primarily based on the cases. The rules alone would not be enough to generate solutions.

Indexing cases and storing them in the knowledge base for quick recall is the other fundamental problem in constructing case-based systems. Many case-based systems use a representation scheme similar to frames or scripts called MOPS, *m*emory *o*rganization *p*ackages (Schank, 1982). MOPS are arranged in hierarchies with inheritance so that searching the case library can go from the general to the specific. Like frames, MOPS have slots, called *norms*. The parent class of a MOP is called its *abstraction,* and children of a MOP are called its *specializations*. In a pest management problem, the norms of a MOP called green bug problem might include the green bug density, the phenological stage of the plant, and the visual signs of damage to the plant. The green bug problem MOP could have an abstraction called aphid problem, which in turn would be a specialization of the general MOP for insect pest problems. Cases are included as instances of particular MOPS. This framework allows a system to classify a new problem even if it cannot find a specific case with the same norm values.

8.7.1 JUDGE and PC-JUDGE—case-based sentencing

We look briefly at a modification of the JUDGE system (Bain, 1986; Riesbeck and Schank, 1989), a case-based reasoner designed to hand down sentences for injurious assault crimes based on previous cases in criminal law. Our system will treat pest control decisions by a farmer as if the farmer were sentencing the pest for some crime, damaging a crop. We will address only the single species problem.

The JUDGE system goes through four stages in determining a sentence: (1) a rule-based interpretation (indexing) phase, (2) a case-retrieval and comparison phase, (3) an adaptation phase, and (4) a generalization phase. The initial phase uses rules to infer the motivations of the actors in the assault. From a description of the facts of the case, the program needs to decide whether the injury was provoked by actions of the victim (i.e., was the victim the defender or the aggressor?), whether the use of force was justified, whether the victim or the offender escalated the level of violence of the interaction, etc. These determinations are then used in case retrieval as indexes. The last phase, generalization, allows the JUDGE system to convert frequently occurring cases into rules.

Our pest management system, which we call PC-Judge, would examine sampling data from a crop field and would first determine such things as the following: Did pest numbers increase suddenly or gradually? Did pest populations show a response to any recent field treatment or cultural practice? Were pest population changes predictable from external factors, e.g., spring emergence or factors affecting nearby fields? The PC-Judge system, given the hypothetical situation that a large number of *Helicoverpa* spp. eggs were observed in a cotton field during early boll formation, would proceed as follows: The interpretation phase would use rule-based reasoning to determine that (1) the infestation is sudden; (2) the infestation could be a response to a recent thunderstorm that struck only a few fields; (3) *Helicoverpa* spp. has been observed in the field before, but at low numbers; (4) no pesticides have been sprayed on the field recently; and (5) there seems to be an average population of beneficial insects in the field. Further inferences are (1) the situation is urgent because large populations of larvae could damage the crop at that stage; (2) spraying is dangerous because beneficial insects would be eliminated and the crop would be susceptible to further damage for some time; and (3) the field might continue to attract more *Helicoverpa* spp. than other fields because of its potential for lush growth.

This interpreted example could then be classified within a MOPS hierarchy. It could be an instance of the MOP called SUDDEN-INFESTATIONS-AFTER-RAIN, for example, which would have the abstraction, SUDDEN-INFESTATIONS. The norms of a MOP fairly high in the hierarchy would include those indicating the number of eggs sampled, the number of beneficials, the stage of the crop, and the urgency. Sudden infestations would be described by norms that describe the cause of the infestation.

The rule-based inferences made above would be used as indexes in selecting related cases from the system's case library. Assume that the following case is then recalled: A large number of *Helicoverpa* spp. eggs were suddenly observed during the later stages of boll maturation, after a rain and in a field that had not been sprayed previously but that had a large population of beneficial insects. In that previous case, no spray was recommended, but there was a sense of urgency about the situation. Sampling in two days was recommended to determine whether the beneficial insects were controlling the population. If large numbers of small larvae were found in the next sample, a spray at that time was indicated.

Once a similar case has been found, the system looks for differences between the recalled case and the current problem. The recalled case is similar in many ways; however, in the recalled case the number of beneficial insects was higher and the crop was at a later phenological stage. The program determines that the urgency of the situation is higher because the bolls are at a more susceptible stage. Furthermore, the situation is more urgent because the number of beneficial insects is lower than in the recalled case. However, the danger from using pesticides is also greater because the crop will be susceptible to damage longer than in the recalled case.

The PC-Judge would then determine that the situation is worse than in the previous case but that the danger of spraying is worse as well. Since the only harsher possible decision would be to recommend a spray, the system decides in the adaptation phase that the pest control recommendation should be the same as in the recalled case. Notice that the system's rules for determining its recommendation only help determine whether a recommendation should be more or less harsh. They could not be used without cases to invent a recommendation from scratch.

Generalization of this type of situation into a rule would occur only after the system has accumulated several examples of *Helicoverpa* spp. infestations after rains and during the boll period for which the resulting recommendation was the same. After some threshold number of repetitions is reached, the system would use the similar features of all the cases as antecedents in a rule. This would certainly make sense for the current situation, in which there seems to be a balance. Factors that increase the urgency of the situation also increase the danger of spraying too soon. Since the result is always the same, a rule will be more efficient.

Case-based reasoning seems to be a good model of how people solve some kinds of problems, particularly problems for which rules do not exist. In many cases, experts clearly rely on their experience to solve problems, and an indexed case library seems like a good representation for such situations. Obviously, case-based reasoning still relies on rule-based reasoning, both in its basic implementation and in generalizing commonly occurring cases. However, case-based reasoning may be better for systems in which experience plays a key role or in which judgment and adaptation are required.

8.8 Discussion

The principal theme underlying this chapter is that the pure rule-based expert system, developed in Chaps. 2 and 3 and applied in the succeeding chapters to various problems, is inadequate in some situations. What are those situations and how do the mechanisms of knowledge representation developed here address these inadequacies? One problem was recognized and described in Chap. 2: organizing and maintaining a large rule base. A rule base has no structure or organization imposed on its rules except by their location in the sequence or their occurrence. The logical relationships among the rules is often not evident from an inspection of the rule base. This can lead to serious bugs within the rule base and to the inadvertent introduction of bugs when the rule base is modified. The frame provides a natural means of organization for production rules; this organizational power was first effectively exploited in the CENTAUR system (Aikins, 1983) and has been used in several other expert systems since that time.

A second problem encountered in Chap. 3 is the problem of encoding attribute values in assertions. For example, in the animal identification problem, suppose one of the distinguishing features of the animal is its color. Within the propositional logic system developed in Chap. 2, each possible color of each specific part of the animal must be identified as a separate assertion. This presents no theoretical problems but can cause practical problems of high memory requirements and a confusing rule base. The predicate logic system introduced in this chapter permits the attachment of variable values such as color to attributes such as parts of the body. This can result in a considerable economy in the structure of the rule base. [See Woolley and Stone (1987) for an example of the application of this concept.] The idea of variable values becomes even more important if numerical values and associated numerical relationships are to be used. For example, in a propositional logic system it would be impossible to compare a wingspan of 12 inches to a test value of 10 inches and conclude that the former is greater. The assertion "The wingspan is 12 inches" is legitimate but carries no intrinsic meaning relative to the assertion "The wingspan is 10 inches." Such meaning can, however, be inferred if the number 12 is a value attached, or, to use the AI jargon, "bound," to the attribute *wingspan*. Recognizing that wingspan is a numerically valued attribute makes possible the comparison *wingspan* > 10.

A third problem with the basic inference engine design is that it provides no means for the integration of knowledge among different rule bases. While in principle several rule bases for different topics could be merged into one large rule base, this gives rise to all the practical difficulties associated with very large, amorphous rule bases, including memory requirements and difficulty of maintenance. The blackboard concept addresses this problem by providing a location on which to write shared information and a means of determining which knowledge bases can make use of this information.

A fourth problem is that of synthesis. The rule-based expert system was originally developed as a diagnostic device and its primary strength is in analysis, i.e., the determination of which among several competing hypotheses is the correct one. The rule-based expert system is much less adept at synthesis, i.e., the development of a detailed plan of response to the diagnosed condition. The script, a device for encoding a sequence of stereotypic actions in response to a well-defined situation, is well suited for this purpose. The guidelines frame of the CALEX system (Plant, 1989) provides an example of this use of the script concept. However, scripts are not by themselves sufficiently powerful to handle the most complex synthesis problems. For these the techniques of the branch of artificial intelligence known as *planning* are required. These are discussed in Chap. 9.

Chapter

9

Planning: Full Season Tactical Management

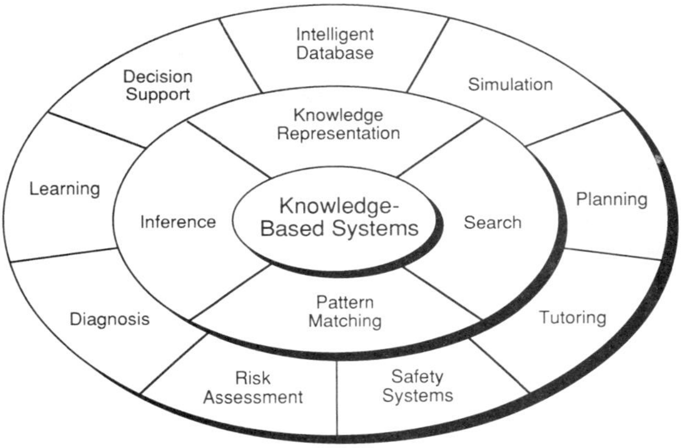

As discussed in Chap. 1, management of an agricultural commodity, or any other kind of management for that matter, involves two aspects, *strategic* management and *tactical* management. Tactical management includes specific "when" and "how much" decisions involving irrigation, pesticides, fertilizers, and so forth. Tactical management often consists of two phases, analysis (also called diagnosis) and synthesis (also called prescription). The analysis phase involves determining the current status of the crop, specifically determining the current threats to the crop's well being, e.g., pest infestation, low phosphorous level, and water stress, or forecasting the date on which such threats will occur. The synthesis phase involves determining a response to the results of the analysis phase.

9.1 The Problem: Irrigation Scheduling

A recurrent theme of this book has been that rule-based expert systems are much better at the analysis phase than at the synthesis phase. In the previous chapter we discussed *scripts,* which can help to overcome some of the deficiencies of rule-based expert systems in the area of synthesis. However, scripts are not the whole answer. They represent a stereotypic response to a well-defined situation. Many synthetic problems of crop management involve situations that are not well defined and, thus, not amenable to the scripts approach. A good example is irrigation scheduling. Chapter 6 described a procedure for integrating a regression model into an expert system to determine the date for the next irrigation of a crop. Given an adequate data set, an expert system using a knowledge base such as this could provide the user with an estimate of the next irrigation date. However, this does not answer all the questions that the user might have about irrigation. Two other important questions are: When should the irrigation date be scheduled if the data are not adequate to provide a regression-based estimate? When should subsequent irrigations take place, i.e., what is a complete irrigation schedule for the crop? It is essential that an expert system in a practical application be able to answer the first question since the farmer must schedule irrigations whether the data are good enough or not. There is less immediate need for an answer to the second question except as it might affect other long-term scheduling. However, the time is not far off when the supply of irrigation water to farmers will be limited, and at that point the exploration of alternative complete irrigation schedules will be an essential part of a crop management decision support system.

A primary alternative method to leaf water potential for scheduling irrigations is the water budget method (Fereres et al., 1981). In this method one first determines the soil moisture level on some reference day. This may be done directly be measuring the soil water potential with a potentiometer or indirectly by estimating the soil moisture at the end of the last irrigation or at the start of the season if no irrigations have yet been applied. Changes in the soil water content are then estimated on a day-by-day basis. Water is lost from the reserve through evapotranspiration. It is added to the reserve through rainfall and, during the early season, by the roots' growth into new regions of the soil where moisture is still plentiful. The gains and losses to the water budget are tallied on a daily basis. Water content is usually measured as a percent of the *field capacity,* which is the maximum amount of water that can be held by the field. The *depletion level* is the relative difference between the actual water content and the field capacity. When this depletion reaches the maximum allowable level, it is time to irrigate.

A third method of irrigation scheduling is simply to look up the irrigation date in a table. This is often used to schedule the first and/or last irrigation dates of a growing season. For example, Grimes and El-Zik (1982) give a table of optimal irrigation dates based on soil type for the first and last irrigation dates for cotton grown in the San Joaquin Valley, California.

In this chapter we will describe a knowledge-based system for scheduling irrigations based on a combination of the leaf water potential method and the water budget method. This system is a simplified version of the CALEX/Cotton irrigation scheduling module (Plant et al., 1988; Plant, 1989; Goodell et al., 1990). The preferred scheduling method for the next irrigation is by measuring the leaf water potential. In CALEX/Cotton the water budget method is used for two purposes: to schedule the next irrigation if the leaf water potential method fails and to schedule all subsequent irrigations. However, some crops are not amenable to pressure bomb measurement, and so in some cases methods based on the soil moisture content, such as the water budget method, are the only reasonable alternative. The table method is used to schedule the first and last irrigations.

9.2 Computer Generation of Optimal Irrigation Schedules

The idea of using a computer to generate an irrigation schedule is not new. In principle, the methods of stochastic optimal control may be used to generate an irrigation schedule for a crop, and this schedule should, in some sense, be optimal. The way optimal control methods work is quite different from the methods we use in this chapter, and they provide a useful contrast. They are easiest to understand if we begin with a *deterministic* optimal control problem, i.e., one in which there is no randomness. A deterministic mathematical formulation of the optimal control problem as applied to irrigation scheduling has the following form.

For convenience we will use a daily time step. In practice we could also use degree-days, as described in Chap. 6, as the time unit. Let u_k denote the amount of water applied to the crop on day k. In optimal control theory parlance this is called the *control*. Suppose that the duration of the growing season is D days. Let $\mathbf{x}_k$ represent a vector whose components represent the values on day k of the components of a simulation model for the crop-soil-water system, and assume that the net economic return on day k of the growing season from applying an amount u_k of irrigation water can be represented by the function $g_k(\mathbf{x}_k,u_k)$. Then the optimal control problem is to choose a sequence $\{u_1, u_2, u_3,\ldots, u_{D-1}\}$ of control vectors in order to maximize an *objective function* denoted J, which is given by

$$J = \sum_{k=0}^{D-1} g_k(\mathbf{x}_k,u_k)$$

This equation simply states that the total profit from irrigation is the sum of the profits gained from irrigation on each day.

In order to solve the problem of maximizing J, it is necessary to know how the crop responds to irrigation. This information is contained in the vector $\mathbf{x}_k$, and the behavior of this vector is governed by the crop-soil-water simulation model. This model can be represented abstractly by the equation

$$\mathbf{x}_{k+1} = \Phi_k(\mathbf{x}_k, u_k)$$

$$\mathbf{x}_0 = \mathbf{a}$$

where k ranges from 0 to $D - 1$. The main components of this model are the *state vector* $\mathbf{x}_k$, the *control* u_k, and *system dynamics* Φ_k. The state vector contains all the information necessary to completely describe the state of the system, in this case, the crop, on day k. The control term u_k describes the control action taken, in this case, the amount of water applied, on day k. The vector function Φ describes the system dynamics, i.e., the evolution of the state of the system from time k to $k + 1$. The mathematical theory describing how the sequence $\{u_k\}$ should be chosen is very well developed and is described, for example, in Bryson and Ho (1969) and in Dreyfus and Law (1977).

A major assumption associated with the model given in the previous paragraph is that the system is deterministic. This assumption means that if the system is started with a given initial state vector $\mathbf{a}$ and a given control sequence $\{u_0, u_1, \ldots, u_k\}$, then the value of each component of the state vector $\mathbf{x}_{k+1}$ can be exactly determined. This is not a very good representation of the actual irrigation scheduling problem since there are any number of factors, e.g., precipitation, temperature, pest infestation level, and so forth, that influence the crop's water status and are highly stochastic in nature. This is important because the optimal irrigation schedule would be different if it is known that there will be a certain level of precipitation in the next few days than it would be if there is a possibility of no precipitation at all. In cases like this where uncertainty is a factor, a more appropriate system model has the following form:

$$\mathbf{x}_{k+1} = \Phi_k(\mathbf{x}_k, u_k, \boldsymbol{\eta}_k)$$

$$\mathbf{x}_0 = \mathbf{a} + \boldsymbol{\omega}$$

where η is a vector-valued random process representing the system "noise" and ω is a vector representing the uncertainty in the initial conditions. Moreover, it is often the case that the components of the state vector $\mathbf{x}$ cannot be precisely measured. In such cases a more accurate representation of the system observations is a noise corrupted version of $\mathbf{x}_k$, represented by

$$\mathbf{y}_k = \mathbf{h}(\mathbf{x}_k) + \boldsymbol{\nu}_k$$

where ν is a random process representing measurement inaccuracy. Often only a few of the components of a system can ordinarily be observed, in which case the dimension of the vector $\mathbf{y}$ is less than that of $\mathbf{x}$. The optimal control problem in this case is to measure $\mathbf{y}_k$ and on the basis of this measurement to select the sequence $\{u_k\}$, where $k = 0,\ldots, D - 1$, so as to maximize either the expected value of the objective function J or some utility function based on J (e.g., Bertsekas, 1976).

There is one final level of complication to be added to the problem. In many cases, the parameters of the system model are not known with precision. We may represent this mathematically by replacing the equation above with

$$\mathbf{x}_{k+1} = \Phi_k(\mathbf{p},\mathbf{x}_k,u_k,\boldsymbol{\eta}_k)$$

$$\mathbf{x}_0 = \mathbf{a} + \boldsymbol{\omega}$$

$$\mathbf{y}_k = \mathbf{h}(\mathbf{x}_k) + \boldsymbol{\nu}_k$$

where $\mathbf{p}$ is a vector of parameters. If the values of some or all of the components of $\mathbf{p}$ are not known with precision but are only estimated using some probability distribution, then the control problem is actually a twofold one. We must both try to optimize the process given the current knowledge about $\mathbf{p}$ and at the same time try to gain a greater precision of knowledge of $\mathbf{p}$ so that controls may be used more effectively. At some times, particularly early in the process, the optimal strategy is to use values of the control variable u_k that test the system response and, thereby, improve the estimates for the components of $\mathbf{p}$. This process is called *adaptive control theory* (Walters, 1986). In the case of irrigation scheduling, it is certainly true that many of the parameter values of the model are only known with great uncertainty. In this example a policy of "exploration" such as this would be interpreted as applying more or less water, or water at different times, to determine the crop's response to the water. It is unlikely that such a policy would ever be used in actual practice.

9.3 Methods of Solution of the Problem and Their Limitations

As in the deterministic case, there exists an extensive mathematical theory for dealing with stochastic optimal control problems of the form given above. If the system dynamics are linear and the noise is gaussian, then we can ordinarily make use of a device called the Kalman filter to analyze the problem. In addition, if the cost function is quadratic, then we can obtain a complete analytical solution. This theory is described by Gelb (1974). In the case of nonlinear systems, we must ordinarily obtain a solution to the problem through stochastic dynamic programming. This theory is described by Bertsekas (1976).

The theories of deterministic and stochastic optimal control were originally developed primarily with engineering applications in mind. Since their development, however, a great deal of effort has been devoted to applying these techniques to problems in other areas. In particular, a number of theoretical studies have been made of the application of these techniques and techniques from the related field of decision analysis (Raiffa and Schlaffer, 1961) to problems in agriculture and biological resource management (e.g., Mann, 1971; Goh et al., 1974; Getz, 1982; Shoemaker and Onstad, 1983; Plant et al., 1985). Considerable research has been done on the application of optimization techniques to irrigation scheduling (e.g., Hall and Butcher, 1968; Dudley et al., 1971; Rhenals and Bras, 1981; Yakowitz, 1982). Given this level of theoretical activity, the number of successful implementations, i.e., implementations that are actually used by real decision makers to aid them in the decision making process, in agriculture has been disappointing. Getz and Gutierrez (1982) discuss this matter in some detail. Some of the problems associated with the use of models such as these in agriculture are the following.

1. *The system model (represented by Φ) may not be sufficiently accurate.* The function Φ in the equations above must contain all of the information necessary to describe how the state variable $\mathbf{x}$ evolves from time k to $k + 1$. For example, in the irrigation of a field crop the model must include at a minimum the development of the plant, the movement of water in the soil, and climatological factors influencing the agroecosystem such as rain and solar radiation. Moreover, other factors such as the level of insect pest infestation and nutrient status may affect the crop's demand for water, in which case a complete model of the agroecosystem is necessary. For some engineering problems, such as controlling the trajectory of a guided missile, the processes associated with the problem are well understood and can be described with accuracy and precision through mathematical models. However, there is no crop system of economic significance where an accurate mathematical description of the complete agroecosystem has been constructed. It is unlikely that any crop system exists where a model has been constructed having sufficient accuracy to provide a basis for optimal control of irrigation.

2. *The parameters of the model for individual cases are difficult to estimate.* As the development of crop and pest models progresses, the time may soon come when reasonably accurate complete mathematical models for some agroecosystems are available. Indeed, such models may be close at hand for some systems (e.g., Baker et al., 1983). Even if such a model was available, however, its parameters would have to be specified for each implementation. Since these parameters may vary considerably in different locations even within the same general geographic area, this parameter estimation represents a major undertaking. This implies that the system model must always be of the form in which the parameters are imprecisely known, and this is the most difficult form to deal with mathematically.

3. *Parameters characterizing system noise may not be known.* A major assumption in the analysis of stochastic models represented by the above equations is that the statistical properties of the system noise, represented by the random vector processes $\boldsymbol{\eta}$, $\boldsymbol{\omega}$, and $\boldsymbol{\nu}$, and the parameter estimates for $\mathbf{p}$ in the adaptive control model, are completely characterized. In actual agricultural crop systems virtually nothing is known about the system noise and very little about the parameter estimates. One must assume that the statistics can be characterized by some probability distribution (usually the normal distribution or some other distribution with mathematically convenient properties) and then estimate the parameters of this distribution. These estimates are usually extremely crude and often reflect more concern for mathematical tractability than for realism.

4. *Stochastic dynamic programming is very computationally intensive.* Even assuming that all of the problems characterizing the mathematical description of the problem could be solved, there remains the matter of actually computing the solution. In the case of a stochastic optimal control problem this solution is ordinarily obtained by the use of stochastic dynamic programming (e.g., Bertsekas, 1976). This technique has extremely high computational resource requirements. This is in part due to a phenomenon known as the *curse of dimensionality* (Bellman and Dreyfus, 1962), i.e., the number of computations necessary for a solution rises geometrically with the number of states and the number of time steps. The curse of dimensionality holds even for deterministic problems. For stochastic problems the computational requirements are even greater because several conditional expectations must be computed at each step of the solution. Thus, in practice the formulation of the stochastic dynamic programming problem must be simplified considerably to make it computationally tractable. This simplification entails the substantial risk of undercutting

the accuracy of the model and is in direct conflict with the requirement of sufficient complexity to accurately simulate the system.

In the case of adaptive control theory, which is probably the most pertinent for agricultural management, the situation is even more difficult. Real adaptive control problems can only be solved approximately, and even these approximate solutions are more computationally difficult than those of the already hard to solve, nonadaptive stochastic dynamic programming problems. Moreover, the nature of these problems eliminates the use of some otherwise attractive approximation techniques such as differential dynamic programming that are used in other areas (e.g., Yakowitz, 1982). In summary, the stochastic optimal control approach to the irrigation scheduling problem, while theoretically attractive, faces severe practical problems. The knowledge-based approach, by making effective use of heuristic knowledge and rules of thumb, can readily deal with system complexity and uncertainty. This approach is potentially very attractive for computer-based generation of irrigation schedules and solution of similar planning and scheduling problems.

9.4 Knowledge-Based Planning and Scheduling

9.4.1 Planning and scheduling

Our discussion of the knowledge-based approach to irrigation scheduling is based on the branch of artificial intelligence known as *planning*. Langlotz and Shortliffe (1989) provide a concise statement of the role of planning in the artificial intelligence context: "Planning consists of devising a course of action that conforms as well as possible to a set of goals." For a review of the subject see Hendler et al. (1990). Winston (1984) provides a very convenient simple example of a planning problem, which he calls the "blocks world." The problem is this: Given an initial configuration of blocks on the surface of a table (such as block A is on top of block B, which is on the table, and block C is on the table) move the blocks to achieve a desired configuration (such as block C is on top of block B and block B is on top of block A, which is on the table) (Fig. 9.1). There may be some constraints, e.g., the lifting

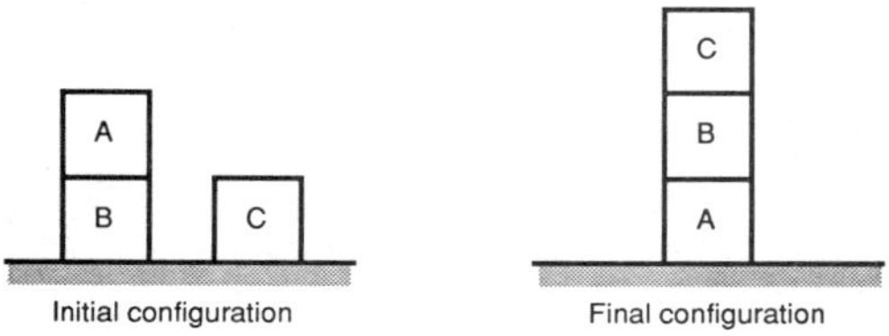

Figure 9.1 Initial and final configurations of an example from the blocks world (Winston, 1984). The problem is to plan a sequence of steps that will take the blocks from the initial to the final configuration.

of a block when another block is on it may be forbidden. This restriction would, in the example of Fig. 9.1, prevent consideration of a plan that included picking up block B until block A had been first removed.

Given an initial and final configuration of blocks, the planner must produce a sequence of actions that will take the blocks from the initial to the desired configuration without violating the restrictions. In the example shown above, the planner might return the following plan: (1) Pick up block A and put it on the table; (2) pick up block B and put it on block A; and (3) pick up block C and put it on block B. Other, more complex plans are of course possible, but it is implicit that the planner will return a plan that is in some sense, such as minimizing the number of steps, optimal.

Most planners work by constructing a sequence of successively longer and more elaborate subplans until the final plan is attained. Each of the subplans involves a portion of the full plan. People plan in a similar way; for example, to plan a trip to a location in a different city, the major portion of the the trip, which may be by car, rail, or air, is planned first. If it is to be by air, then a flight is booked; the ground transportation to and from the airport at each end of the trip is then arranged. Each of these planned stages of the trip are subplans of the complete trip plan. The earliest planners, such as ABSTRIPS (Sacerdoti, 1975), started at the beginning and formed a tentative plan until they came to an infeasible step; they then backtracked through the sequence of steps of the plan to a point where they could start over again. For example, in planning a trip this would entail first planning the ground transportation from home to the airport, then the flight, and finally the ground transportation at the destination. This is clearly inefficient since the ground transportation from home might put the traveler at the airport at an inconvenient time for catching a plane.

The most successful planning systems have generally been those which avoid early commitment to plans that later prove infeasible or very poor. These systems do this by initially forming the full plan in very abstract terms and gradually reducing the level of abstraction. In the traveling example given above the initial phase would be to decide the modes of transportation for each leg of the trip, and more detailed planning would follow from there. The introduction of this notion is generally credited to the NOAH system (Sacerdoti, 1975). Another important concept introduced by the NOAH system is the use of *critics,* which are subprograms that review the plan in its current state to search for suboptimal, infeasible, or otherwise undesirable portions of the plan. Critics will be used below in the generation of irrigation schedules. NOAH did not include any provision for backtracking from a plan once commitment to that plan had been made. It is frequently advantageous for a planner to have the capability to backtrack, i.e., to form a tentative plan that may have to be adjusted later. The planner NONLIN (Tate, 1977) is based on the abstraction and critic concept of NOAH but includes a backtracking capability.

The planners described above were created and tested primarily to solve problems of the type represented by the blocks world problem. Crop management has a number of features that differ significantly from the blocks world and similar problems. One difference is that most planners work with a well-defined starting state of the system and a well-defined goal state. In the blocks world, for example, the starting and final configuration of the blocks is expressly given. In the case of crop management, on the other hand, the ultimate goal may be vaguely stated as permitting the grower to produce a crop with the greatest economic yield possible while minimizing unnecessary inputs. However, specific, well-defined goals cannot usually be given. A second difference is that the state of the system, the crop, has its own internal dynamics, which are highly stochastic and imprecisely known. A third difference between crop management and the blocks world (although possibly not other domains) is the level of knowledge that must go into the formulation of the plan. In the moving of a block, and in any plan of this type, the basic actions, e.g., picking up a block or putting it down, are very simple and require no particular knowledge base to determine which action is needed; all of the knowledge is encoded in the set of constraints, e.g., that a block cannot be moved with another block on top of it. In crop management, by contrast, for each activity the decision of which actions to take and when to take them are based on a complex analysis performed by an expert system. The planner must therefore have access to an expert system.

Finally, there is one way in which the crop management problem domain is, fortunately, similar to the simple blocks world domain: the manner in which time is incorporated. In the artificial intelligence literature (e.g., Vere, 1983), there is a distinction between a *scheduler* and a *planner*. A planner has no reference to the absolute time at which an action occurs, only the relative reference that it occurs before or after other actions. For example, no value is given for the time at which a specific block in the blocks world is picked up; all that is specified is the order in which the blocks are picked up and moved. In contrast, a scheduler keeps track of the actual time at which actions are scheduled. The more traditional schedulers employ a complex method of understanding time relations (Allen, 1983), e.g., Vere's work is concerned with the optimal scheduling of activities in a satellite. This level of complexity has been found to be unnecessary in scheduling crop management actions (Plant, 1989). Like traditional planners, the only aspect of time necessary in constructing a crop management schedule is whether actions occur before or after or simultaneously with one another. The way a crop management scheduler differs from traditional planners is that it must maintain an absolute time, i.e., the date, at which an event is to occur in addition to the before and after relationships. However, this maintenance of absolute time does not impose significantly greater complexity on the representation of time than that contained in traditional planners. Because of this, we will use the terms *planner* and *scheduler* synonymously in this chapter.

9.4.2 Irrigation scheduling as a planning process

An irrigation schedule consists of a sequence of dates on which water is to be applied to the crop and, for each date, an amount of water to apply. The planner has at its disposal three fundamental methods for determining the appropriate date of the next irrigation: the pressure bomb, water budget, and table methods. The planner must accomplish the following tasks: (1) determine which of the three methods is most appropriate for generating the next irrigation date, (2) use the procedure associated with this method to determine the irrigation date, and (3) determine the amount of water to apply on that date. Before the season begins, the iterative scheduling process must start with the first irrigation. As the season progresses the scheduler must take into account irrigations that have already been applied as well as changes in weather and other factors that affect the irrigation schedule. Finally, the irrigation scheduler must be able to determine when to stop the iteration because no further irrigations are appropriate before harvest.

If followed precisely, the NOAH planning method described above would first derive a plan at a very abstract level, perhaps simply consisting of a determination of the appropriate method for scheduling each irrigation in sequence. The planner would then gradually make the irrigation dates more precise in a stepwise manner by decreasing the level of abstraction and, thereby, increasing the degree of precision with which the actual date is determined. That is, the NOAH planning algorithm initially would attempt to determine a globally optimal irrigation schedule for the entire season, with the actual irrigates initially represented only imprecisely. It would then, using search techniques developed in Chap. 7, gradually refine and make more precise this globally optimal schedule.

The CALEX planning system works on a sequential basis in which each irrigation is scheduled precisely before any attempt is made to schedule the following one. The reason for this is that many of the rules for determining the most appropriate method for scheduling a future irrigation depend on the time in the season when the actual date is scheduled. Therefore, a planner that gradually decreases its level of abstraction uniformly over the whole plan might be placed in a vicious circle. The planner cannot determine the appropriate method for scheduling an irrigation because it does not know the time in the season when the irrigation occurs, and it is not able to determine when the irrigation occurs because, lacking knowledge of the appropriate method, it cannot accurately schedule the next irrigation. While this can be avoided with the appropriate planning and search technique, the solution might become very complex. The sequential planning algorithm does not guarantee a global optimum because it does not change the dates of earlier irrigations based on information about later ones. However, it does provide a solution to the planning problem that is both conceptually and computationally simple and provides an adequate schedule.

The first step in the scheduling process is to decide on the appropriate method. The decision rules are these:

1. If the irrigation to be scheduled is either the first or the last of the season, find the date in the appropriate table.
2. If rule 1 does not apply *and* if the irrigation to be scheduled is the next one *and* the appropriate regression analysis and extrapolation can be computed, then use the leaf water potential method.
3. If rules 1 and 2 do not apply, then use the water budget method.

In summary, the water budget method is used if (1) the irrigation is neither the first nor the last of the season and (2) either it is not the next irrigation or, if it is, an extrapolation of leaf water potential cannot be computed. This decision sequence can be represented in a flowchart similar to that shown in Fig. 9.2.

In developing a knowledge-based architecture for selecting from among alternative methods such as these, one must adhere to a number of design criteria. Fundamentally, the system should not eliminate the advantages conferred by the use of rule-based expert systems as opposed to some other programming scheme. What are these advantages? Four of the most important advantages are:

Domain independence. The expert system is able to provide the same level of inference to any domain irrespective of the specific content of the knowledge characterizing that domain.

Simplicity. The inference process in a rule-based system is very simple and, partly as a result of this simplicity, very general.

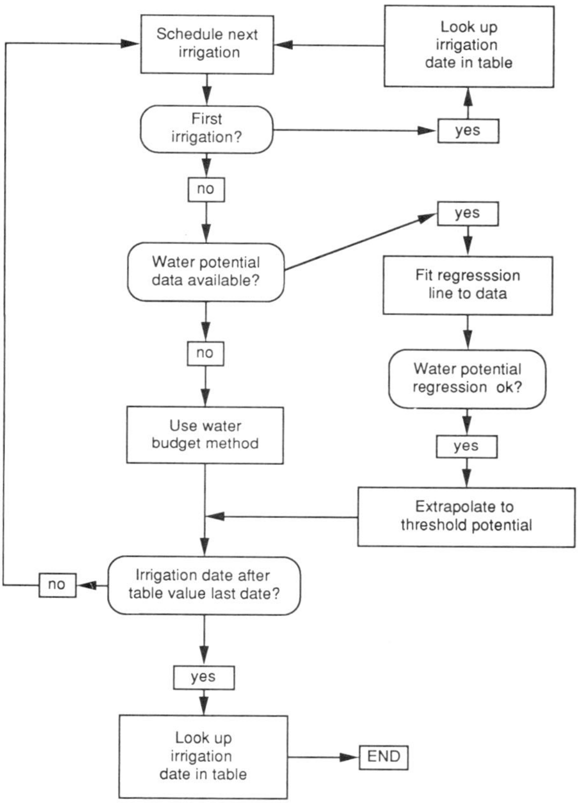

Figure 9.2 Flowchart for an irrigation scheduling algorithm based on a simplified version of the algorithm used in CALEX/Cotton.

Intuitiveness. The function of a rule base and the logic of an individual rule is fairly straightforward and easy to understand.

Explanation. An explanation facility that can describe the process that generated the plan is automatically available to the user.

As we discussed in Chap. 8, the planning involved for any action, such as an irrigation scheduling, may be broken into two phases. The first phase consists of an analysis of the current state of the system relative to that action, and the second phase consists of the synthesis of a response to that system state. In the case of an irrigation scheduler, the state of the system represents the moisture level of the crop-soil complex, and when this state reaches a critical level, the response to the current state consists of scheduling an irrigation. This process is repeated again and again. Having broken the planning into two phases, let us consider the first phase, the analysis. The second phase, the synthesis, can be handled through the use of guideline scripts as described in Chap. 8.

It is evident from the flowchart of Fig. 9.2 that each of the three basic methods of irrigation scheduling can in principle be implemented using a rule base. However, the iterative process cannot be easily implemented with a simple rule-based system. A structure is needed that will permit more than one rule base to be incorporated into a knowledge base. Also, a means must be provided for moving between rule bases and deciding which rule base to invoke next. The CALEX activity frame discussed in Chap. 8 provides a logical choice for such a structure.

One of the slots of the CALEX activity frame is called the *methods* slot. This slot is used to house the rule bases. The methods slot contains a sequence of one or more methods, each of which is itself a frame. A method frame has two principal slots, the *rule base* slot and the *calling states* slot. The rule base slot contains the knowledge necessary to analyze the current status of the crop relative to the activity. For example, in irrigation scheduling the method frame associated with looking up the date in a table contains rules that determine whether or not the irrigation is either the first or last of the season. This rule base also contains rules that permit the irrigation date to be determined from the table. The method frame associated with the extrapolation of leaf water potential contains rules that determine whether a regression analysis can successfully be constructed based on the available pressure bomb data and, if possible, do the extrapolation. The method frame associated with the water budget method contains rules that permit the next irrigation to be scheduled based on a water budget analysis. In summary, the rule base contains the knowledge necessary to schedule the next irrigation.

The activity frame must also contain a mechanism for determining which method to invoke next. The individual rule bases cannot contain this knowledge. If they did then the rule bases would have a direct connection to the knowledge contained in the other methods. This would defeat the purpose of organizing the rule bases in the frame structure. Therefore, the rule bases must be kept in isolation from each other and must concern themselves only with the direct implementation of their particular phase of the planning process.

While the information necessary to implement the method selection flowchart of Fig. 9.2 must not be contained in the rule base of any method, it is desirable that this information be directly contained in that method frame rather than in some location external to all the methods. Maintaining this information within the method frame retains the modularity of the system and permits methods to be added, changed, or removed with less effect on the other methods. Therefore, the information is housed in the other principal slot of the method frame, the calling states slot. This information determines which method is invoked, or "called," in attempting to solve the current problem.

The decision of which method to call next is made by a device called the *method level critic*. This uses the *calling states* slot of the method frame to decide which method should next be invoked. The method level critic must be able to select methods based on the same philosophy that a human expert would use in determining which of several available methods to use in attacking a large problem. The domain independence requirement of the expert system restricts the type of information available to the method level critic. The information that the critic can use is knowledge of which method was previously called and whether or not that method was successful in verifying a hypothesis in its rule base. These two pieces of information taken together comprise the *state* of the program.

Figure 9.3 shows a flowchart of the algorithm used by the method level critic to determine which method to select next. Figure 9.4 shows the organization of the method frames in the irrigation scheduling system. When the processing begins, and no method has yet been called, the state of the program is defined to be *Start*. Accordingly, one and only one method must have this as a calling state. For example,

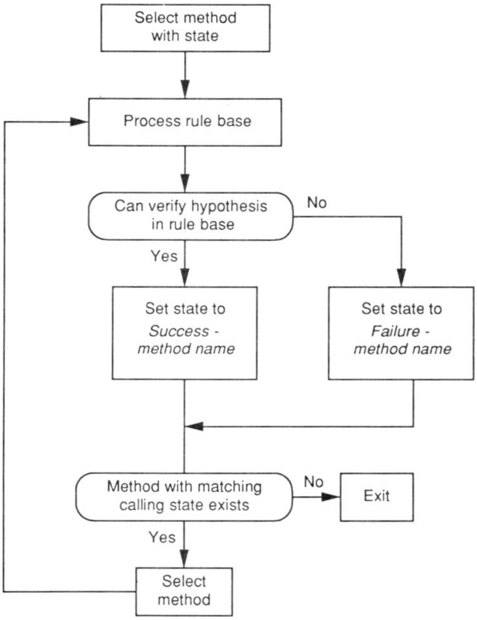

Figure 9.3 Flowchart of the algorithm used by the CALEX method level critic.

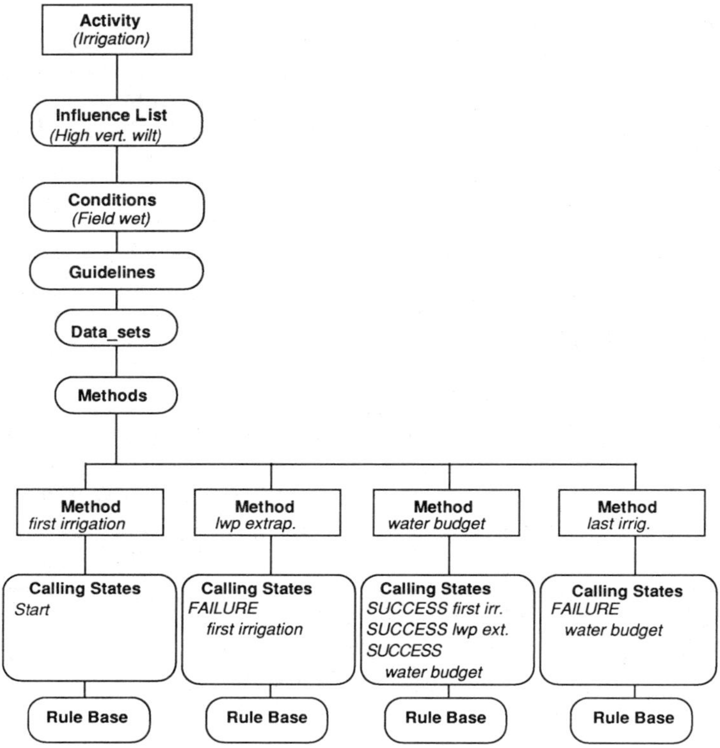

Figure 9.4 Structure of the irrigation activity frame showing the method subframes and their calling states.

the calling state of the *first irrigation* method in the irrigation system is *Start.* Therefore, this is the first method called when processing begins. When a method is called, it performs the inference process on its rule base. This consists of analyzing the available data, and possibly questioning the user to determine if it can verify a hypothesis in the method's rule base. For example, in the first irrigation method the one hypothesis is *irrigate via table.* This hypothesis is verified if the rule base determines that the irrigation to be scheduled is either the first or the last of the season.

If the method determines that it can verify a hypothesis in its rule base, then it has completed the analysis phase. The method calls upon the guideline specifier described in Chap. 8 to complete the synthesis phase, and it sets the current state of the program to *Success-<name>*, where *<name>* is the name of the method. If the expert system determines that the method cannot verify a hypothesis, the program state is set to *Failure-< name >* . This completes one cycle of the loop, and the next cycle is begun. The method critic examines each of the available methods to see if any of them have the current state as a calling state. If so, that method is called. Otherwise, the processing of the activity is concluded and control is returned to CALEX.

This system is used to implement the irrigation scheduling algorithm as follows (Fig. 9.4). The four methods are *first irrigation, lwp extrapolation, water budget,* and *last irrigation.* The calling state of the *first irrigation* method is *Start*; the calling state of the *lwp extrapolation* method is *Failure-first irrigation*; the calling states of the *water budget* method are *Success-first irrigation, Success-lwp extrapolation, Failure-lwp extrapolation,* and *Success-water budget*; and the calling state of the *last irrigation* method is *Failure-water budget.* With these calling states the system functions as follows. The first method called is always *first irrigation.* This method's rule base contains only one hypothesis. This hypothesis, which is verified if no irrigation has yet been applied, causes an irrigation to be scheduled on the date listed in the table. Before any irrigations have been applied, this hypothesis will be verified and the state of the program will be changed to *Success-first irrigation.* After the first irrigation this hypothesis will not be verified, and the state of the program will be set to *Failure-first irrigation.* In the first case the next method called is *water budget* while in the second case it is *lwp extrapolation.* If *lwp extrapolation* is called and the appropriate regression line is generated, then the next irrigation date is established by extrapolation of this line and the state of the program is set to *Success-lwp extrapolation.* If this extrapolation cannot be done, then no irrigation is scheduled and the state of the program is set to *Failure-lwp extrapolation.*

At this point no matter which of the three calling states is in effect, the next method called is *water budget.* This method computes the next irrigation date and checks to see if this date is before the optimal last irrigation date specified in the table. If it is, then a hypothesis in the rule base is verified and the state of the program is set to *Success-water budget.* Otherwise, no hypothesis is verified, and the state of the program is set to *Failure-water budget.* If the method is successful, then, since one of the calling states of the *water budget* method is *Success-water budget,* this method is called again. This process continues, and a succession of irrigation dates are computed, until the optimal date of the last irrigation is passed. At this point the *last irrigation* method is called and schedules the last irrigation. Since there is no method with *Success-last irrigation* as its calling state, the process terminates and the complete irrigation schedule is presented to the user.

9.5 Encoding Procedural Knowledge

Unlike other agricultural management tasks discussed so far in this book, the knowledge required to schedule an irrigation by the water budget method is almost entirely procedural. The soil moisture level is repeatedly estimated by subtracting the loss to evapotranspiration and adding the effect of any precipitation. The estimated moisture depletion must then be checked against the allowable depletion level for that time of year, and an irrigation is scheduled when the allowable depletion is exceeded. A flowchart for this process is shown in Fig. 9.5. The process involves the sort of repeated iteration that is much more common in conventional programming than in a rule-based expert system.

Although the water budget algorithm could be programmed into the system by means of a demon of the type discussed in Chap. 5, it is not hard to modify the chaining mechanism of an expert system's infer-

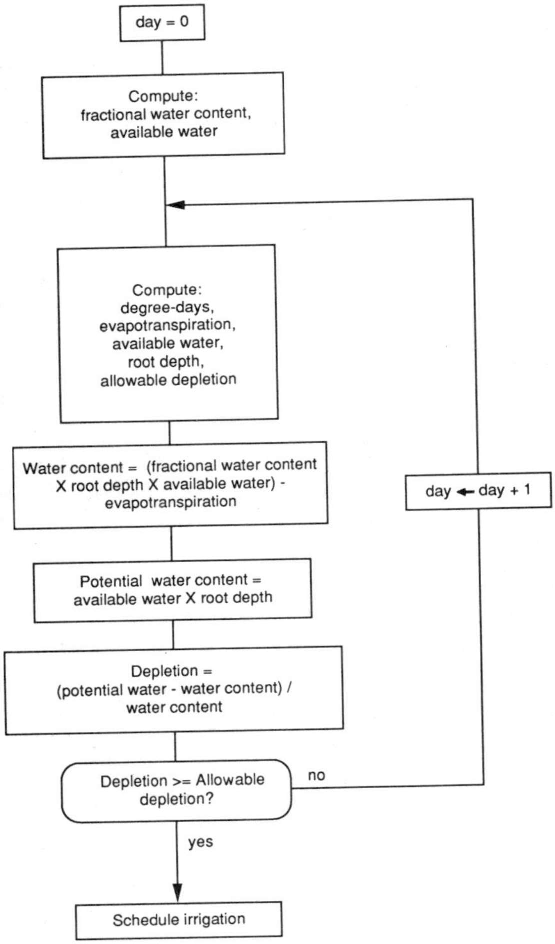

Figure 9.5 Flowchart for the water budget irrigation scheduling method.

ence engine to accomplish the same thing. There are some very practical reasons for doing this. One of the most important is the relative simplicity of the expert system's rule base structure. The primary structural component of the rule base is the production rule, and this structure is relatively simple to understand, even for the nonprogrammer. A second reason is the presence of the automatic explanation facility. A user who wishes to see an explanation for the scheduling of an irrigation on a particular date can, by seeing the rules that went into the generation of this date, determine whether the reasoning is acceptable.

How can the inference process be converted from the linear, branching mode of forward and backward chaining to the circular, iterative mode of a procedural system? One simple way is to enforce a continued cycling through the rule base, starting with the first rule and ending with the last on each cycle, until either some hypothesis in the rule base is satisfied or a preassigned maximum number of iterations is completed. Unlike chaining, in this sort of looping the position and order of the rules are critical. In effect, the declarative rule base is turned into a conventional computer program. Figure 9.5 shows a flowchart of the water budget algorithm. Each computation in the flowchart is accomplished by a rule or part of a rule. For example, the step in which total depletion is set equal to the difference between the potential water content and the actual water content divided by the water content may be accomplished by a rule of the form

IF ()
THEN depletion = (potential_water − water_content) / water_content

The parentheses in the IF part of the rule indicate that no antecedent needs to be verified; the rule is always instantiated. The last rule of the loop compares the allowable depletion with the actual depletion. This rule has the form

IF depletion ≥ allowable_depletion
THEN schedule an irrigation

Here *schedule an irrigation* is a hypothesis in the rule base. Once this hypothesis is satisfied, execution of the looping stops.

What is the advantage of implementing a procedural sequence such as this in a rule base? Isn't it like attempting to squeeze a square peg into a round hole? Yes it is, nevertheless, there are advantages to doing it. Perhaps the greatest advantage is that it makes available to the user the expert system's explanation facilities. The user can step through the sequence of actions leading to the scheduling of an irrigation date and determine whether this procedure is acceptable or not. This sort of option is usually not available with traditional computer programs.

9.6 Discussion

Two very different methods have been described in this chapter for scheduling a sequence of irrigation dates. The optimal control method treats irrigation scheduling as analogous to the problem of planning the flight of a missile. Indeed, optimal control theory grew out of the need to control guided missiles. An objective function is first established. In controlling a guided missile, the objective function measures the deviation from some planned trajectory; in irrigation scheduling, it measures some level of value associated with the crop, such as the net economic yield. After the objective function has been established, a sequence of control actions is computed. In the case of the guided missile these control actions are levels of thrust of the rocket engine and positions of the control ailerons. In the irrigation schedule the control actions are the days on which water is applied and the amount of water applied. The computation involves choosing the control actions to optimize the objective function. The computation can be either deterministic or stochastic. The stochastic method is clearly the only appropriate choice for irrigation scheduling, but this method carries with it a number of practical pitfalls involving computational complexity and parameter determination.

If the optimal control method is akin to remote guidance of a missile, then the knowledge-based planning method is analogous to driving a car. Indeed, much of the impetus for early planners grew out of the desire of the military for an autonomous land vehicle, i.e., a tank or armored car that could travel and maneuver without a driver on board. Guiding a car, unlike guiding a missile, involves many small adjustments to stay on a road, e.g., stop at stop lights, avoid traffic, and so forth. This process is far too complex to be tractable with optimal control methods, so the heuristic methods of artificial intelligence are more appropriate. The difference can be fully appreciated by thinking of programming a rocket to fly to the next city. (No warlike intent need be imagined; suppose a packet of mail is to be dropped by parachute.) The flight could be pretty well preprogrammed, with small corrections made from a remote controller to correct for deviations from the planned trajectory due to wind gusts and so forth. The trajectory could under these circumstances be optimized to minimize fuel consumption, time of flight, or some other performance parameter.

Now imagine driving the same mail packet to the next city. The route could be planned in advance, but no remote controller currently available could be counted on to safely drive a car from one city to the next. The task is too complex and full of too many uncertainties. Moreover, the route chosen may be a generally good one in the sense that it approximately minimizes the time of travel, but one does not ordinarily attempt to compute in detail the precisely optimal route, or to stick to this route at all costs. One simply chooses a reasonably good route. The problem of scheduling a sequence of irrigations suffers from the same complexity as driving a car, and the solution is, like the route between cities, only required to be reasonably good, not precisely optimal. Therefore, the knowledge-based methodology seems most appropriate.

With this methodology the computer functions like a human expert with a hand calculator and a set of tables. The expert has several methods available and chooses the one most appropriate for the circumstances. Most of these methods are not in themselves particularly advanced from a technological standpoint; indeed, they have been available for years. The novelty lies in the use of the computer to decide which method to use and in the programming of these methods in a rule-based form. The advantage of using the computer is that it permits the farmer to use the methods without requiring the mathematical sophistication necessary to do the computations. The advantages of the rule-based format are the same ones we have encountered before, i.e., relative conceptual simplicity, availability of an explanation facility, ease of modification, and so forth. The planner described in this chapter links several rule bases together into a coherent knowledge base and provides a mechanism for moving from one rule base to another. The planner retains the domain independence of the expert system by focusing on domain independent information, namely, the identity of the rule base used previously and whether or not one of its hypotheses was successfully verified.

What other management activities are amenable to this approach? Experience with the CALEX system indicates that it is very powerful. Consider, for example, the case of insect pest management. Unlike irrigation scheduling, in insect pest management we do not, or should not, generate a schedule of pesticide applications. Rather, we scout the insect population repeatedly and treat only when the population exceeds an action threshold (Stern et al., 1959). Therefore, a pest management advisory program should first be capable of determining whether or not to apply a pesticide by comparing the observed infestation level with the action threshold. If the level is above the threshold, then a pesticide application is recommended. If it is below the threshold, then the system determines the next date to scout the field and presents this information to the user. Therefore, the first recommendation can follow one of two branches.

If the second branch is followed and a future scouting is recommended, then the consultation is complete for the present. However, if a pesticide application is recommended, then a date should still be provided on which the field is to be scouted again. This date should take into account reentry restrictions for the pesticide and the expected time until the pest population rebounds. Therefore the expert system must generate either one or two recommendations, depending on the nature of the first recommendation. The first method, called *check threshold,* has only one hypothesis, *over threshold.* This hypothesis is verified if the infestation level is indeed over the action threshold. A guideline can then be generated recommending a pesticide application. In this case the state of the system is set to *Success-over threshold.* If the infestation level is not over the action threshold, then the hypothesis is not verified, no guideline is generated, and the state of the system is set to *Failure-over threshold.* The second method *determine scouting date* has both of these states as calling states and is therefore invoked to generate a scouting date no matter what the pesticide status. If a pesticide application has been recommended, then this information can be put on the blackboard for use by the *determine scouting date* method.

The scheduling method developed in this section is *sequential*; each action is planned precisely before work begins on the next one. This method offers the advantage of simplicity but does not guarantee a globally optimal plan. The problem can be easily understood by considering the scheduling of the last two irrigations in a season. It may happen that after scheduling the next-to-last irrigation, it is found that the last irrigation must be scheduled one day later because of constraints on drying the crop. Clearly this should influence the scheduling of the next-to-last irrigation, but the algorithm as presented here precludes this. The actual CALEX algorithm has the ability to adjust the next-to-last irrigation based on this eventuality, but it cannot adjust the whole schedule. A planner based on the NOAH algorithm does have this ability. Such a planner would follow a heuristic search procedure of the kind described in Chap. 7 to search for a globally optimal irrigation schedule in a manner analogous to the search for a globally optimal path through a network. The disadvantage of this approach, and the reason that it was not used in CALEX, is that it does not mesh as well with the rule-based expert system architecture used in the rest of the program and that it is more conceptually difficult and domain dependent.

Chapter

10

Object-Oriented Models

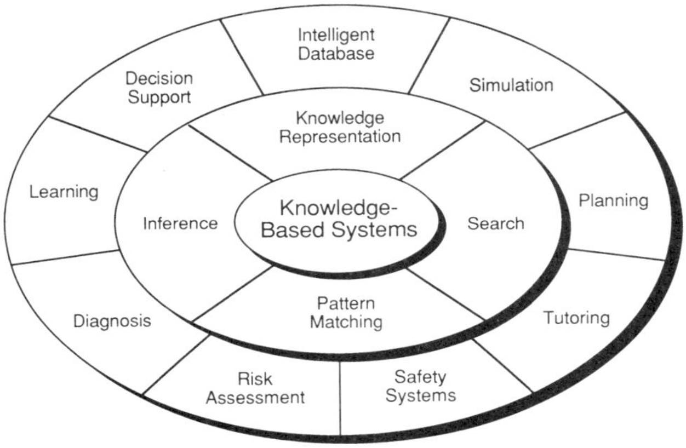

This chapter discusses behavioral modeling, event-driven simulation, frames, and object-oriented programming. AI simulation techniques are explored using the example of simulating a host-parasite system. AI techniques, which were developed initially to mimic human intelligence and behavior, are applied instead to animals in order to model animal population dynamics as the sum of the motivations and behaviors of the individual animals.

Object-oriented programming has caused a revolution in the computing world and has been embraced wholeheartedly by AI practitioners. In agriculture there has been much less enthusiasm, but that is probably only because agricultural scientists are typically slow to adopt a new technology. Also, programming efficiency will not likely ever be an exciting topic in agricultural science. Rather, new techniques that help solve problems are sought.

Therefore, the key question that we will address is: What does object-oriented programming allow us to do that could not be done before? In a sense, since object-oriented programming is only a style of programming, it is wrong to say that it expands the capabilities of the computer. At a very low level, an object-oriented program, like any program, is converted back to assembly language. Whatever can be done in an object-oriented language, therefore, can also be done in assembler or most languages that provide hooks to low-level commands, like C. However, just as the LISP language made it possible for programmers to start thinking in terms of words and relationships, object-oriented languages let programmers build systems that conform much better to the programmer's concept of the real world.

Object-oriented programming has been widely adopted by AI developers because of its applicability to simulation and blackboard systems, i.e., systems with multiple information sources (see Chaps. 9 and 12). Here we will look at its application to simulation, in particular, individual-level simulation as opposed to population-level simulation.

10.1 Problem: Predator-Prey Population Dynamics

The interaction of two species, one a predator or parasite of the other, has been a favorite modeling example in disciplines from mathematical ecology to economic entomology throughout the twentieth century (Hassell, 1978). The basic issues are: How do both species survive? Why doesn't the predator eventually consume all the prey, thereby causing its own extinction? How can we predict what the outcome of a predator-prey or parasite-host interaction will be? From a mathematical standpoint, the relative simplicity of the system has been very appealing, particularly the finding that highly complex and even chaotic population dynamics can be derived from very simple mathematical models. On the other hand, agricultural researchers would very much like to understand why some beneficial insects, parasites and predators of insect pests, do such a good job regulating their hosts' densities while other species fail to reduce their hosts' populations.

Two approaches to modeling have emerged from these two perspectives on the problem, though the two overlap considerably. The mathematical ecologists have employed an analytical approach, using differential equations and their discrete analogs, difference equations, to model the two-species interaction. These equations have been manipulated in countless ways over the years, but the general form has remained constant from Lotka's work (1925) to the modern formulations as exemplified by Hassell's (1978) difference equations for host-parasite interaction:

$$N_{t+1} = rN_t f(N_t, P_t)$$

$$P_{t+1} = \phi N_t (1 - f(N_t, P_t))$$

where N is the population density of the host, P is the population density of the parasite, and $f(N,P)$ describes the proportion of hosts surviving parasitization or predation in one generation. The constant ϕ converts the kill rate into a birth rate for parasites. Volumes, literally, have been written about the behavior of these equations and the system they represent, given different forms of $f(N,P)$.

The applied ecologists, meanwhile, have concentrated on developing more realistic and complex simulation models of the interactions between parasites and their hosts (e.g., Gutierrez et al., 1984). These models include far more detail about the biologies of the two species. Generally, the species' developmental rates and age structures are included, and natality and mortality are modeled explicitly, as are the relationships between these processes and abiotic factors like temperature and rainfall.

One of the fundamental characteristics of both the analytical modeling approach and the simulation approach is that the goal is to describe the change of population numbers over time. In the classical analytical models, only the total number of predators P or prey N appear in the equations. In the simulation approach, these become arrays indexed by age class, e.g., $N = \{n_1, n_2, n_3, \ldots, n_k\}$, but the elements of the array still describe a population number or density. Essentially, the species interaction is modeled as if there were some object in the system called a population density with a predictable behavior. In fact, of course, populations are not true objects. Population dynamics is the sum of the individual actions, and population-level effects are really emergent properties of many independent or coordinated individual behaviors. Even when population-level models adequately describe population trends and dynamics, they only summarize the integrated actions of individuals; they do not simulate the basic processes responsible for population change.

This focus of classical host-parasite modeling at the population level becomes a problem when one is interested in the basic behaviors and characteristics of individuals that are responsible for population-level effects. Let us take the problem of resource utilization as an example. In host parasite systems, resource utilization by the parasite is called the functional response. It is the relationship between the number of hosts attacked and the number of hosts available. In reality, this relationship is itself highly dependent on the age of the organisms involved and the environment in which the hosts are to be found. However, if we assume that we are dealing with a single age-class, adult parasites for instance, and a reasonably homogeneous environment like a monoculture, these difficulties can be overcome.

We can develop a very simple model of the functional response for parasites by assuming that each parasite searches its environment at random and independently. Assuming that the probability of finding a host by a parasite in 1 second of searching is very small and the amount of time, i.e., number of seconds, spent searching is very large, the probability that a host will be encountered a given number of times is described by a Poisson distribution

$$P(x) = \frac{\mu^x e^{-\mu}}{x!}$$

where x is the number of encounters and μ is the mean frequency of occurrence, i.e., the product of the probability ρ of finding a host per trial, i.e., per second, and the number η of trials, i.e., number of seconds in the experiment. The zero term of the Poisson tells us the probability that a host will remain undiscovered and is the term of interest. Thus,

$$P(0) = e^{-\mu}$$

and the probability that a given host will be attacked at least once is

$$P(x \geq 1) = 1 - e^{-\mu}$$

The number of hosts attacked N_a in the experiment can then be calculated as the probability of a host being attacked multiplied by the number of hosts available, that is,

$$N_a = N(1 - e^{-\mu})$$

This is essentially the same model proposed by Thompson (1924) and by Nicholson and Bailey (1935), except that we have not represented μ as a function of N and P. The mean frequency of occurrence of a host being found can be defined as the total number of host-parasite encounters N_e divided by the total number of hosts available N. Nicholson and Bailey assumed that the number of encounters per parasite would be proportional to the number of hosts, that is,

$$\frac{N_e}{P} = aN \quad \text{or} \quad \frac{N_e}{N} = aP$$

Their functional response model can now be derived as

$$N_a = N(1 - e^{-N_e/N}) = N(1 - e^{-aP})$$

They called the constant a the parasite's "area of discovery." By using this name, Nicholson and Bailey implied that it is some observable characteristic of a parasite. In fact, this is not the case. The parameter must be estimated from experimental data by curve fitting.

Holling (1959) proposed a more complicated model for the functional response based on some assumptions about individual parasite search behavior. He began with the assumption that parasites have a characteristic search rate a' and that there is some fixed amount of time available for search T_s. This meant that the total number of encounters per parasite is equal to

$$\frac{N_e}{P} = a'T_sN$$

He further assumed that at each encounter between parasite and host, some time is taken in manipulating and parasitizing, i.e., there is some handling time T_h. The time available for search is then the total time less the total handling time.

$$T_s = T - T_h\frac{N_e}{P}$$

These two expressions, combined and rearranged, give us Holling's "disc equation."

$$\frac{N_e}{N} = \frac{a'PT}{1 + a'NT_h}$$

This equation's primary contribution seems to have been that, quoting Hassell (1978), it "pinpointed the essential biological ingredient" that is responsible for the plateau in host numbers attacked per parasite as host density increases, i.e., the handling time associated with parasitization. However, the equation has not been such a practical success. Both search rate and handling time have been shown to depend on prey density, feeding patterns and recent history, and environmental characteristics. Furthermore, although a' and T_h have conceptually clear meanings, they cannot be estimated by direct observation, they must be calculated by curve fits to experimental data. That is, they really are population-level summary statistics, not necessarily representative of individual behavior.

Clearly, parasite searching efficiency is fundamentally dependent on the individual behavior of the parasite and the location of hosts in a complex environment. The traditional modeling approaches discussed above have never proven to be of much practical value. Often too, when a population model was seen to deviate from observed data, new parameters were added to the model, justified as the result of some previously unaccounted for behavior.

For example, Hassell and Varley (1969) noticed that for many species interactions, the search efficiency decreased with increasing parasite density (Fig. 10.1). They proposed a modification to the Holling

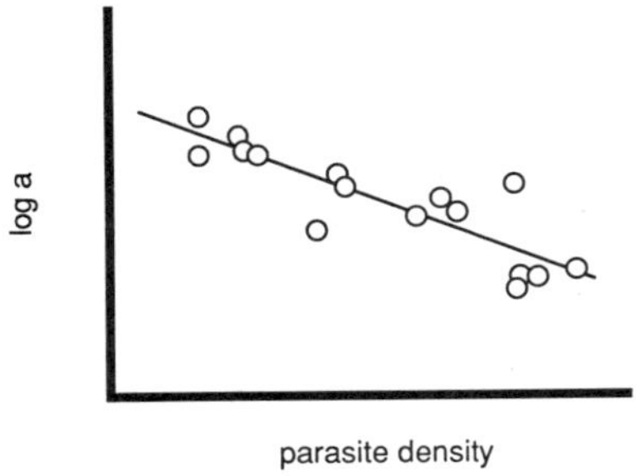

Figure 10.1 Relationship between the parasite's searching efficiency and parasite density that led to the proposal for a mutual interference constant.

functional response model, letting the search rate a become a log-linear function of parasite density P_t, that is,

$$a = QP_t^{-m}$$

or

$$\log a = \log Q - m \log P_t$$

where m is the mutual interference constant.

This mutual interference constant was supposed to describe the result of parasites physically interacting with one another at high densities, thus, reducing their searching efficiency. But it was later discovered (Free et al., 1977) that the same effect can be produced by an entirely different underlying mechanism, parasite aggregation in areas with high prey density. If higher than expected numbers of parasites are in patches of high prey density, then their overall host attack rates will be lower than expected due to more competition, not due to physical interference. This has been called *pseudointerference.*

The example of real versus pseudointerference points out the problem of making individual-level assumptions based on population-level phenomena. Until it can be verified by observation that individual parasites physically interfere with one another, the inclusion of a mutual interference correction parameter has no more basis than a statistical curve fit. A population-level parameter may help describe the dynamics of populations, but it does not explain the cause. The underlying mechanism must be explained at the individual level.

When the influence of individual behavior or environmental heterogeneity on population dynamics is strong, as it seems to be in host-parasite interactions, it would be helpful to build a model more from first principles. The behavior of individual insects is often observable; at the least it is easy to make hypotheses about how individuals behave. If we can build models of population processes that are based on a true description of individual behaviors and interactions, then testing hypotheses about how behavioral changes affect population processes will be much simpler and straightforward.

10.2 Individual and Behavior Modeling

We need to distinguish between modeling animal behavior and modeling animal populations by describing individuals. The field of ethology, while it provides the basis for individual-level modeling of populations, is more concerned with the patterns of animal behavior than with its demographic consequences. Ethologists study the basis of animal behavior, while individual-level modelers might be satisfied with purely statistical descriptions of behavioral choices and actions.

However, a knowledge-based approach to individual-level modeling clearly implies an explicit representation of causes of individual actions and behaviors. In fact, the field of artificial intelligence is devoted to this very same goal, making computers behave intelligently. Although artificial intelligence research has not yet produced a computer that truly mimics human intelligence, its products are much more capable of mimicking the intelligence of some animals, particularly the so-called lower animals like insects.

Before we describe a knowledge-based approach to modeling animal behavior, we will briefly develop a general model of animal behavior and explore some of the problems with classical modeling approaches along the way. The following discussion draws from a book chapter by Packard et al. (1990).

10.2.1 A structure for behavior

Behavior is usually described as a series of stimuli and corresponding responses. A stimulus can be associated with a single particular response, as is the case with reflexes. On the other hand, some stimuli will elicit different responses at different times, depending on the internal state of the animal, other environmental stimuli, etc. The processes responsible for changing the response of an animal to a given stimulus are motivation and learning (Staddon, 1983; Toates, 1986).

Figure 10.2 shows the overall model of behavior. Stimuli, shown as messages from the external environment to an animal, are filtered by the animal's perception. Many examples of this filtering or selective detection of messages from the environment are evident in the world around us. Insects, for example, see ultraviolet light while humans cannot. The appearance of flowers to pollinating insects is often very different from our perception of flowers as a result. Likewise, noctuid

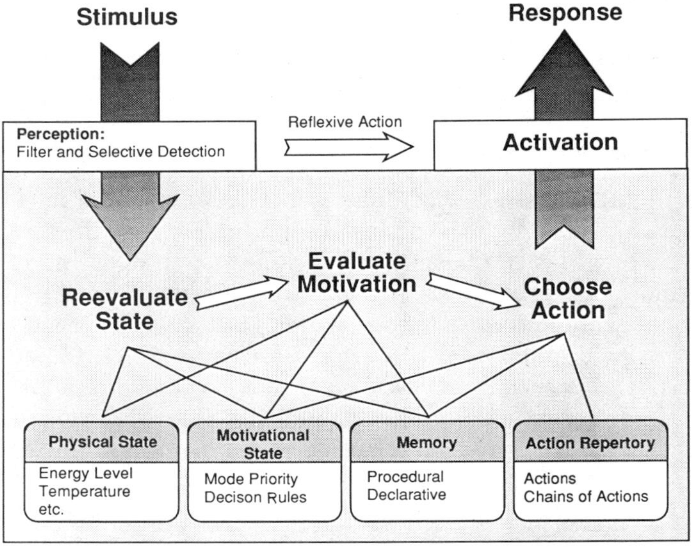

Figure 10.2 A very basic model of an animal's behavioral system, showing a hierarchical decision process, and its relation to observable stimuli and responses. (*Modified from Packard et al., 1990, Figs. 1 and 2.*)

moths' hearing is tuned to frequency ranges used by bats for navigation and prey location.

The effect of a perceived stimulus in an animal is some change in state. Stimuli can have direct reflexive effects on an animal, or when combined with memory, stimuli can give the animal new information about its status. A chemical stimulus may tell the animal that it has encountered food or an enemy. A stimulus from within the animal may indicate that it's stomach is nearly empty or that its muscles are fatigued. We do not make any distinction between stimuli from the external environment and stimuli from within the animal. Both types of stimuli act on the brain, although the physiological mechanisms may be quite different.

The rules for choosing behavior based on a stimulus are thought to be grouped into hierarchical sets (Dawkins, 1976). First an animal determines its motivational mode. These might include ingestion, agonism, dispersal, reproduction, etc. Each mode can be thought of as competing for control of the animal's choice of actions. The selection of a motivational mode is made, therefore, based on priorities. For example, Kovac and Davis (1980) have shown that in the sea slug, escape always receives top priority. Oviposition has a higher priority than mating, and feeding generally outweighs moving away from a tactile stimulus.

Once a behavioral mode has been chosen, animals then choose actions from their action repertory. The rules for choosing actions may depend on several factors. Sometimes actions are more likely to occur in particular sequences, so that the probability of an action being chosen depends on recent previous actions and their order. For instance, a digger wasp after capturing a prey, say a grasshopper, carries it to its nest. It then places the grasshopper at the entrance to its hole and enters the hole where it has previously laid an egg. The wasp then comes out of the hole and drags the grasshopper down into the egg chamber. However, if the grasshopper is moved while the wasp is in its hole, the wasp will emerge and repeat the entire process. It moves the grasshopper back to the entrance, enters the hole without the prey and returns to drag the prey down. By repeatedly moving the grasshopper, we can keep the wasp in an endless loop.

In other cases, actions seem to be chosen in a goal-oriented fashion. Feeding behavior and thermoregulation are good examples. Animals behave as if they are trying to maintain some "set point" (Staddon, 1983). When their physical state differs enough from the set point, they initiate actions with the goal of reducing the difference. Moving along temperature gradients or searching for prey more actively when hungry are examples of goal-oriented behavior.

Memory and learning are clearly important elements of behavioral choice. Some animals seem capable of recording procedural knowledge like sequences of actions that produced a desired result. This tends to increase the action repertory by including sequences of actions as well as individual actions themselves. Other animals are capable of storing declarative knowledge, e.g., knowledge about how actions are related to effects. Such animals may be able, like humans, to choose behaviors based on an internal expectation of the results of its actions.

Learning is closely related to memory. Learning is often broken down into associative learning, as in the conditioning of a rat to press a bar to get food, and nonassociative learning, for instance, the process of becoming more or less attuned to a particular stimulus. Learning can also be fast or slow and seems to be genetically facilitated in many cases. For instance, a blue jay learns to avoid eating monarch butterflies after only one unpleasant encounter with a poisonous one. It is as if there are preexisting memory locations set aside for experiences relating to prey capture and stomach reaction.

10.2.2 Modeling approaches

Many modeling approaches have been used in ethology, from Lorenz's mechanical models of motivation (Nisbett, 1976) to optimization and game theory. We will not review all these modeling approaches here except to say that they generally treat important aspects of behavior but not the complex interactions within populations. Also, most of the quantitative models assume stable and/or homogeneous environments. Thus, game theory allows the modeler to decide whether particular behaviors are more or less adaptive and whether behaviors are evolutionarily stable. Optimal foraging models permit a researcher to predict expected patterns of resource utilization in a patchy environment, assuming the patchiness is an inherent and unchanging characteristic of the environment.

Environments are seldom stable in their properties. Agricultural environments, in particular, are characterized by disruption and almost continual change in evolutionary time. The classical modeling approaches that have served behavioral theorists so well are not well suited to practical applications in agriculture because of things like environmental heterogeneity in space and time, individual animals with incomplete or inexact knowledge of their environment, animals whose behavior depends on their particular experience, and animals whose behavior is goal oriented or mediated by competing motivational systems.

Our goal in knowledge-based, individual-level modeling is to describe the environment and the individuals interacting within it in ways that take account of individual variation, spatial heterogeneity, and motivational, goal-oriented behavior. There are really two problems here. The first is how to represent all the relevant actors in a model in a consistent way and a way that can be incorporated into a knowledge-based system. The second is how to model behavior using knowledge-based tools.

10.3 Object-Oriented Programming

Any modeling approach will become bogged down when trying to account for all the variability and heterogeneity in the real world. Mathematical approaches become particularly unwieldy. Numerical methods like computer simulation are therefore required. However, traditional computer simulation methods are much better at dealing with and manipulating arrays of state variables, like the number of individuals aged 10 days, than data structures describing individuals. Object-oriented programs, on the other hand, are specifically designed to represent individuals or specific objects found in an environment. Computer memory will still limit the number of individuals that a simulation model can include, but the model's structure can remain extremely simple.

We will begin with some background on object-oriented programming (OOP), comparing it with procedural programming, and finally extending it with access-oriented programming. Interested readers should also see Goldberg and Robson (1983) and Stefik and Bobrow (1986).

10.3.1 The basics of OOP

Classical programming languages and many rule-based systems use variables and their values to document the state of the system. The number of prey alive in age-class k would be represented by the value of a variable $N[k]$. In a Fortran or C model, we perform operations on our data by calling subroutines or functions and passing the appropriate data to them. Thus, a function to compute the proportion of hosts killed during the current time-step of the model could be written in C as

```
for (i = 0; i < max_age; i++)
{
   number_attacked [i] =
      functional_response(N [i], P [i], a_prime,
        hand_time, delta_T)
}
```

where the definition of the functional response function would include the code for determining the number attacked and would be found elsewhere in the code, e.g.,

```
functional_response(N, P, a, Th, T)
int N, P;
float a, Th, T;
{
   int number_attacked;
   code here;
   return (number_attacked)
}
```

Object-oriented systems describe all data and information in the system in framelike objects. An object is a data structure that maintains a local state, described by local variables, as well as all the procedures that operate on the local variables. Objects, therefore, have both state and behavior. For example, a patchy environment might be described using objects to represent each patch. A patch object could then include descriptive information, like the number of hosts within it. If we wanted to calculate the number of hosts attacked in our environment in a way that is similar to what we do in a classical program, we would first create our patch object as

```
Object subclass: Patch                    Patch is a subclass
  variables:                              of the generic 'object'
    numberHosts                           class.
    numberParasites
  methods:
    calculateNumberAttacked
      | numberAttacked |                  Name of method.
      code using object and local         Bars enclose local
      variables only                      variables used only in
      ^numberAttacked                     the method.
```

Next we would send the object the message, `calculateNumberAttacked` with no parameters. The `patch` would return the value it calculates for `numberAttacked` to whatever object sent the message.

It doesn't matter whether we are programming a model of parasitization in a patchy environment or calculating the mean and standard deviation of a set of numbers. All the elements in an object-oriented program must be represented as objects, all behaviors as methods, and all actions must be triggered by methods passed to objects. If we were writing code to take the average of a set of numbers, the objects would be the numbers, the numbers would have methods that allowed them to behave properly to messages like 'add' and 'divide.' The expression

```
c = a + b
```

which normally we might interpret as "Assign the sum of two numbers *a* and *b* to the variable *c*" would be interpreted in an object-oriented language as "Send object `a` the message `plus` with the parameter `b`, and assign the result to `c`" and would be written as

```
c := a plus: b
```

Of course, most object-oriented languages allow the programmer to substitute the familiar syntax for numerical operations, but the syntax above is more correct. Messages are sent between objects. Each message invokes a locally defined method of the same name within the receiving object. A message may include parameters, as in the case of the `plus:` method above, and methods may return values to their caller.

Returning to our patchy environment, what have we gained from the object-oriented style? Not much yet. However, consider what happens when there are two different kinds of patches in our environment. We also make the assumption that the parasite's search rate a' is different in the two patch types. In our C program, we now have to create an array of patches and modify our iterative function call. The value of the patch array entries is patch type 0 or 1, and the code looks like

```
for (i = 0; i < max_age; i++) {
   number_attacked [i] = 0;
   for (j=0; j < max_number_patches; j++) {
      if (patch [j] == 0)
        a_prime = 0.05
      else
        a_prime = 0.003;
      number_attacked [i] +=
        functional_response(N [i], P [i], a_prime,
           hand_time, delta_T)
   }
}
```

Our object-oriented model requires almost no changes because of the principle of inheritance. As with frames, objects are defined in terms of object classes. Each class represents a type of object. Specific objects in the system are then created as instances of the class. In our example above, `Patch` is really an object class. Each patch in nature would be represented in our model as a specific instance of the patch class. Each instance of a class inherits its variables and methods from its parent class. Classes, therefore, contain the program code, but instances of classes, true objects, contain all the specific data and perform all the actions in a simulation.

In order to have two different types of patches, we can do either of two things. First, we could simply include the variables, `patchType` and `parasiteSearchRate` in the definition of the `Patch` class. Then, when specific instances of patches are created, the appropriate value would be given to patches of each type by the `instanceCreation` method in the `patch` class, as shown below:

```
Object subclass: Patch
  variables:
    patchType
    numberHosts
    numberParasites
    parasiteSearchRate
  methods:
    calculateNumberAttacked
       | numberAttacked |
       code here
       ^numberAttacked
    setType: anInteger
       patchType := anInteger
       parasiteSearchRate :=
          (patchType = 0
             ifTrue: [0.05]
             ifFalse: [0.003])
  class methods:
    createType: anInteger
       ^super new setType: anInteger
```

Given this definition of the `Patch` class, we could create a set of patches and then compute the total number of hosts attacked with the following messages:

```
attacked := 0.
patches := Set new.                          "Create a set of patches"
10 timesRepeat:                              "Add 10 of type 0"
   [newPatch := Patch createType: 0.
   patches add: newPatch].
5 timesRepeat:                               "Add 5 of type 1"
   [newPatch := Patch createType: 1.
   patches add: newPatch].
patches do:                                  "Send each patch the"
   [:eachPatch |                             "same message and sum"
   attacked :=
     attacked + eachPatch calculateNumberAttacked]
```

The second way of creating different patch types is to create subclasses of the `Patch` class. Like instances, subclasses inherit their parent classes variables and methods. However, they can also add their own variables and methods or override those of their parents. So, if we have two or three different kinds of patches, each one can be a subclass of the `Patch` class. Each subclass could redefine a **`setUp`** message, and the `Patch` class could be simplified as

```
Object subclass: Patch
  variables:
    numberHosts
    numberParasites
    parasiteSearchRate
  methods:
    calculateNumberAttacked
      | numberAttacked |
      code here
      ^numberAttacked
    setUp
      (subclass implementation needed)        "The subclass(es) must
                                               define this method"
  class methods:
    new
      ^super new setUp

Object: PatchA
parentclass: Patch
  variables:
  methods:
    setUp
      parasiteSearchRate := 0.05

Object: PatchB
parentclass: Patch
  variables:
  methods:
    setUp
      parasiteSearchRate := 0.003

Object: PatchC
parentclass: Patch
  variables:
  methods:
    setUp
      parasiteSearchRate := 0.01
```

Now, creating patches and computing hosts attacked looks like

```
attacked := 0.
patches := Bag new.                       "Create a set of patches"
10 timesRepeat:                             "Add 10 of type A"
  [newPatch := PatchA new.
  patches add: newPatch].
5 timesRepeat:                            "Add 5 of type B..."
  [newPatch := PatchB new.
  patches add: newPatch].
2 timesRepeat:                            "...and some type C"
  [newPatch := PatchC new.
  patches add: newPatch].
patches do:                               "Send each patch the"
  [:eachPatch |                             "same message and sum"
  attacked :=
    attacked + eachPatch calculateNumberAttacked]
```

Notice we've now eliminated the need for any conditional statements, and the creation of new patch types is elementary. As we make the situation more and more realistic, the advantages of the object-oriented design become even more evident. Consider, for instance, what would happen if there were several different types of parasites as well as three different kinds of patches. The C program becomes too complicated to show. However, we can modify our object-oriented model so that there is a class called parasite with several subclasses and the object-oriented model is still tractable.

In this example, we have seen several of the basic principles of OOP at work. First of all, the process of programming involves representing the key types of actors in the system being modeled as object classes. Whenever possible, these classes are structured in a hierarchy to take advantage of code inheritance. Once the object classes are determined, each object class's behavior is encoded as methods. This step requires thinking about the interactions among objects in the physical system as messages sent back and forth among objects. An object-oriented program is run by instantiating objects, i.e., creating copies of the object classes that will actually send and receive messages in the model. The objects are then sent into action by sending one or more of them a message or set of messages.

Object-oriented programs are said to exhibit *information hiding* and *data abstraction* because the sender of a message has no access to and does not need to make any assumptions about the internal representation of data types in the receiver. When we request each patch to compute the number of hosts attacked, all we needed to say is `calculateNumberAttacked`. It is possible, for example, to change the method that responds to that message without changing the message that invokes it.

Object-oriented programs also exhibit *polymorphism,* i.e., two different objects can respond to the same message in different ways. In our example, the different classes of patch responded to the same message in different ways because of the different values of a local variable. However, one patch could reimplement the method to use a Nicholson-Bailey type of functional response while another patch type uses the Holling model.

10.3.2 OOP, frames, and access-oriented programming

Recall from Chap. 1 that the development of knowledge-based systems can be seen as the inclusion in representation schemes and reasoning systems of more and more associations. Rules represented the basic building blocks of a knowledge-based system. Frames allowed knowledge engineers to describe objects in the real world with data structures that grouped related information. Essentially, frame-based programs extended the *variable* concept, in which names are associated with some changeable value, to the *object-attribute-value* concept, in which attributes are first associated with some object and then given a value. Frame-based systems also include associations among frames through inheritance.

Object-oriented programming uses a framelike representation scheme. That is, information is stored in object-attribute-value format, and classes of objects comprise a knowledge representation scheme employing objects, hierarchies, and inheritance. However, while frames are used primarily to structure knowledge, objects in object-oriented systems also include all the procedural knowledge of the system in their methods. The distinction is evident when one compares an object-oriented reasoning system with a frame-based system.

As we saw in Chaps. 8 and 9, a frame-based system uses frames to represent its knowledge while a separate inference engine acts on that knowledge. For example, a cotton pest management system might contain a frame called date-of-first-flower, with a slot that holds the date, another that holds a default value, or perhaps another that holds a procedure to be called to calculate the value if it is missing or unknown. The power of the frame system is activated by the inference engine when it needs to know what is the value of the date-slot.

A major difference between frames and objects is that the information within a frame is meant to be referenced externally. Thus, a rule could refer to the pest density of cotton field A. Information within objects, on the other hand, is inaccessible outside of the object itself. If we had an object class called `CottonField` with a variable called `pestDensity`, the only way another object could query the pest density of an instance of cotton fields is if there is a method in the `CottonField` class that returns the value of `pestDensity`, e.g.,

```
methods:
  pestDensity
  "Answer the pest density of the receiving instance"
      ^pestDensity
```

How then can we use rule-based reasoning with an object-oriented representation? Most often, as in the case of the planning system described in Chap. 9, systems that include both rule-based and object-oriented paradigms are really frame-based. They allow the slot values of frames to be referenced directly from external frames. In this type of system, it is the object-oriented paradigm that is compromised. Frames are allowed to contain method slots that are fired when messages are received by the instances of the frames. However, the notion of information hiding is lost.

It is possible to implement rule-based reasoning within a purely object-oriented environment. The Humble system, an extension to

Smalltalk-80, is one example. For the remainder of this chapter, we will follow the syntax and structure of Smalltalk but we will assume that rule-based reasoning can also be performed by special rule-object classes that are allowed to access the values of variables in other objects directly.

Another characteristic of frame-based systems that has been incorporated into some object-oriented systems is the ability to include associations with slots. Some frame-based systems allow slots to have not only values, but also facets of the slots. So, in our hypothetical cotton pest management system, a single frame could be constructed called a cotton field, and one of its slots could be the date of first flower. That slot could also have facets, such as a default value, a certainty factor, a value class that limits the possible values allowable, a cardinality that tells how many values may be put into the slot at the same time, etc.

Some object-oriented systems allow similar associations to be attached to object variables. For example, a variable may be associated with a *when-changed* method which is activated any time the value of the variable is changed. Other associations might include a *when-accessed* or *if-condition-then* to allow for actions to take place when the system detects appropriate conditions. This style of adding methods to variables has been dubbed *access-oriented programming* (Bobrow and Stefik, 1986) and is very useful in building graphical interfaces to object-oriented systems; however, access-oriented programming is also useful in designing blackboard systems, for example.

10.4 Application to the Problem

We will develop a simple predator-prey model using a combination of object-oriented programming and rule-based reasoning. The model will explicitly represent each individual within an environment and will begin to implement the model of animal behavior developed above. This model is based on more fully developed models for host-parasite (Makela et al., 1988) and predator-prey (Stone, 1990) behavior.

The first step in model development is to describe the system in terms of objects. Each object is the member of some class, and all relevant information about the system must be contained in the variables of the objects represented. We have two main types of actors, predators and prey. For simplicity, we assume that they are elementary versions of the same kind of insect or mite, such as the predatory and phytophagous mites found in citrus groves. Because both are similar, they have common biologies and share many attributes and behaviors, allowing us to take advantage of OOP's inheritance capability.

We also need to represent their environment. We assume that the environment is a network of patches suitable for the prey's development. These patches must have food for the prey and must be linked in some kind of network, so that prey can move from one patch to another. Again for simplicity, we put the patches into a closed grid format, in which movement is possible only between neighboring patches. We need two object classes to describe this, an Environment class that defines the spatial network of patches and a Patch class whose instances are the patches themselves.

10.4.1 Representation using objects

Patches. Our class-object hierarchy will look like Fig. 10.3. We will now begin to flesh out this elementary hierarchy, starting with the simplest class, `Patch`. We assume that each patch has a food source for the prey and a maximum capacity for prey. It also needs a grid location, here described by a cartesian coordinate pair, and the name of the environment object containing its spatial arrangement.

```
Object subclass: Patch
  variables:
    food                "a real number of food quantity"
    capacity            "an integer"
    gridLocation        "an x-y point"
    environs            "the Environment that contains
                        this Patch"
```

The patch must be able to interact with the prey by providing them food when they send a message that they are eating, and the patch should also replenish the food at some rate. We put the replenishment behavior into a method called `act`, assuming that each location will be called upon to update its food resources at each time-step of the model. Also, we will include a method that lets prey ask how much

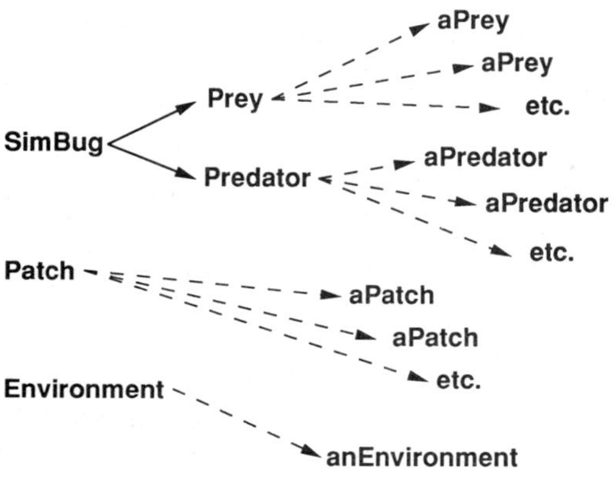

Figure 10.3 Class and object hierarchical relationships in the predator-prey model developed in the text.

food is in the patch. We add the following to the `Patch` class description:

```
variables:
  foodGrowthRate                        "a real number"
methods:
  act
    food := food + foodGrowthRate
  food
    ^food
  provideFood: amountRequested          "Return the amount of food requested or
                                        the amount available, whichever is lower
                                        and decrease the value of the food vari-
                                        ables"
    | eaten |                           "temporary variable"
    food > amountRequested
      ifTrue: [eaten := amountRequested]
      ifFalse: [eaten := food].
    food := food - eaten.
    ^eaten                              "Return the amount eaten to the
                                        sender of the message"
```

It is also important for the patches to keep track of any predators or prey it contains. That way, if we need to know how many predators are in a patch, or what the patch's capacity is, we can ask the patch to tell us. We set up two lists, one for predators and one for prey. The patch must also be able to keep these lists updated as the insects come and go. Notice below how simply the updating methods can be written because of inheritance. We can write one general method for adding insects and one general method for removing insects.

```
variables:
  predators                             "a list of all the predators"
  prey                                  "a list of the prey"
methods:
  addBug: aBug
    (aBug isKindOf: Prey)
      ifTrue: [prey add: aBug]
      ifFalse: [predators add: aBug]
  removeBug: aBug
    (aBug isKindOf: Prey)
      ifTrue: [prey remove: aBug]
      ifFalse: [predators remove: aBug]
  numberOfPrey
    ^prey size                          "Return the size of the prey list"
  numberOfPredators
    ^predators size
  capacity
    ^capacity
  prey
    ^prey                               "Return the list of prey"
  predators
    ^predators                          "Return the list of predators"
```

Before we leave the patch class, we add a constructor method, i.e., a way to create new instances of the patch class and to initialize their variables. This is a special kind of method that is sent to the class, not to an instance of a class, and it is called a *class method*.

```
class methods:
  newAt: aGridLocation in: anEnvironment
     ^super new setupAt: aGridLocation in: anEnvironment
```

The constructor uses the message `new` to create a new object, then sends the new object the message, `setupAt: in:`, which needs to be included as a regular method in the `Patch` class.

```
methods:
  setupAt: aGridLocation in: anEnvironment
    environs := anEnvironment.
    gridLocation := aGridLocation.
    prey := List new.
    predators := List new.
    food := 10.0.
    capacity := 10
```

Now, an instance of the `Environment` class will be able to create a set of patches at specific locations by sending the class `Patch` messages like `Patch newAt: (2,5) in: self` where the pair (2,5) is a point on the grid and *self* is the specific instance of the `Environment` object sending the message.

Bugs. Now we turn to the `Bugs` class. All bugs have some similar characteristics: an age, sex, a stomach or food reserve, a location, and an environment. We ignore males entirely in this discussion and so we further assume all bugs have the capability to produce and lay eggs at the rate of about one per day. Shared bug behaviors include moving, aging, dying, reproducing, eating, and metabolizing reserves. Our first pass at the representation of the `Bugs` class is now

```
Object subclass: Bugs
  variables:
    environs                     "The specific environment"
    location                     "The bug's current Patch"
    age                          "An integer"
    eggsReady                    "An integer"
    reserves                     "Amount of energy left"
  methods:
    ageOneDay
      age := age + 1
    moveTo: aPatch
      location removeBug: self.  "Tell current patch I am leaving"
      location := aPatch.        "Reset current patch"
      location addBug: self      "Announce arrival in patch"
    die
      location removeBug: self.
      environs removeBug: self
    metabolize
      reserves := reserves
         - (reserves min: (self class reserveLossRate))
    eat
      (subclass implementation needed)
    reproduce
      (subclass implementation needed)
```

Notice that the last two methods are not implemented at this level. This means that all `Bugs` respond to the method reproduce, but the method used has to be coded in all subclasses of `Bugs`. The `Bugs` class itself is not intended to have any instances, only subclasses. It is therefore called an *abstract* class. It also does not need a constructor method since no instances will ever be created.

All bugs also share a simple version of our motivational model of behavioral choice. They all need the variable `motivationalMode` and the methods `chooseMode` and `chooseAction` also implemented at the subclass level. We can now add action to our bugs by defining their `act` method, i.e., the daily cycle of actions and behaviors each bug will go through at each time step.

```
act
  | nearlyDead |                                  "Temporary variable"
  self ageOneDay.
  nearlyDead := (reserves = 0.0).                 "Bug might starve"
  age > (self class maxAge)
    ifTrue: [self die]                            "Die of old age"
    ifFalse:
      [self metabolize.                           "Use some reserves"
      self chooseMode.
      self chooseAction.
      (motivationalMode includes: #Ingestions)
        ifTrue: [self eat].
      (nearlyDead and: [reserves = 0.0]           "Still no reserves?"
        ifTrue: [self die]                        "Die of starvation"
        ifFalse:
          [(motivationalMode includes: #Reproduction)
            ifTrue: [self reproduce]]]
```

If, at the beginning of a day, a bug has no food reserves left, it is declared nearly dead. If after its actions that day it still has no food reserves, it dies of starvation. Bugs also die if they have aged beyond their class's maximum age (defined below). Assuming they do not die, all bugs deplete their food reserves by some class-specific metabolic loss rate; they then choose their daily action by sending themselves the messages, `chooseMode` and `chooseAction`. Bugs then eat or try to eat if they want to, and finally, all bugs in reproductive mode can attempt to reproduce or lay eggs.

Even though we have not implemented some of these methods, all bugs share them. Predators and prey respond differently when asked to eat, of course. This is reflected in very different implementations of the eat method in the two subclasses and is an example of polymorphism.

Predators and prey. The predator and prey classes are now written as subclasses of the `Bugs` class. All variables and methods described above for bugs are inherited by these classes. Here, for the first time, we will use class variables. These are variables whose values are shared by all instances of the class. `Predators` and `Prey` classes include class variables for the maximum age attainable, the metabolic use rate, and the age at reproductive maturity. We also need to include the class methods to allow access to the class variables, as used above in the `act` method in `Bugs`, and to create new instances, i.e., eggs, of predators and prey.

```
Bugs subclass: Prey
  Class variables:
    AgeAtReproduction
    MaximumAge
    ReserveUseRate
  Class methods:
    initialize
      AgeAtReproduction := 10.
      MaximumAge := 20.
      ReserveUseRate := 0.05
    newAt: aPatch in: anEnvironment
      ^super new setupAt: aPatch in: anEnvironment
    reserveLossRate
      ^ReserveUseRate
    maxAge
      ^maximumAge
  variables:
    "In addition to those defined in Bugs"
    feedingRate
  methods:
    setupAt: aPatch in: anEnvironment
      environs := anEnvironment.
      location := aPatch.
      age := 0.
      reserves := 1.0.
      eggsReady := 0.
      feedingRate := 0.1.
      location addBug: self.
      environs addBug: self
    age
      ^age
    eat
      reserves < 0.9
        ifTrue: [reserves :=
          reserves + (location provideFood: feedingRate)
    reproduce
    "Create new instances of Prey in the current Patch and the
    current object's Environment as long as there are eggs ready and
    suitable places to oviposit"
      | places |
      eggsReady := eggsReady + 1.
      places := location capacity - location numberPrey.
      [places > 0 and: [eggsReady > 0]]
         whileTrue:
           [self class newAt: location in: environs.
           places := places - 1.
           eggsReady := eggsReady - 1]
```

Notice below that predators live longer than prey, have a higher metabolic rate, and mature later. Predators differ also in their `eat` and `reproduce` methods. Also notice that the predator `eat` method needs to be able to tell the age of specific prey. We include the method age above to supply this information.

```
Bugs subclass: Predators
  Class variables:
    AgeAtReproduction
    MaximumAge
    ReserveUseRate
  Class methods:
    initialize
      AgeAtReproduction := 20.
      MaximumAge := 30.
      ReserveUseRate := 0.15
    newAt: aPatch in: anEnvironment
      ^super new setupAt: aPatch in: anEnvironment
    reserveLossRate
      ^ReserveUseRate
    maxAge
      ^maximumAge
  variables:
    "None beyond those defined in Bugs"
  methods:
    setupAt: aPatch in: anEnvironment
      environs := anEnvironment.
      location := aPatch.
      age := 0.
      reserves := 1.0.
      eggsReady := 0.
      location addBug:self.
      environs addBug: self
    eat
    "If not full and prey are available in the current Patch, select
    an appropriate prey and eat it. Predators can kill any prey up
    to their own age plus 5."
      | preyAvailable killAge kill |
      preyAvailable := location prey copy.
      (reserves ≥ 0.9 or: [availablePrey isEmpty])
        ifFalse:
           [killAge := age + 5.
           kill := nil.
           "Select the first prey in the list of available prey that
           is young enough to kill."
           kill := preyAvailable
                select: [:prey|prey age < killAge].
           kill isNil
             ifFalse:
                [reserves := 1.0 min:
                   reserves + 0.1 + kill age * 0.05.
                kill die]]
  reproduce
  "Create a new instance on the class Predator in the current Patch
  and the current object's Environment. Lay eggs as long as there
  are more prey than predators in the Patch or if there are more
  than 3 eggs ready."
    | places |
    eggsReady := eggsReady + 1.
    places := location numberOfPrey
       - location numberOfPredators.
    [[places > 0 and: [eggsReady > 0]]
      or: [eggsReady > 3]]
        whileTrue:
           [self class newAt: location in: environs.
```

```
            places := places - 1.
            eggsReady := eggsReady - 1]
```

At this point, we have bugs, both predators and prey, that can move from patch to patch, grow older, reproduce, eat, and die. Predators can kill prey, as well. All we need now are the decision rules in the methods `chooseMode` and `chooseAction` that let them decide for themselves, as it were, what to do based on a simplified version of our motivational system of animal behavior.

10.4.2 Motivational model implementation

The stimuli affecting bugs are internal factors like age, eggs ready to be laid, and the extent of reserves. External factors are also stimuli, like the number and type of other bugs in the bug's current patch. The presence of predators, for instance, should elicit an escape mode in prey. We will endow our predatory bugs with four motivational modes, #Ingestion, #Hunting, #Reproduction, and #Rest, and our prey with five modes, #Ingestion, #Escape, #Dispersal, #Reproduction, and #Rest. We can now write rules for determining the mode or modes that are active at any time. For example, for prey the following rules determine the set of competing motivational modes:

IF age ≥ AgeAtReproduction
THEN goals add: #Reproduction

IF age ≥ AgeAtReproduction AND
environs predatorsNearby: location
THEN goals add: #Escape

IF age ≥ AgeAtReproduction AND
location numberOfPrey = location capacity
THEN goals add: #Dispersal

IF age < AgeAtReproduction AND
location numberOfPredators > 0
THEN goals add: #Escape

IF reserves < 0.2
THEN goals add: #Ingestion

IF goals isEmpty
THEN goals add: #Rest

For predators, the rules would look somewhat different. Recall that each predator will be allowed to try to eat at each time-step, even if the predator's goal is to rest. Resting, in this context, really means remaining in the current location as opposed to hunting.

IF age ≥ AgeAtReproduction
THEN goals add: #Reproduction

```
IF      not: [goals includes: #Ingestion] AND
        location numberOfPrey > location numberOfPredators
THEN    goals add: #Rest

IF      not: [goals includes: #Ingestion]AND
        location numberOfPrey < location numberOfPredators
THEN    goals add: #Hunt

IF      reserves < 0.3
THEN    goals add: #Ingestion

IF      goals isEmpty
THEN    goals add: #Rest
```

Choosing actions depends on another set of rules. For the prey, Fig. 10.4 shows the logic represented by the chooseAction method. Notice the motivational mode priority scheme. Reproduction will occur (in the act method) whenever possible once the bug has reached adulthood. Ingestion takes precedence over dispersal, and escape has priority over ingestion.

As we can see from the flowchart, predators and prey need to be able to get information from their environment and their current patch. Accessing this type of information occurs by sending messages to the appropriate object. If a predator is trying to sense the direction toward a patch with prey, e.g., as would a predator that is relying on the concentration of a pheromone, it is sampling its environment. We implement that as a message to its environment, e.g., environs targetDirectionToPrey. If a prey's motivational mode depends on the number of predators in its current patch, it senses this by locally sampling its patch, e.g., for pheromones. We implement that type of sampling by a message to its patch, like, location numberOfPredators.

The implementation of the `Environment` class is left to the imagination of the reader. Keep in mind that it must contain methods for (1) setting up a grid of patches, each with a specific x-y coordinate, (2) controlling the simulation by telling each `Patch`, `Predator`, and `Prey` object to `act`, (3) keeping track of what predators and prey are in the system, (4) returning the `Patches` that are adjacent to a given `Patch`, (5) determining if any predators are in or adjacent to a given `Patch`, (6) returning a direction toward areas of high prey concentration, and (7) returning a direction toward areas favorable for dispersal.

Output from a somewhat more detailed model (Stone, 1990) initialized with 50 prey and 10 predators is shown in Fig. 10.5*a*. Notice that the time traces of population densities are very rough, much like field or experimental data and very much unlike the classical predator-prey population traces produced from population-level models (Fig. 10.5*b*). The roughness in this case does not come from any randomness in the model. As you may have noticed, although we created our

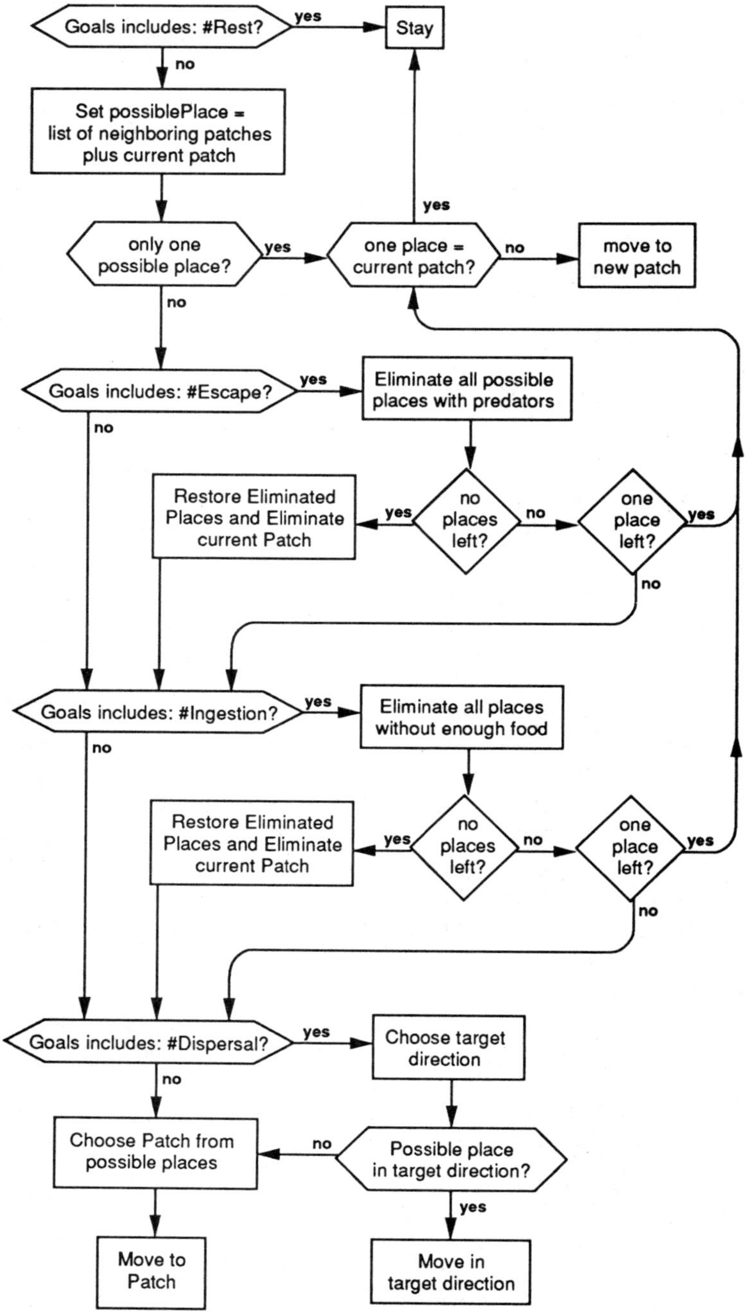

Figure 10.4 Flowchart representing the chooseAction method for Prey. This flowchart can be implemented in rules or procedural code, or in some combination.

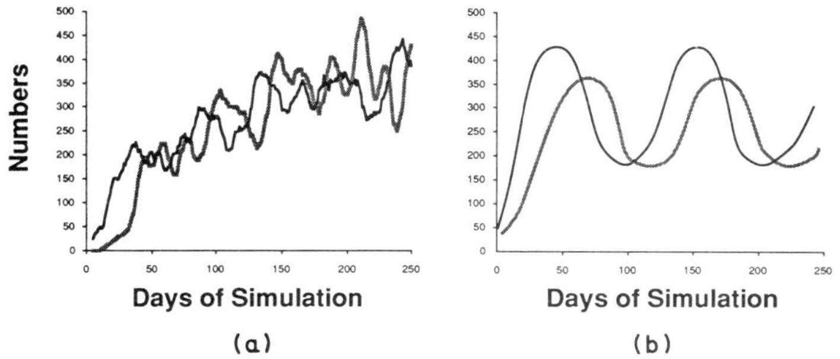

Figure 10.5 Comparison of time series plots of predator (patterned line) and prey (solid line) densities in (*a*) the individual-based model and (*b*) a Lotka-Volterra predator-prey model.

classes so that individual bugs and locations could vary in their initial parameters, no variation was actually included. All prey are initially identical, as are all locations and predators. The apparent variability in the output is an emergent property of the system, a result of the accumulated experiences of several thousand individuals. Figure 10.6, for instance, shows the functional response sampled from this interaction. Notice the word sampled. Since this model does not specify the relationship between prey density and predator search efficiency, we have simply sampled the simulation at different times to determine the relationship. The same is true for the survivorship and natality functions shown in Fig. 10.7. Even though no age-specific mortality was included in the model (other than death by old age), the results show type II to III survivorship functions, typical of insects (Pearl, 1928). These are purely the result of the interaction of individual behaviors and the environment. They are not built-in functions based on population-level parameters.

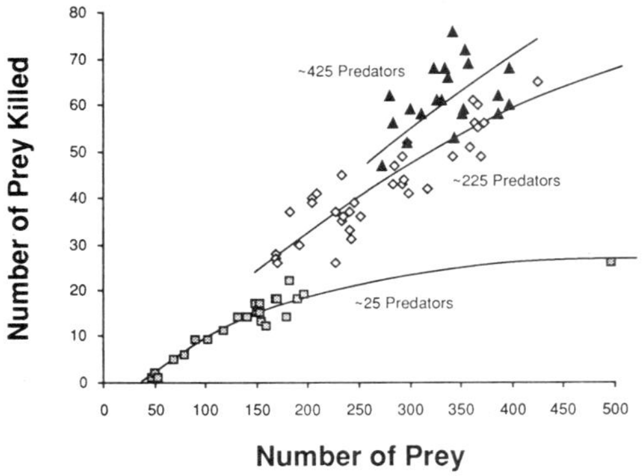

Figure 10.6 Functional response showing the number of prey killed as a function of prey density for three different predator densities.

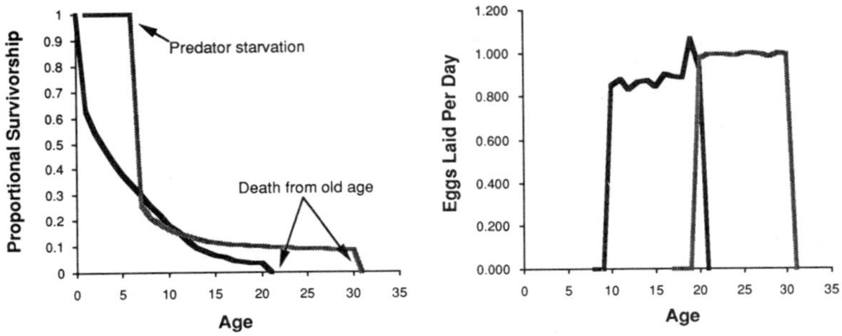

Figure 10.7 Survivorship and natality functions as sampled from runs of the simulation model developed in this chapter. Prey functions are shown as solid lines, predator functions as patterned lines.

10.5 Conclusions

Modeling at the individual level appears to provide at least three advantages to classical population-level modeling, and at least one disadvantage. On the positive side,

1. Individual-level models are conceptually closer to the natural system. They do not involve a great deal of abstraction since actors in nature can be represented by objects and classes in the model. An animal's behavior can be easily modeled in accordance with accepted ethological theory by using goal-oriented inference schemes, rules, and an object-message model of motivational, stimulus-based behavioral mechanisms.

2. Population-level phenomena can be observed as an indirect result of individual behavior. This provides researchers with a key linkage between their models, or hypotheses about how animals behave, and their data, which tend to show only population changes. One can actually vary an individual's behavior directly and observe the population-level result as an emergent property. This is, for example, much preferable to approaches like the ones we observed in the case of mutual interference.

3. Output from individual-level models indicate that a considerable amount of the variation we commonly see in field and laboratory data is explainable in terms of individual behavior and experience, not necessarily heterogeneity among individuals. Classical modeling approaches can only recreate the noise in field data by including random variables in their models.

On the opposite side, individual models are memory hogs. They stress computers to their limits. A simulation model like the one developed here might take 2 to 3 minutes per day to run with 500 to 1000 individuals in the simulation. Classical models are much faster and, therefore, more useful in analyses like optimization, which can require many thousands of runs in a short time span. Luckily, the shortcomings of this approach are being lessened by the advent of multiprocessing computers, parallel processing computers, increased speed of microprocessors, and computers with massive amounts of random access memory.

Chapter

11

Qualitative Models and Model-Based Reasoning

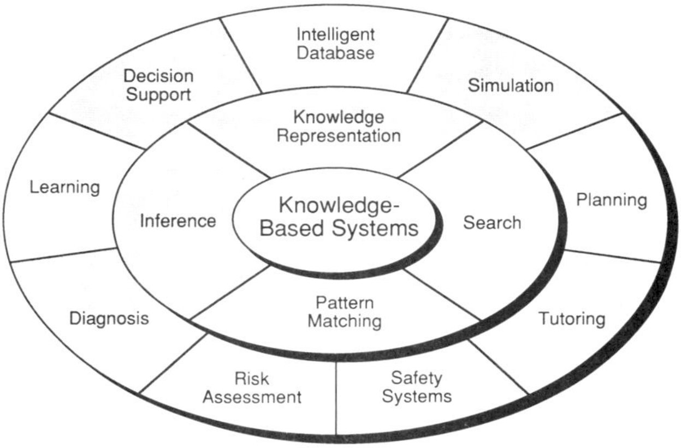

This chapter describes the representation and use of deep knowledge, i.e., the fundamental principles underlying observed behavior. The development of a knowledge-based system that uses fundamental knowledge requires the integration of three fundamental concepts described in this chapter. These are causality, counterfactual reasoning, and qualitative models. Causality is the specification or determination of the causal relationships among a collection of events. Counterfactual reasoning is reasoning from hypotheses that are not true, i.e., determining what would be the consequences if the hypotheses were true. Qualitative simulation models are computer models that provide a coarser, qualitative level of resolution than do traditional computer simulation models.

11.1 Introduction

Most of this book has been concerned with the development and implementation of expert systems, and most of the expert systems in the book have been rule-based. Some have used other forms of knowledge representation such as frames or semantic nets, but these share with rule-based expert systems a common feature: They make use of a "shallow" level of knowledge. Shallow knowledge of a domain is knowledge that is only based on observable features of that domain (Kuipers, 1984). "Deep" knowledge, on the other hand, is knowledge about a domain in which observations of domain phenomena are interpreted based on an understanding of the underlying mechanisms of the domain that cause the observed behavior.

The ability to reason based on shallow knowledge and heuristics has always been considered one of the great advantages of the rule-based expert system. In many domains shallow knowledge seems to be the only form of knowledge available. Human experts frequently make judgments based on such shallow knowledge as observational experience and hunches, and an expert system is uniquely capable of representing and using this form of knowledge. However, an expert system that relies solely on shallow knowledge does not truly represent the reasoning of human expert. For example, the expert system for animal identification first discussed in Chap. 2 is shallow. Identifications are based on a classification of observable features of the specimens. An actual human taxonomist might construct an identification key, or rule-based expert system, using only observable features of the specimens, but in doing this construction the taxonomist would, at least implicitly, use knowledge of the underlying phylogeny of the system. This knowledge would include an understanding of subjects such as population genetics, paleoecology, and geography. Through this understanding, the taxonomist can determine the relationships of present day species by tracing their lineage back in time to common ancestors. The expert uses fundamental, or deep, knowledge of this sort to readily incorporate new information into the compendium that constitutes the expert's personal knowledge base. For example, upon the discovery of a new species, the taxonomist can, using fundamental knowledge, more readily classify this species and relate it to those already known. The discovery of new fossil evidence may lead to a reorganization of the phylogentic tree. At a deeper level, new theories of how species arise and evolve may be developed. The taxonomist uses all of this knowledge in constructing an identification key, even if none of it is actually apparent in the key itself.

Some deep knowledge may be represented in a rule base. Indeed, this representation is what gives a rule base its "depth." For example, in Chap. 2 we discussed the incorporation of classes of animals, such as mammal, carnivore, and so forth, into the rule base. When a new species is discovered, it is easier to modify the rule base to include that species if such a classification hierarchy is used. However, while such a hierarchical rule base represents domain knowledge, it does not represent the way an expert actually *thinks* about that domain. A rule base more closely resembles a set of instructions that an expert would give to a novice. An identification key is a good example of such a set of instructions. The expert, in thinking about the same domain, makes use of knowledge of how its components interact. Such thinking is sometimes called reasoning from *first principles*. Because it reasons with the help of a model of the domain, an expert system that incorporates reasoning from first principles is sometimes called a *model-based* expert system.

Model-based expert systems have many potential advantages over rule-based expert systems. Three of these are pointed out by Whitehead and Roach (1990). One is that rule-based expert systems cannot handle unanticipated situations. In a rule-based expert system, if the knowledge base does not contain rules covering the situation at hand, then the expert system has no knowledge of it whatsoever. Rule-based expert systems are sometimes called *idiot savants* since they have knowledge of a precise and constrained domain. In this sense rule-based expert systems themselves do not display graceful degradation (see Chap. 2). A model-based expert system can often determine the cause of an unanticipated situation by determining how its model of the domain must be modified to produce behavior similar to the observed situation.

A second potential advantage of model-based expert systems is that they are more easily adapted to slightly different domains without the intervention of a human expert. For example, the irrigation scheduling expert system of Chap. 9 was developed for cotton in California. Transferring this knowledge to another crop or location would require the cooperation of an expert on irrigation under the new conditions. If irrigation has never been done before under these new conditions, then such an expert would not exist. An expert system able to reason from first principles about plant-soil-water relations might be able to extrapolate its knowledge to the new system without the aid of an expert.

A third advantage is that a model-based system can deal with changes in the domain parameters. For example, in the irrigation scheduling system a new variety might have different demands for water. A rule-based expert system would require reprogramming of the rules to accommodate this. A model-based expert system might make these changes based simply on a knowledge of the new parameter values in the model. This ability of the model-based system to make small leaps of generalization can make the problem of maintenance of a rule base in a changing environment considerably simpler.

All of the advantages described above stem from the use of a model that "knows" how individual components of a process behave in isolation and how their interaction causes the behavior of the overall process. Such an expert system can deal with novel situations provided that the novelty can be described in terms of the components of the model or their interaction. For example, in diagnosing the cause of a problem in a complex system, the expert system makes use of a model of how the system is *supposed* to work and compares the behavior of this model with the observed behavior of the faulty system. In principle, rules describing the cause of the problem are unnecessary because the expert system can determine this cause by comparing the behavior of the model with that of the real system. In practice such a *deep* model of system behavior may be supplemented by shallow diagnostic rules to cover portions of the domain where the expert's knowledge is insufficiently precise to permit the construction of a model.

People generally think of computer models as being composed of differential equations or difference equations, but in fact a model-based expert system need not be constructed this way. The model may even consist of a set of rules. Clancey (1989) describes the use of production rules in a model of the domain and Starfield et al. (1989) gives a rule-based model of an estuary system. The important distinction between a model-based expert system and a rule-based expert system is that the knowledge base of the former has at its core a model for the fundamental structure of the domain while the latter has a knowledge base consisting of rules for directly diagnosing the state of the domain. The model itself may consist of rules, differential equations, or some other construct such as will be described later in this chapter.

In a complex problem, such as crop management, it is unreasonable to expect that a deep model would replace the rule bases associated with the expert system; rather, the model would complement these rule bases and possibly serve as a reference for them. Current crop management expert decision support systems often contain a crop model. For example, three major cotton systems, CALEX/Cotton (Goodell et al., 1990), GOSSYM-COMAX (Lemmon, 1986), and COTFLEX (Stone et al., 1986), each contain a simulation model for the cotton plant. However, a model of the type described in this chapter may be very different from the traditional crop simulation model. In determining what form the crop model associated with a model-based expert system for crop management should take, we must first determine the functions of that model. These functions may be divided into two broad categories, predictive and postdictive.

Predictive functions of the model include the prediction of yield and the predictions of the time and nature of management actions and inputs that will be required for the crop. These include irrigations, chemical applications, and so forth. The term *postdiction* is a neologism apparently coined by artificial intelligence researchers. It is used to mean the determination of the cause of a behavior deviating from normal by comparison of that behavior with the behavior of the model. There are at least two reasons why a crop model for a crop management expert system would be called upon to provide postdiction. The first is in the diagnosis of problems in the crop, such as loss of fruit, poor color, wilting, and so forth. The second is to provide an adaptive feedback system for the model itself. In any real crop management situation, the behavior of the crop will almost certainly deviate over the course of the season from that predicted by the model. This deviation is caused by factors in the real system that are not represented or inaccurately represented in the model. For example, insect pressure, variability in soil properties, a slight inaccuracy in the portion of the model that deals with photosynthesis, or a combination of these could set the model off on a different course than that of the real crop. If this occurs, then the model must be brought back onto course through corrective feedback. There is in engineering a complete theory of feedback control (e.g., Chen, 1970). However, this theory has not proved very applicable to crop simulation models. The models and the processes that they simulate are too complex, and their properties too poorly understood, for engineering feedback control methods to be of much use. Therefore, the crop model must use a combination of predictive and postdictive analysis to make adaptive corrections to itself based on observed differences between its behavior and that of the real crop.

The foregoing discussion has consistently referred to the crop model in the singular, as if there could only be a single one. With traditional crop models this is indeed true. Even in this case, however, the crop model is actually a collection of interacting submodels, one for water and nutrient assimilation by the roots, one for carbohydrate partitioning, one for dry matter accumulation, and so forth. These submodels may be written at different levels of resolution, reflecting different demands on the model as well as different degrees of precision of knowledge about the process being modeled. In a model-based expert system, the model may in reality be made up of a loosely knit collection of individual submodels for different processes. The submodels may have completely different forms, some being regression models of the type described in Chap. 6, rule-based models, differential equation models, and qualitative models of the type described later in this chapter. Moreover, there may be multiple models of the same process, with each model serving a different function. For example, the model used for prediction might be different from that used for postdiction, or the model used for determining the response to nitrogen in irrigation scheduling might be different from that used for a similar purpose in pest management.

Crop models have for many years been composed of differential or difference equations and have been dynamic in nature. A dynamic model is one that provides, through the solution of the differential equations of the model, a simulation of the behavior of the crop as a function of time. Static models, which only describe the crop at one time, are also possible and potentially useful. A static model for a complex entity such as a biological organism can be used to describe how the components of the organism interact to produce the observed behavior of the whole organism. Often in complex systems this behavior cannot easily be determined simply from a knowledge of the behaviors of the individual components. In a similar way, a static model of a complex structural system such as a bridge or building can be used to determine how the individual structural components, the beams, trusses, and so forth, interact to bear the load of the structure.

In their discussions of the philosophy of modeling in agriculture, France and Thornley (1984) and Penning de Vries and van Laar (1982) point out that there are two fundamentally different applications for a model: research and management. In research applications the model is used as a mathematical statement of a hypothesis. The advantage of a mathematical model in this context is the precision with which its predictions can be stated. The purpose of the model is to aid in *increasing* knowledge about the system being modeled. In agricultural research the scientist uses a crop model to help increase the scientist's understanding of the crop and its physiology, e.g., to better understand the effect of different nutrient application strategies on the pattern of plant growth. The simulation model is executed under varying conditions, and the results of these simulations are compared with data collected from the real system under conditions similar to those simulated. The concern in such studies is ordinarily with the behavior of a generic or "typical" crop plant, rather than plants that are a part of any specific crop at a particular location.

Management applications are concerned not with increasing knowledge about a system but with *using* knowledge already gathered. The model is used to reproduce the behavior of the system with sufficient accuracy and precision that the output can be used to make management decisions. In crop management, for example, one might want to forecast the yield of a particular farmer's crop under various alternative management scenarios. Management applications generally require less speculative models with properties that are firmly rooted in widely accepted principles. Within management applications there is a distinction between strategic and tactical uses of the knowledge. As described in Chap. 1, strategic applications involve broad policy questions, e.g., is an early season nitrogen application likely to increase crop yield in general? Tactical applications involve the application of a strategy to specific individual crops, e.g., how much nitrogen should a particular farmer apply and precisely when should it be applied? Like research models, strategic management models generally are intended to represent the behavior of a *typical* instance of the system, e.g., of a typical crop plant. Tactical models, on the other hand, must usually provide accurate forecasts of the behavior of a *specific* instance of a system, e.g., of the specific collection of plants that makes up a particular farmer's crop.

In addition to the distinctions between prediction and postdiction and between research and management, there is a similar, but separate, distinction between primary uses for a model. One use is to obtain precise estimates of certain parameter values. For example, in a research application one might want to know if the logistic growth model (e.g., Emlen, 1984) provides a good fit to the observed growth of a certain population or whether some other mathematical form such as the Gompertz equation (France and Thornley, 1984) would be better. To determine which model provides a better fit to the data, an attempt is made to determine the parameter values of the model, solve the equations of the model either analytically or numerically, and test their fit to the data. The second possible use for a model is to determine emergent behaviors of a complex system of interacting components. Emergent behaviors are those that can occur because of the interaction of the components and cannot be predicted based on the study of the components in isolation. This is the sort of behavior discussed in Chap. 10. The problem discussed in this chapter, determining what is causing a plant to wilt, involves precisely this sort of emergent behavior. The wilting of leaves is the visible manifestation of the interaction between the plant's leaves, shoots and roots, the atmosphere, and the soil.

11.2 The Problem: Plant Water Use and Turgor

11.2.1 Description of the problem

The problem discussed in this chapter involves the maintenance of turgor pressure in a plant sufficient to avoid wilting. Virtually any home gardener, upon seeing a wilted plant, will respond by watering it. The gardener might, in explaining to a small child how to care for plants, provide a simple rule such as "If the plant is wilted, then it needs water applied to the soil." However, anyone with at least a high school level knowledge of biology has a more fundamental picture of the interaction between the water, the soil, and the plant. In terms of water potential, discussed in Chap. 6, water moves from regions of higher potential to regions of lower potential. The turgor pressure, one component of the water potential, in a well-watered plant is sufficient to maintain its structural integrity. Thus, if the plant is wilted (low water potential) and the soil surrounding the roots is moistened (high water potential), then the water will move from the soil to the roots and restore the plant's turgor pressure. Aside from water stress caused by dry soil, however, there are a number of other potential causes of wilting, including disease, damage to roots, high temperatures, and insect attack. If a plant is wilted and the soil surrounding it is moist, then some problem such as one of these must be the cause. An individual with some knowledge of plant physiology would be able to formulate several initial guesses as to the cause of wilting in well-watered soil and could then do further testing based on these initial guesses to determine the actual cause.

The problem as we have posed it is a purely postdictive one. There are also predictive functions associated with plant-water relations, the most common being the prediction of the next irrigation date in an irrigated crop. Postdiction is easier, however, because it is not necessary to include dynamics. A second simplification is to focus purely on the diagnostic aspect of the management problem. In an actual decision support system, diagnostic capability is only one of two components, the other being the synthesis of the recommended solution. The theory of model-based expert systems is still sufficiently undeveloped that the linkage of the model to decision support systems has not been considered in any detail. However, this linkage can in principle be made by linking the model to a guideline script system or planner of the type described in the previous chapters.

11.2.2 Rule-based expert system approach to the diagnosis of wilt

Let us first examine a rule-based expert system approach to diagnosing the cause of wilt. This will permit us to identify some of the shortcomings that we hope the model-based system will correct. The reader who has read this book from the beginning should by now have a fairly clear idea of how a rule base for diagnosing the cause of wilting would be constructed. The single rule given above by the gardener to the child is one example. Let us make things more precise by constructing a rule base with only two rules:

Rule 1: IF The plant is wilted
THEN The water potential is too low

Rule 2: IF The water potential is too low
THEN Water the soil

This rule base illustrates the concept of shallowness described above.

It short circuits all the intervening biology and physics by relating one observation, wilting, to another, dry soil. The disadvantages given by Whitehead and Roach (1990) of the shallow reasoning system can be illustrated in the context of this rule base. The first is the inability to deal with unanticipated situations. Suppose, for example, that the soil is watered and the plant does not respond. As described above, an experienced gardener might begin to suspect a problem with the roots. However, the rule base does not have any information about roots and is at a loss to continue.

The second problem is the need for an expert to have made each of the observations on which the rules are based. In the context of the simple rule base this means that researchers must have already done experiments or made field observations leading to the conclusion that increased soil moisture cures wilting. This particular fact has, of course, probably been known since prehistorical times. However, it is not hard to conceive of novel situations that have not been studied to an extent sufficient to provide a diagnosis for an observed problem. Suppose, for instance, that a new chemical for regulating plant growth is developed. When this chemical is applied to plants of a particular species, wilting is observed even in moist soil. The experts, in this case plant physiologists, have not yet studied the chemical sufficiently to determine why this should happen. However, based on their knowledge of the chemical properties of the plant growth regulator and the physiology of the plant, they might be able to develop a strong hypothesis for the cause or even to explain it with confidence. If this is the case, they can integrate this information directly into their body of personal knowledge and use it to make irrigation recommendations. However, a rule-based expert system cannot make this sort of extrapolation. Since each relation must be explicitly encoded, the rule base cannot make the sort of automatic extrapolation that a human expert can.

The third problem, maintenance in the face of small changes in domain parameters, is illustrated by the following example. Some plants, when grown in a sufficiently hot, arid environment, will display a small amount of reversible wilting in the heat of the day. Suppose the rule base for the plant in question was developed in a cool environment where this phenomenon was not observed. When the same plant is grown in a hot, dry environment, wilting is observed during the hottest part of the day no matter how much the soil is watered. The shallow rule base cannot deal with this. It must be explicitly modified for the new situation.

The way to get around these problems while maintaining the rule-based format is to make the rules deeper. This process is discussed in Chap. 2 in the context of animal identification, and the principle is the same for wilting diagnosis. The rule base must include a representation of the intervening links in the logical chain from wilted leaves to dry soil. We can, for example, construct a set of rules based on water potential. An initial construction of such a diagnostic rule base might look like this:

RULE 1: IF The leaves are wilted
THEN The leaf water potential is too low

RULE 2: IF The leaf water potential is too low
THEN The stem water potential is too low

RULE 3: IF The stem water potential is too low
THEN The root water potential is too low

RULE 4: IF The root water potential is too low
THEN The soil water potential is too low

RULE 5: IF The soil water potential is too low
THEN Water the soil

Suppose that we now wish to include the possibility of a plant disease as a cause of wilt. In a real knowledge base for diagnosing wilting in a specific crop, we could write rules containing statements about specific diseases. For our simple model system, however, we will stick to a more generic approach.

Since there are now more than one possible cause for the observed symptom, the straightforward chain of reasoning given above will no longer work. The problem is that there is more than one possible cause of the observed symptom; in other words, the precise cause is uncertain. One method for incorporating multiple causes is to use certainty factors as described in Chap. 3. For example, we could replace Rules 4 and 5 with a set of rules such as

RULE 4: IF The root water potential is too low
THEN The soil water potential is too low (CF = 0.75)
The roots are infected with a disease (CF = 0.25)

RULE 5a: IF The soil water potential is too low
THEN Water the soil

RULE 5b: IF The roots are infected with a disease
THEN Treat the disease

The process of inference in a rule base such as this would work by chaining rules in the manner described in Chap. 2. In backward chaining, dry soil and disease would be hypotheses, and a backward chain from Rule 5a or 5b to Rule 1 would result in the presentation of a prompt to the user asking whether the leaves are wilted. In forward chaining the user would initially be presented with this prompt, and an affirmative answer would start a chain from Rule 1 to Rule 5a or 5b. The rule base could then be strengthened with other rules and prompts designed to tighten up the conclusion. For example, the rule base could contain rules permitting the system to narrow the list of hypotheses by asking further questions of the user.

What is wrong with such a rule base? The problems are exactly the same ones that troubled the simple single-rule rule base that says "If the leaves are wilted, then water the soil"; they have just been pushed back a bit. The more powerful rule base still cannot deal with unanticipated causes of wilt, e.g., hot, dry weather. This rule base is still founded on the association of specific symptoms to specific causes, and it still must be maintained in the face of changes to the system, e.g., the introduction of a disease that causes stem discoloration but not wilting. From a theoretical perspective, the more powerful rule base is no better than the original one, although it may be better in practical use.

There is a second fundamental feature of the diagnostic rule base that is of little practical importance in rule-based diagnostic expert systems but will become crucial to an understanding of the distinction between a rule-based and a model-based system. This is that in the rules, causality is in the wrong direction. In other words, if Rule 3, for example, is read without knowing that it is part of an expert system rule base, we might get the impression that low stem water potential causes low root water potential, rather than the other way around. The reason for this seeming paradox is that each rule represents a *production,* i.e., a diagnosis moving from symptom to cause rather than a logical statement of implication or causality. This issue is discussed in more detail below.

In any case, our final conclusion seems to be that although adding depth to a rule base can be of great practical value, in the end this

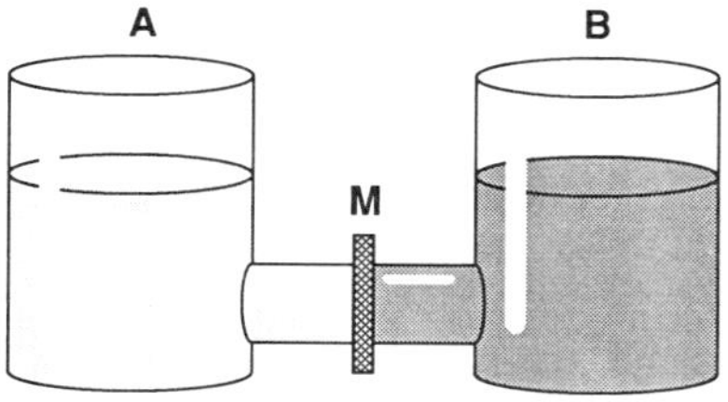

Figure 11.1 Diagram of a simple experiment. Chambers A and B initially contain equal volumes of liquid. They are separated by a membrane M that is permeable to water but not to dissolved sugar. Chamber B contains a sugar solution, while chamber A contains pure water. Water moves through membrane M into chamber B, expanding the volume of water in chamber B until the osmotic pressures balance.

procedure does not remove the fundamental problems with rule-based expert systems discussed in the previous section. It only reduces the impact of these problems.

11.2.3 Representing the problem with a simple experiment

Rather than pursue in detail the diagnosis of plant wilt symptoms directly, we will examine in detail a simple schematic model that has many of the same physical properties as the plant-soil-atmosphere system. The model, shown in Fig. 11.1, is an experiment that people who have studied plant-water relations or physical chemistry will be familiar with. It consists of two cylindrical tanks, open at the top, filled with water and connected by a semipermeable membrane, i.e., impermeable to dissolved sugar. Cylinder A contains pure water while sugar is added to cylinder B to form a solution. Both cylinders are at atmospheric pressure and initially contain equal volumes.

A person who has never done the experiment but has some knowledge of physics can describe, based on observational experience, the outcome of such an experiment. It is even possible to sketch a graph of the volumes $V_A(t)$ and $V_B(t)$ of the cylinders as functions of time. Such a graph is shown in Fig. 11.2. It is sketched freehand, without the use of a differential equation model, but it captures the essential features that experience and intuition indicate would be present in the solution of such a mathematical model. These features are the following: (1) The volume of fluid in cylinder B increases until the water potential in the two cylinders is balanced; (2) because matter is conserved, the volume in cylinder A declines in such a way that the sum of the two volumes remains constant; and (3) the magnitude of initial rate of change of volumes is high, and as time increases, the system gradually approaches a dynamic equilibrium.

We can also formulate a simulation model for the experiment. This model is based on standard physical chemistry found, for example, in Glasstone (1946) and on nonequilibrium thermodynamics found, for

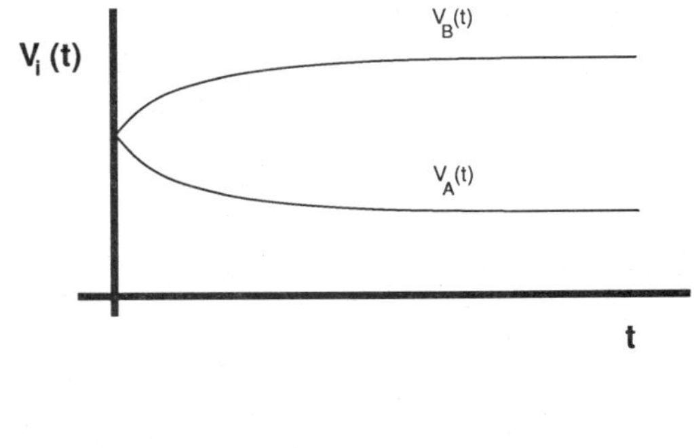

Figure 11.2 Sketch of the qualitative behavior of the simple experiment shown in Fig. 11.1. The volume in chamber A gradually decreases while the volume in chamber B increases at the same rate. The two volumes eventually approach an equilibrium distribution.

example, in Katchalsky and Curran (1974). Recall from Chap. 6 that the water potential Ψ, if matric and gravitational potential are neglected, is equal to the difference between the pressure potential P and the osmotic pressure Π, that is,

$$\Psi = P - \Pi$$

Since both cylinders are at the same elevation and have no internal matrix structure, the only contributions to water potential in this experiment will be from solution potential and pressure potential. Let Ψ_A denote the water potential, Π_A the solution potential, and P_A the pressure potential of cylinder A (the same quantities for cylinder B are denoted similarly). All six of these quantities are functions of time. The flow rate Q across the membrane between the cells is

$$Q = K\Delta\Psi$$

where

$$\Delta\Psi = \Psi_B - \Psi_A$$

Here K is the hydraulic conductivity of the membrane, and $\Delta\Psi$ is the water potential difference across the membrane. Since the water in cylinder A is pure, the solution potential in cylinder A is identically zero, that is, $\Pi_A = 0$. For a dilute solution the osmotic pressure of cylinder B is approximately proportional to the concentration. The Onsager equation for the flux across the membrane is

$$J = L_P \Delta P + L_{PD} \Delta \Pi$$

together with a second equation for the velocity of the solute relative to the solvent (Katchalsky and Curran, 1974).

Assume that both cylinders have unit cross sectional area, so the volume of fluid is proportional to the height H in the tank. Then the concentration in cylinder B is M/H_B, where M is the mass of solute in the tank. On the other hand, since the cylinders have unit area, the pressure potential on each side of the membrane is equal to the weight of water in each cylinder, which is proportional to the height of the water on each side. Therefore, the potential difference across the membrane is

$$\Delta\Psi = L_P\rho(H_B - H_A) - \frac{L_{PD}M}{H_B}$$

Substituting the equation for Q and noting that $dV_B/dt = Q$ leads to

$$\frac{dH_B}{dt} = K\left[L_P\rho(H_B - H_A) - \frac{L_{PD}M}{H_B}\right]$$

This equation, together with the relationship

$$H_A + H_B = H_0$$

where H_0 is the initial height of the fluid in the tanks, provides a complete model for the system.

Suppose now that we set up the experiment, and the behavior does not conform to Fig. 11.2. Suppose, for the sake of definiteness, that the volumes do not change with time. The problem is not, in this case, that our intuition is incorrect, since this intuition was confirmed by the differential equation model. Rather, the actual experiment must in some way not conform to the properties of the model. We might suspect, for example, that the membrane is not permeable to water or that it is permeable to both the solute and water. Other possibilities are that the fluid in cylinder B is really pure water or that the fluid in cylinder A is a sugar solution. An example of a rule base to cover the first two possibilities is

Rule 1: IF The volume of fluid in cylinder B does not rise
THEN The membrane is impermeable (certainty = 0.7)

Rule 2: IF The volume of fluid in cylinder B does not rise
THEN The membrane is permeable to the solute (certainty = 0.3)

In the present context the three fundamental flaws of a rule-based expert system given by Whitehead and Roach (1990) manifest themselves as follows. The first flaw, narrowness, is reflected in the fact that a rule base constructed to diagnose the situation in which the volumes $V_A(t)$ and $V_B(t)$ remain constant is incapable of dealing with the situation, for example, in which the sum of $V_A(t) + V_B(t)$ does not remain constant or in which $V_A(t)$ increases and $V_B(t)$ decreases. The second flaw manifests itself in the need for a human expert to develop plausible rules for these new situations. The third flaw is that the rules would have to be completely reworked if the domain were changed, for example, if chamber B were pressurized.

11.3 Causality and Counterfactual Reasoning

11.3.1 A theory of causality

A human expert, in seeking to determine why the experiment of Fig. 11.1 failed to produce the behavior predicted in Fig. 11.2, would seek to determine the *cause* of this failure. The *consequents* of the rules in the diagnostic rule base contain two possible causes of this behavior. As has already been pointed out, however, the present rules cannot be considered as a model of the system because the direction of implication does not match the direction of causality. It would be more appropriate to write Rule 1 in the form "If the membrane is impermeable, then the water in cylinder B does not rise." Considered as a sentence in propositional logic, is this an acceptable model of causality? Simon and Rescher (1966) have addressed this question and demonstrated that it is not: there is no relation between logical implication and causality. The reason is simple, and it is because of the effect of the application of the logical principle of the contrapositive (Manna and Waldinger, 1985), i.e., if the statement "If X, then Y" is true, then the statement "If not Y, then not X" is also true.

Suppose that logical implication were treated as equivalent to causality. Then the statement "If the membrane is impermeable, then the water in cylinder B does not rise" would be considered equivalent to "The membrane's impermeability causes the water in cylinder B not to rise." But by the principle of the contrapositive, which says that the statements "If A, then B" and "If NOT B, then NOT A" are logically equivalent, the logical statement is equivalent to "If the water in cylinder B does rise, then the membrane is permeable." Since implication is being considered equivalent to causality, this means that we must accept the statement "The rising of the water level in cylinder B causes the membrane to be permeable." This latter statement clearly does not meet the common sense notion of causality. Therefore, we may not be able to use the standard propositional logic and inference mechanism of the type used in an expert system rule base as a straightforward model for cause and effect even if the direction of implication is reversed.

There is an engineering definition of causality associated with differential equation models (Zadeh and Desoer, 1970). This is usually phrased in the form of a complex integral equation involving a quantity called the impulse response function. The engineering definition of causality equates it with the property of being *nonanticipatory*. This means that if an input is applied to a system at time t, then that input can only affect the behavior of the system at times later than t, i.e., the system cannot anticipate the input. However, this definition is not useful for our diagnostic purpose. For example, the differential equation model given above for membrane flux is nonanticipatory. However, it is not directly apparent from the model which are the causal agents. Indeed, one way of viewing the model, which most closely approximates the conditions of the experiment, is to view the water potential difference $\Delta\Psi$ across the membrane and the membrane conductivity K as the causal agents and the flux Q as the effect. However, if the membrane contained an active transport mechanism so that Q and

K were predetermined, then these would be the "causal" agents and the potential difference $\Delta\Psi$ would be the effect. The differential equations of the model are exactly the same for both situations.

There is also a statistical treatment of causality associated with path analysis (Draper and Smith, 1981). This is an attempt to infer causality from regression analysis data that indicates correlation. However, path analysis is not directly appropriate for our application. Our problem is not to infer causality from data, it is to represent causal relationships that have already been determined or to infer them from models that have already been constructed. Path analysis may well have been a component in the construction of the models themselves.

The problem of inferring causality has been studied by a number of workers in a variety of disciplines, including economics, philosophy, and artificial intelligence. A number of criteria have been proposed. We will follow the work of Simon and Rescher (1966) in our discussion. They introduce a mechanism based on linear algebra for assigning causality.

Five equations are first written that completely characterize, in an abstract way, the conditions at the start of the membrane experiment. The first equation is

$$Q = K\Delta\Psi$$

which describes the physical law relating membrane flux to conductivity and water potential difference. The second equation is

$$\Delta\Psi = f(\Delta P, \Delta\Pi)$$

which defines abstractly the physical law relating water potential to osmotic and pressure potential differences.

The next three equations describe the conditions established by the experimenter. The first of these is

$$K = \overline{K}$$

which states that the experimenter assigns a conductivity $\overline{K}$ to the membrane. The next is

$$\Delta P = \overline{\Delta P}$$

which states that the experimenter assigns a pressure difference $\overline{P}$ at the start of the experiment. The last is

$$\Delta\Pi = \overline{\Delta\Pi}$$

which describes the osmotic potential difference established at the start of the experiment.

The next step is to establish a table of 0s and 1s called a *linear structure* that has as its rows and columns the five parameters present in the equations. The rows of the table correspond to the equations. In each row of the table a one is placed in each column if that equation contains the corresponding variable. The table thus looks like that shown in Table 11.1.

TABLE 11.1 Linear Structure for the Physical Model of the Experiment Shown in Fig. 11.1

Q	$\Delta\Psi$	K	ΔP	$\Delta\Pi$
1	1	1	0	0
0	1	0	1	1
0	0	1	0	0
0	0	0	1	0
0	0	0	0	1

The next step is to note that these five equations can be reduced to two by performing the elementary row operations of gaussian elimination (e.g., Noble and Daniel, 1973). This simply "solves" the last three equations for their variables K, ΔP, and ΔΠ, and substitutes these into the first two equations. We are then left with a reduced system of two equations. The linear structure for this system looks like that shown in Table 11.2. The third step is to repeat the process, "solving" the second equation for $\Delta\Psi$ and substituting this into the first equation. In this way, we are left with a single equation in the variable Q.

TABLE 11.2 Reduced Linear Structure after One Reduction Operation from Table 11.1

Q	$\Delta\Psi$
1	1
0	1

This three-step process defines a hierarchy of causality represented in Fig. 11.3.

The experimentally controlled parameters K, ΔP, and $\Delta\Pi$ *cause* $\Delta\Psi$, which in turn *causes* the flux Q. This causality hierarchy conforms to our common sense idea of causality in the experiment. It can be inter-

$$\left.\begin{matrix} K \\ \Delta P \\ \Delta \Pi \end{matrix}\right\} \rightarrow \Delta\Psi \rightarrow Q$$

Figure 11.3 Causality hierarchy derived from Tables 11.1 and 11.2.

preted in a manner relevant to our intended use of the model: The diagnosis of the cause of failure of the experiment to conform to the behavior predicted by the model. In this context position in the hierarchy is a function of the strength of belief in the validity of the expression defining that parameter (Rescher, 1964; Roach et al., 1990). That is, the relation $Q = K\Delta\Psi$ is a physical law and as such is the least likely to be invalid. Thus, our confidence in the relationship determining Q is very high. At the other end of the scale, the values of K, ΔP, and $\Delta\Pi$ are fixed by the experimenter and, thus, might be considered the most subject to error. The relation determining $\Delta\Psi$ occupies an intermediate position: It is defined by the physical properties of the system.

Now that we have established a hierarchy of causality for the simple model of the experiment, how do we use this to diagnose the failure of the experiment to conform to the predicted behavior? This failure might be expressed as follows: According to the model, the statement "The water in cylinder B rises" is true. In fact, the observation is that the water does not rise. The statement is thus a *counterfactual,* a statement contrary to fact, at least as the fact is stated in the model. Since the model is internally consistent, the observed behavior is inconsistent with the model. There are a number of possible ways to restore internal consistency to the model, i.e., to make the failure of the water in cylinder B to rise consistent with the assumptions of the model. Finding the correct way to change the model to restore internal consistency is equivalent to diagnosing the cause of the failure of the experiment. The diagnosis problem is therefore a matter of rapidly determining the most reasonable or likely alterations to the assumptions of the model to restore its internal consistency.

Standard propositional logic as described in Chap. 8 is of no use in dealing with counterfactuals. In the propositional statement "If p, then q," if the assertion p is false, then the statement is logically consistent no matter whether q is true or not. For example, the statement "If the moon is made of green cheese, then pigs can fly" is logically consistent. Therefore, some other method must be found for dealing with counterfactuals. Several methods have been proposed and discussed (e.g., Rescher, 1964; Simon and Rescher, 1966; Lewis, 1973; Ginsberg, 1986; Roach et al., 1990). Our discussion will follow that of Roach et al. (1990) and Whitehead and Roach (1990). This work is an extension of that of Rescher (1964) and Simon and Rescher (1966).

The procedure is best illustrated through a simple example described by Simon and Rescher (1966). Consider the statement "Striking a dry match in the presence of oxygen causes it to ignite." This statement can be expressed in terms of a simple model consisting of four boolean, i.e., having values *true* or *false,* variables together with a rule. The variables are S, struck; D, dry; O, oxygen present; and I, ignited. The rule is S & D & O → I, i.e., if S, D, and O are all true then I is also true. Roach et al. (1990) refer to the rule, together with the three auxiliary relations $S = \bar{S}$, $D = \bar{D}$, and $O = \bar{O}$ as a *world*; in our terminology, it is a model. We will stick with our terminology and denote the model by M. By constructing a linear structure it is easy to show that the causality hierarchy is that shown in Fig. 11.4.

$$\left.\begin{array}{l} S \\ D \\ O \end{array}\right\} \rightarrow I$$

Figure 11.4 Causal hierarchy of the simple match-striking model.

Consider a specific instance of the model denoted M_0 in which the values of the parameters are S = *false,* D = *true,* O = *true,* and I = *false*; that is, in our model the match is not struck. The model M_0 is internally consistent. Now suppose we are presented with the counterfactual "The match is struck," i.e., S = *true.* We need to determine how to modify our model parameter values to restore internal consistency. Obviously we must eliminate the statement S = *false* and replace it with S = *true.* However, this is not sufficient to restore consistency to the model. Roach et al. (1990) describe an efficient algorithm, called WIF, for *w*hat *if,* for restoring consistency in the presence of a counterfactual. The algorithm is described below.

For every counterfactual C introduced into a model M:

1. If *not* C is also a counterfactual, then end and announce failure.
2. Remove *not* C from M.
3. For every rule R in M, if C is on the left side of R or *not* C is on the right side of R, then do step 4.

4. If the left side of R is in M and the right side of R is negated in M, then recursively invoke step 1 with a model M′ consisting successively of the model M augmented by
   4.1. The right side of R
   4.2. The negation of each item on the left side of R
   4.3. The negation of the rule R

This algorithm is applied to the match model as follows. The counterfactual C is the statement S = *true*. Therefore, we first apply step 1 to *not* S = *true*, which gives S = *false*. This is not a counterfactual, so go to step 2. In this step, the statement S = *false* is removed from the model. Moving to step 3, there is only one rule in M, and it has C, that is, S = *true*, on the left side. Therefore step 4 is executed. In this step each of the substeps is tested, with priority going to the first one. If this alternative is inconsistent with other rules in the model, then substep 4.2 is tested, and if 4.2 is also inconsistent, then 4.3 is tested.

For the match example, the alternative steps to be tested are as follows:

Step 4.1		Step 4.2a		Step 4.2b		Step 4.3	
Retain	Replace	Retain	Replace	Retain	Replace	Retain	Replace
D = *true* O = *true*	I = *false*	D = *true* I = *false*	O = *true*	O = *true* I = *false*	D = *true*	D = *true* O = *true* I = *false*	S&D&O → I

These have the following interpretations:

4.1. "If the match is struck, it will ignite."
4.2a. "If the match is struck, it will not be dry."
4.2b. "If the match is struck, oxygen is not present."
4.3. "If the match is struck, it is dry, and oxygen is present, it will not ignite."

Clearly alternative 4.1 is the correct one, and by the consistency restoration algorithm it is invoked. This algorithm can be explained through the process of *modal categorization* (Rescher, 1964; Simon and Rescher, 1966), which is an ordering of model parameters in terms of their firmness of acceptance. In its simplest form it says that parameter values should be rejected first for those parameters that are last in the causal hierarchy. In the match striking example, we have indicated in Fig. 11.4 that I is in the lowest position, so its value is rejected. This is equivalent to accepting alternative 4.1, the correct one. The reason this procedure works is because of the equivalence between the ranking of the modal categorization and the direction of causality. The fact that the value of S is *false* and the values of O and D are *true* cause the value of I to be *false*. Therefore, when one of the three values, the value of S, is changed to *true*, resulting in a breakdown of internal consistency, the resolution of this breakdown is to alter the truth value of the variable that *caused* the change. If this alternative is somehow inconsistent with other parts of the model, then the values of O and D are tested. The physical law represented by the rule is considered most firmly believed and is, therefore, the last alternative tested.

11.4 An Assessment of Models of Causality

In the previous section we developed an inference method for fault diagnosis using model-based reasoning. The method is based on introducing a counterfactual representing the observed fault into the model's database. This database represents the state of the model, or "world," and must be internally consistent. We, therefore, determine adjustments to the model that are necessary to restore internal consistency. These adjustments provide the diagnosis of the cause of the fault. In the simple match example of the previous section, the knowledge comprising the model was represented by a set of boolean variables together with a simple rule. This method of knowledge representation is probably inadequate for a more complex model, such as would be used in the diagnosis of the cause of wilting. In this section we discuss a method by which such a model may be constructed.

We begin by examining in greater detail the two previously discussed methods of knowledge representation for the simple experiment of Fig. 11.1, rules and differential equations. This simple problem has the same characteristics as a more complex problem such as representing the internal hydraulics of a plant. The objective of the discussion is to identify the strengths and weaknesses of these two methods, with the hope of motivating the development of a method that incorporates the strengths to the greatest possible extent and the weaknesses to the least possible extent. We shall always maintain the assumption that the ultimate application is management rather than research, i.e., the primary objective for the model is not to increase the level of knowledge of the system but rather to use the currently held body of knowledge about the system to predict its behavior.

In developing a causal rule base for the experiment we will use the notation $\rightarrow$ to indicate causality. The rule base then becomes

Rule C1: The membrane is impermeable
→ The volume of fluid in cylinder B does not rise

Rule C2: The membrane is permeable to the solution
→ The volume of fluid in cylinder B does not rise

There is an immediately apparent problem associated with these rules. This is, the rules say nothing about the behavior of the system if the membrane is functioning as it is supposed to. We might propose a third rule that says "If the membrane is permeable to water and not to sugar, then the water in cylinder B does rise." Unfortunately, this is not quite correct since there might be other problems; for example, the water in cylinder B might not contain any dissolved sugar. To cover all possibilities it would be necessary to construct a rule that completely describes the proper functioning of the experiment. However, this would defeat the purpose of the causal model since it would subject the model to the same limitations as the rule-based expert system.

The difficulty with this approach is that merely ensuring that causality is in the same direction as implication does not actually model the causality of the system. That is, the fact that the membrane is semipermeable does not *directly* cause the water in cylinder B to rise; the effect is indirect. The membrane's semipermeability causes sugar not to flow across it. The water potential difference between cylinders A and B, together with the membrane's permeability to water, causes water to flow across the membrane. The flow of water causes the level of water in cylinder B to rise. The causal rules shown above are not *localized,* i.e., they extend across several links in a chain of causality. A more appropriate collection of rules to describe the flow of water across the membrane might have the form

Rule 1: The water potential in cylinder A exceeds that in cylinder B
AND The membrane is permeable to water
→ Water will flow from cylinder A to cylinder B

Of course an analogous rule would be needed for flow in the other direction, together with a rule to specify no flow if the water potential in the two cylinders is equal and a rule to specify no flow if the membrane is not permeable to water.

Sadly, we must point out that there is still a serious flaw. A collection of rules to cover all possibilities would be so unwieldy that it would be virtually useless. It would be almost impossible to determine whether or not the rule base was complete in the sense that it covered all possibilities, including both normal and abnormal behavior.

The problem is that we are trying to describe a dynamic functional relationship with a set of IF-THEN rules, when some other form of function would be more natural. Therefore, let us turn to the other alternative, the differential equation model. Here we represent the behavior of the model with the differential equation

$$\frac{dH_B}{dt} = K\left[L_{P\rho}(H_B - H_A) - \frac{L_{PD}M}{H_B}\right]$$

supplemented by a modal categorization of causality. Confronted with a failure of the water level in cylinder B to rise, we search, using the modal categorization as a guide of which parameters to search first, for a modification of the parameter values or equations of the model that mimics the observed behavior, for example, by setting the value of $K = 0$.

This approach might actually be feasible in the description of a simple experiment such as that represented in Fig. 11.1. Unfortunately, it is probably not feasible in a real problem such as the diagnosis of plant wilting. Traditional plant growth models present a number of obstacles in management applications, particularly in dealing with the emergent behavior of complex systems. Some of the most significant problems are

1. *The model may require too much data.* Anyone who has worked with plant growth models in a management application is familiar with the hunger of these models for data. The models have many parameters, each of which must be assigned a value. Often these values can only be estimated or are expensive to determine even in a laboratory setting.

2. *The model may give a solution that is needlessly precise.* In a knowledge-using context, such as crop management, we often need only information with a very coarse resolution from a model. For example, it is often sufficient to know whether something is increasing or decreasing, or to have a rough estimate of its value. However, a traditional plant growth model cannot take advantage of this lack of required precision. It must represent each of its components precisely in order to get the interactions right, and, therefore, it produces a precise solution.

3. *Differential equations have too much structure to control their behavior.* Even very simple differential equations or difference equation models can often have solutions that are complex or depend on the parameters in a complex way (e.g., May and Oster, 1976). Often the solutions are unpredictable or counterintuitive. In addition, small changes in parameter values may lead to drastic, and possibly incorrect, changes in model behavior. In a research application this can be a very useful property since it can point out areas of incomplete understanding and suggest new experiments. In a management applica-

tion, however, such unpredictable behavior is usually undesirable. The objective in constructing this type of model is to capture information that is already known, so the behavior of the model should be unsurprising to the individual or individuals whose knowledge went into the model's construction. Even if the behavior is unexpected, which may be the case if the model describes a complex system, the causes of this behavior should be readily understood and accepted by the experts.

In summary, there are serious problems affecting the use of both rules and traditional dynamic models as the basis for model-based expert systems. Recognition of these problems has led expert system research to explore alternatives, and one of the most promising has been *qualitative* models, whose theory originated as a means of understanding how people reason about physical problems. These models are described in the next section.

11.5 Qualitative Models

The discussion of rule-based models and traditional dynamic models given in the previous section pointed out serious weaknesses if these are to be used in a deep expert system model. The basic problem is that neither method of knowledge representation really captures the way that an expert thinks about a problem. The rule-based system represents the way that an expert would write a description of the system, and the dynamic model is an attempt to simulate the dynamics of the system rather than to represent fundamental knowledge of its behavior. However, there are other forms of modeling that do attempt to capture the expert's fundamental knowledge. These are often called qualitative models.

The development of qualitative models was motivated by artificial intelligence researchers' recognition of deficiencies in traditional forms of knowledge representation such as rules, frames, and semantic nets. The researchers have attempted to develop new forms of knowledge representation that focus on the understanding of underlying mechanisms rather than the shallow knowledge of observable behavior. Much of their effort has been focused on an area that has come to be known as *qualitative physics* (de Kleer and Brown, 1984; Forbus, 1984; Kuipers, 1984, 1986). Qualitative physics attempts to describe how people, both ordinary individuals and physicists, reason about the physical world. The field is still immature, and there is no uniform theory. Indeed, the appropriateness of the discipline itself is controversial (de Kleer and Brown, 1986; Iwasaki and Simon, 1986a, 1986b). We will not attempt to review the entire literature; the interested reader is referred to Weld and de Kleer (1989). Instead we focus on using some of the more common ideas of the field to describe the simple experiment of Fig. 11.1 and indicate how these concepts could be extended to a real problem such as the diagnosis of wilting.

The fundamental idea of qualitative modeling is that differential equation models may in some circumstances provide too fine a resolution for the objectives of a simulation. These circumstances call for a coarser resolution, in which the time course of the behavior of the system is not simulated precisely. Instead, the behavior of the system is simulated in descriptive terms, so that in the case of the experiment of Fig. 11.1, for instance, we might simply say that the level of water in cylinder B first rises and then equilibrates, without attempting to describe precisely the rate of increase or the final equilibrium level. Qualitative modeling is suitable for the application of object-oriented programming techniques as described in Chap. 10, and our discussion will be phrased in this context.

Figure 11.5 shows a schematic diagram of a five component qualitative model for the experiment of Fig. 11.1. One of the fundamental principles of qualitative simulation is the principle of *locality*. This says that any component of a qualitative model can only influence directly the behavior of components connected directly to it. Thus the membrane in Fig. 11.5 can only influence the state of the fluid pipe on either side of it, and each pipe can only influence the states of the membrane and the accumulator connected to it. Each element of the qualitative model has a particular behavior, and these behaviors together with the description of how the components are connected completely specify the model. The description of connections is called the *topology* of the model.

The qualitative dynamics of compartmental models has been studied by Nicolosi and Leaning (1989). Our discussion of the behavior of the model will be limited to its steady state behavior. Implicit in this discussion is the assumption that such a steady state exists, e.g., the

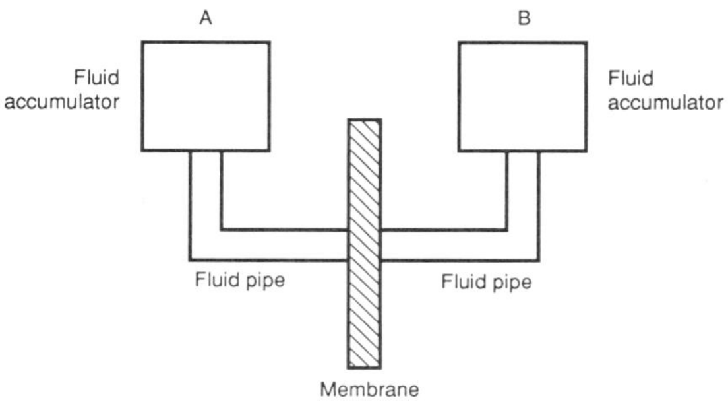

Figure 11.5 Schematic diagram of a five-component model for the experiment of Fig. 11.1.

system does not oscillate. More complex models can explicitly include dynamics (e.g., de Kleer and Brown, 1984; Kuipers, 1986). The behavior of each component in our model is specified in a manner very similar to that of Karnopp and Rosenberg (1975) in their formalization of systems analysis. The membrane is analogous to a resistance in their formalism. Its physical properties are described by two parameters: permeability to water and permeability to solute. Each of these is a binary variable taking on the values *permeable* and *impermeable*. Its state is specified by the direction of flow of fluid through it, which can take on three possible values: *positive* (flow from A to B), *negative* (flow from B to A), and *zero*. The value of the flow of water is determined by the relation

$$K_w[\Psi_A - \Psi_B] \rightarrow Q_w$$

where Ψ_i is the potential on side i of the membrane. The flow of solute is given by

$$K_s Q_w \rightarrow Q_s$$

These relations must be interpreted as causal statements. The interpretations are

IF $\Psi_A - \Psi_B$ is $\begin{cases} \text{positive} \\ \text{zero} \\ \text{negative} \end{cases}$

AND K_w is permeable

THEN Q_w is $\begin{cases} \text{positive} \\ \text{zero} \\ \text{negative} \end{cases}$

and

IF Q_w is $\begin{cases} \text{positive} \\ \text{zero} \\ \text{negative} \end{cases}$

AND K_s is permeable

THEN Q_s is $\begin{cases} \text{positive} \\ \text{zero} \\ \text{negative} \end{cases}$

and

IF K_w is impermeable THEN Q_w is zero

and

IF K_s is impermeable THEN Q_s is zero

The fluid pipes are pure connectors. Their state is specified by the flow Q_i through them and by the potential Ψ_i inside them. They have the property that the potential at one end of the pipe equals the potential at the other end, i.e., no resistance to fluid flow, and the flow rate of fluid into the pipe at one end equals the flow rate out of the pipe at the other end. The state of the fluid accumulators is specified by their potential Ψ_i. This is determined by the pressure P_i at the accumulator mouth, which depends on the volume, or height, of fluid in the accumulator and on the osmotic pressure Π_i, which depends on the solute concentration. Each of these is a binary variable taking on the values *high* and *low*. Thus,

$$P_i - \Pi_i \rightarrow \Psi_i$$

This equation has a causal interpretation similar to that given earlier describing the membrane flow.

In addition to the five components of the model, the intrinsic state of the fluid in each component must be specified. The intrinsic state of the fluid consists of two variables: pressure and solute concentration. Solute concentration is related to volume of water and volume of solute according to Table 11.3.

TABLE 11.3 Final Concentration of Solute for All Possible Values of Solute and Solvent

	Solute amount		
Solvent volume	High	Low	Zero
High	Moderate	Low	Zero
Moderate	Moderate	Moderate	Zero
Low	High	High	Zero

Table 11.3 has the force of a cause-and-effect rule. The causality associated with Table 11.3 is as follows: solute amount and water volume are the causal agents, and concentration values are the caused agents. Finally, a physical law describes the conservation of matter. This law states that if there is a flow of material across the membrane, then the amount of material on the side in the direction of flow increases and the amount on the other side decreases.

Assume that the initial state of the model is as follows:

Parameters	Cylinder A	Cylinder B	Membrane
Water volume	Moderate	Moderate	
Solute amount	Zero	Low	
K_s			Impermeable
K_w			Permeable

According to the model, the equations for water potential on either side of the membrane at the initial state are

$$\Psi_A = \text{moderate} - \text{zero} = \text{moderate}$$

$$\Psi_B = \text{moderate} - \text{low} = \text{low}$$

Therefore, Ψ_A is greater than Ψ_B and therefore Q_w is positive. According to the conservation of material this implies that at the steady state, the volume of water in cylinder B at steady state is *high* and the volume in cylinder A is *low*. Therefore, the equations for water potential become

$$\Psi_A = \text{low} - \text{zero} = \text{low}$$

$$\Psi_B = \text{high} - \text{moderate} = \text{low}$$

The water potentials balance, as required at steady state.

The counterfactual

at steady state, $H_A = H_B = \textit{moderate}$

is now introduced into the system. According to the WIF resolution algorithm of Whitehead and Roach (1990), the first step in restoring consistency to the model is to replace the values $H_A = \textit{low}$ and $H_B = \textit{high}$. We next look for rules having water height as a causal agent. These rules involve solution concentration and solution potential. The WIF algorithm must be modified slightly to take into account the nonbinary nature of the values of the model (cf., Roach et al., 1990; Whitehead and Roach, 1990). Each modification of a rule invokes a recursive application of the WIF algorithm on the model variables in that rule. For example, Table 11.3 has water volume as a causal agent. Therefore, this table is tested, first by testing the effect of varying the solute amount and then by discarding the table itself. Through the restoration algorithm, we can enumerate the possible modifications to restore consistency, such as $K_w = \textit{impermeable}$, $K_s = \textit{permeable}$, and so forth. The relative likelihood of these possibilities cannot be determined by the WIF algorithm without further, domain-specific information.

11.6 Discussion

The material and concepts discussed in this chapter are highly speculative, much more so than the contents of other chapters. As this book is written, no model-based expert systems have been implemented in agriculture, although qualitative models have been implemented in physiology (Bratko et al., 1989) and ecology (Starfield et al., 1989). We can only offer guesses as to what form such systems will take or for which problems they will be appropriate. In our discussion we have tried to present a picture of the level of technology that exists in the field at the present time, realizing that, as with many aspects of computer science, the information will be obsolete by the time it is printed. However, it does seem clear that *some* form of model-based or deep reasoning will take place in agricultural decision support systems. The advantages are too great to pass up.

A model-based reasoning system contains two components, the model itself and an inference mechanism capable of resolving internal conflicts and restoring consistency in the presence of counterfactuals. Although these components are in principle independent, the form and function of each influences the form and function of the other. A variety of modeling paradigms, ranging from production rule-based (Clancey, 1989) to qualitative model-based (de Kleer and Brown, 1984; Forbus, 1984; Kuipers, 1986) to differential equation–based (Iwasaki and Simon, 1986a; Sacks, 1990) are possible. Similarly, there are several methods for dealing with counterfactuals (Ginsberg, 1986; Roach et al., 1990). How these various pieces of the puzzle will fit together remains to be seen.

The model discussed in the previous section for the simple experiment is based on the notion that qualitative models should have discrete valued variables (de Kleer and Brown, 1984; Kuipers, 1986). This idea has been questioned by Iwasaki and Simon (1986), who assert that the study of the properties of dynamic equilibriums of classical systems of differential equations provides a simple and meaningful approach to qualitative analysis. A considerable body of theory has been developed over a period of more than 50 years on the qualitative theory of differential equations (e.g., Andronov et al., 1966; Arnold, 1973) and on the application of this theory to the construction of simulation models (e.g., Karnopp and Rosenberg, 1975); this approach may ultimately prove superior to the discrete variable method (cf., Sacks, 1990).

The primary application of model-based reasoning in this chapter has been in diagnosis. However, this is only one of many potential applications for this methodology. Another is in the development of adaptive simulation models. These are models for the behavior of a particular, specific instance of the system being modeled. For example, an adaptive model of a crop is a simulation of the behavior of that crop on one particular farm. These models are used to forecast the future behavior of the crop and to compare the predicted effects of alternative management actions and environmental scenarios. They are a fundamental part of some agricultural decision support systems. Since no fixed model can be accurately preprogrammed with the data for a specific instance of the real system, some degree of adaptability, or "parameter feedback" must be built into the model (e.g., Wilson, 1989). A counterfactual reasoning system provides one method of determining which of several possible parameters must be modified, and by how much, to cause the behavior of the model to more closely match the observed system.

Another application of causal models is in the construction of explanation and interpretation systems. One possibility is the association of a causal model as a "front end" for a classical simulation model. The interpretation of the output from a complex simulation model is a notoriously difficult task, and a causal model may be able to provide an automated explanation and interpretation of such output.

Although the potential of causal models is great, there will probably be an extended gestation period before they play a significant role in implemented, as opposed to research-oriented, agricultural systems. The motivation and development of these systems has primarily been associated with physical applications, especially kinematic mechanisms (e.g., Whitehead and Roach, 1990). There are a number of fundamental differences between models for kinematic mechanisms and models for crops and other biological systems that make it impossible to simply translate the former into the latter. Some primary differences in this context are the following:

1. *Precision of knowledge about fundamental mechanisms.* Since it is a mechanical device designed by humans, the structure and function of every component of a kinematic device is precisely known in every detail. Moreover, the intended function of the interacting components, i.e., their emergent behavior, when the system is functioning correctly is precisely known. Since the system is a complex one, the emergent behavior when one or more components malfunctions is not necessarily obvious, and an intended use of the model is to determine this emergent behavior in the presence of malfunction or, more precisely, to determine possible component malfunctions that could cause an observed emergent behavior.

In the case of a biological system, the situation is completely different. While good guesses are available for the function of many of the components of such a system, the level of knowledge about these components is uneven and very little is known with complete precision. Phloem transport in plants is a good example. Much is known about phloem transport with considerable confidence, for example, virtually everyone would agree that the function of phloem is to transport carbohydrates from the leaves to the rest of the plant. However, the precise mechanism of carbohydrate transport in the phloem is not completely understood although there are several attractive hypotheses (Ray, 1972). Because of this lack of knowledge about the fundamental mechanisms underlying phloem transport, it would be impossible to formulate a model of this component with the same degree of confidence that would be available in the model of, for example, a hydraulic component in a mechanical system. Therefore, one of the requirements of a causal or qualitative plant model that has no counterpart in the mechanism model is that it incorporates components whose function is well understood, or at least can be modeled as if it were well understood, but whose structure is not known with certainty. The model must be able to capture the knowledge of the function of such components well enough to simulate their emergent behavior when interacting both in a properly functioning and a malfunctioning manner, without relying on information that is poorly understood.

2. *Distributed versus localized components.* In a mechanical system, each component, e.g., valve, cylinder, or linkage, exists as a single unit. The number of such units, while large, is small enough that it makes sense to consider including all the units in the model and modeling their interaction directly. Thus, a mechanism model can be formulated as an exact description, component by component, of the mechanism itself. Such is not possible with most models for biological systems such as plant models. Some of the organs of the plant, such as the leaves and fruit, exist in small enough numbers that it is physically possible to model them individually. In some circumstances this may be appropriate and in others it may not. Other organs, such as the roots and stems, are distributed in structure and this may have to be taken into account. Finally, some of the mechanical components of the plant that might be considered analogous to the components of a mechanism, such as the stomata, exist in such large numbers that it makes no sense to consider modeling them individually. Both because of the distributed structure of the plant organs and the large number of its components, the plant could still not be modeled with the same accuracy as the mechanical system even if the structure and function of all aspects of the plant were precisely known.

3. *Emergent population properties at several levels.* Since an individual plant or other biological system consists of a population of organs, e.g., leaves, each of which contains populations of structures, e.g., stomata, emergent properties resulting from population interaction can occur at several levels. It is traditional in plant modeling (e.g., Harper, 1977; Wang et al., 1977) to use population dynamics models in describing individual plants. Thus, while a mechanism model consists of the relatively simple interaction of components at a single level, a plant model has complex interactions at multiple levels. A model for a crop, consisting of a population of plants, takes on still more levels, and a model for an agricultural ecosystem still more.

4. *Dynamic variation of the structure of the system.* A mechanical system has dynamics; it can move its components relative to each other, rotate and translate positions, and so forth. However, the structure of the mechanical system itself ordinarily does not spontaneously change with time. However, the structure of a biological system does change with time due to the internal dynamics of the system. This change in structure can affect the emergent properties of a population of individual plants. For example, as individual plants grow they may compete with each other for light and nutrients, resulting in diminished yield for each individual relative to that obtained if the plants were growing in isolation.

It should be evident from the foregoing discussion that much remains to be done before causal or qualitative models for biological systems reach the status of implementability currently enjoyed by rule-based systems. However, the advantages of model-based systems are such that progress on their implementation seems inevitable.

Chapter 12

Integrated Decision Support Systems

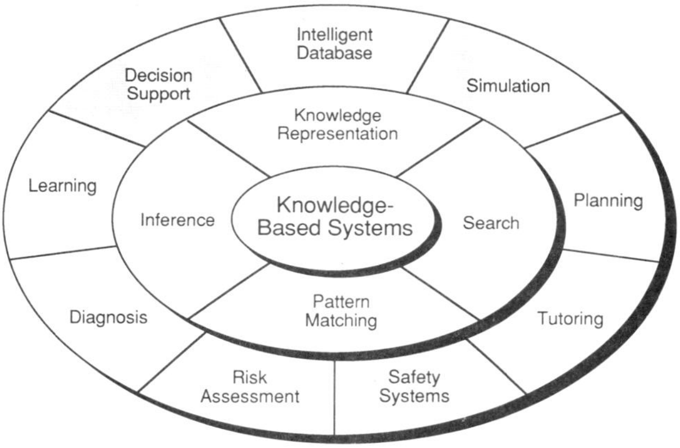

The future of knowledge-based systems on the farm and their integration with database management, prediction, automation, and planning systems are discussed. In this chapter we review the basic issues of intelligent database systems and go into some more depth on blackboard systems for the coordination of multiple knowledge sources. The concept of safety systems is introduced as the next logical step for knowledge-based systems in crop systems, particularly in integrated crop management.

12.1 A Future Farm and the Role of Knowledge-Based Systems

Donald Holt (1985) described a futuristic scenario of what one day on a farm might be like in the computer age. The following is an excerpt:

> The computer has been scanning by telemetry a number of miniature portable weather stations placed in the fields. It scans these every 15 minutes, computes heat unit [= degree-day] and solar radiation accumu-

> lations, and plant available moisture, and summarizes and stores these data, which are forwarded automatically to state, federal, and private weather agencies for further processing....The computer uses weather and soil moisture data, plus short- and long-range weather forecast information from the network, in several simulators to predict the status of farm crops and pests. On this particular day, it anticipates low soil moisture in sands near the river, has actuated the pivot irrigation system in that field, and is monitoring and controlling the pump....One simulator has identified this day as the optimum time in terms of weather conditions and plant growth stage, for dealing with foxtail infestation in the soybean field south of the road....[Another] simulator predicts an infestation of European corn borer in the early-planted corn. After breakfast [farmer] Bob enters the field, carrying a small electronic device consisting of a 7 by 10 centimeter video screen and a small keyboard. Using the eraser end of his pencil, he types IPM...which accesses an IPM expert system on a random access optical disk in the county extension office. The program leads Bob through identification, diagnostic, and decision steps by means of voice, text, and color pictures. When the predicted degree of infestation has been verified by sampling, a least-cost, environmentally safe, biological control procedure is prescribed.

In Holt's scenario, simulation models are being run on a farm-based computer that is getting its input not only from on-farm databases describing the farm's local weather, soil types, topology, and field use, i.e., crops planted in each field, but also from centralized computers connected to the farm via a public network. This network provides the farm computer access to expert systems running at the county agent's office and to weather predictions for the farm's general area. In addition, the farm-based computer is using the output from the simulation models to anticipate problems, insect infestations in corn, and to schedule the application of other inputs, herbicide in soybean.

Technically, there is nothing in this scenario that is impossible with today's computers and the techniques described in this book. However, the extent of integration required among different types of computer programs and even between different computers themselves poses new problems not yet addressed.

12.1.1 Integrated expert systems

Knowledge-based systems working in combination with simulation models and database systems have been called integrated expert systems (Stone et al., 1986). Essentially, such systems attempt to compile all the data, information, and knowledge needed to solve problems in a broad domain and to make it all accessible from a single computer program. Two approaches have been advanced to accomplish this integration. The first focuses on software integration and is exemplified by the COTFLEX and CALEX/Cotton systems, modular programs for cotton management in Texas and California, that include dynamic linkages to simulation models and databases (Plant, 1989; Stone and Toman, 1989). We will discuss these programs, which include their own database systems, specially modified simulation models, and knowledge bases. Another approach is to build a computer program that transparently links together resources across a network of computers, so that different computers can be tapped for information and software modules, like weather predictions, when they are needed. In such systems the focus is on hardware integration to provide the user access to information sources dispersed on a network; however, the user sees all the functions of the multimachine system through a single interface. These systems have been called expert workstations (Faught, 1986) and knowledge-based integrated information systems (Manola, 1990) and have been developed in domains like biotechnology, geology, and aerospace engineering (Fikes and Kehler, 1985). Work is currently under way in forestry (Loh et al., 1988; Saarenmaa, 1988) but not in production agriculture.

Shared access to information. In this book, we have seen how knowledge-based systems can be linked with simulation and statistical models to solve tactical and strategic problems and to schedule the application of repeated inputs to a cropping system, like water or fertilizer. In each case, the information needed by the knowledge-based system has to be entered by the user or has to exist in a specially defined format on the computer running the knowledge-based system.

The scenario described by Holt, above, places much heavier demands on the data handling features of our systems. Data must be collected automatically, stored in local files, sent across networks to files on remote computers, and retrieved in various formats from the same and other centralized computers. The soils data, for example, is likely to be retrieved from a centralized database maintained by state or federal agencies, like the Soil Conservation Service, with which they perform conservation planning analyses. Since data are being collected automatically, the system must be able to detect data that include errors or that indicate failure, for example, of the weather station probes. That is, data must not only be collected, they must be understood and stored appropriately. Data also have to be reused by different decision support modules. The descriptive information on soil types, for example, must be available both to an irrigation scheduler, to a crop growth model, and probably to a farm-level crop rotation planner that will operate in the off-season.

Intelligent database systems not only store and retrieve data, but also apply domain-specific knowledge to data to aid the user in information discovery, error detection, and problem-based querying (Parsaye et al., 1989). These systems can be used to build knowledge-based systems with the functionality required in Holt's scenario.

Planning and practice. Another requirement of the computer system for our future farm is farm-level coordination. The computer should know enough, for example, not to recommend both an irrigation and scouting for potential insect infestations in the same field at the same time. In previous chapters we have examined in isolation various aspects of crop management, such as pest management, irrigation scheduling, and nutrient management. Now we must deal explicitly with the interrelatedness of these problems, so that decisions made regarding one problem influence how decisions are made regarding the others. Moreover, while these activities influence each other in specific ways, they are nevertheless conceptually independent, and their management is often the purview of different sets of individuals. This last comment applies in two different ways. First, the experts whose knowledge goes into the construction of the knowledge base for the different components are ordinarily different people, and second, particularly in large organizations, the individuals who actually implement the controls may be different for different components. For example, on many large farms different employees are responsible for irrigation and pest management.

The construction of plans by the interaction of semiautonomous individuals has been addressed in the planning literature and is called *multiagent planning*. This term was coined by Konolige and Nilsson (1980), who were among the first to discuss the use of the computer to formulate plans in this manner. The concept of multiagent planning has also been discussed by Davis and Smith (1983). Later in this chapter we discuss multiagent planning as it relates to coordinated farm management.

12.1.2 Movement from responsive to preventive systems

One of the most appealing aspects of Holt's futuristic vision of computerized farming comes from a change in emphasis from problem solution to problem prevention. In this book, we have discussed systems that a user can access to help solve a problem. Farmer Bob's computer, however, runs constantly and anticipates problems before they arise. It is the computer that alerts the farmer to problems like a potential pest outbreak in early planted corn or moisture stress in cotton planted in sandy soils near the river. In contrast to the systems discussed in previous chapters, such a program would (1) run in real time, not after the fact; (2) be connected to a vast array of sensors measuring such things as soil moisture, ambient, canopy, and soil temperatures, and crop color; and (3) use simulation models to compare the actual system's performance to expected behavior and to recognize dangers before they become big problems.

Monitoring and automation. Sensors and automated machinery are not the focus of this book, but they will clearly be integral components in the computerized farm of the future. For the most part, the sensor technology exists to measure virtually all the relevant input parameters used in existing simulation models of crops. In addition, the microprocessors and telecommunication devices needed to compile and transmit the data from sensors back to a nearby, central computer are available today. For example, the GOSSYM/COMAX system (McKinion et al., 1989), developed for cotton production, includes a remote weather station at every installation. It is the knowledge required to interpret the data from these sensors that is limiting.

Model-based reasoning. As difficult as it is to simulate the development of a crop that is subject to farmer manipulations, pests, and weather, simulation alone is often insufficient to enable us to reason about the causes of problems that arise. When the natural system diverges from its expected, i.e., simulated, course, the first concern is to determine why. A simulation model can be used in a trial and error mode to come up with an explanation. However, what is really needed is a model-based reasoning system as discussed in the last chapter. There, we discussed the problem of diagnosing an observed disorder: What caused a plant to wilt? Here we want to answer the related question: Do the observed crop conditions indicate any disorder or any need for intervention or further sampling?

As in the previous chapter, we discuss the use of model-based reasoning with the counterfactual approach developed by Whitehead and Roach (1990). That is, we assume that we have a model describing how the crop system behaves, i.e., how growth responds to different combinations of inputs, damage, and weather. How these relationships are represented internally is not important for the moment. The important assumptions are that the model accurately represents the natural system and that the model is written in a form consistent with the counterfactual reasoning system.

We also make one further assumption, namely, that the knowledge-based system has recommended schedules of inputs and field operations for each field based on its predicted growth pattern, planting date, soil type, pest history, etc. These plans are followed unless the sensor and sampling information indicate that the field is not behaving as expected.

Whenever the natural field is sampled or sensor data from the field are received, these data or observations are asserted to the reasoning system. For example, the sample information (thrips_damaged_leaves 10) might be asserted to indicate that 10 percent of the leaves show feeding damage from thrips. The counterfactual inference engine then determines whether or not the new assertion is consistent with all the previous assertions. (Note that the AI term for the set of all known facts and conditions about a particular problem at a point in time is the *state-of-the-world* or more simply, the *world*. Thus, the counterfactual engine determines whether the new assertion is consistent with the current world.) If it is consistent, the system returns *true*. If it is not consistent, i.e., if the new information cannot be explained by the model and what is already known, then the counterfactual engine returns all the potential worlds in which the new information is consistent and changes to the existing world that would cause the new world to exist. For example, the system might say that 10 percent damaged leaves is consistent if there had been an undetected immigration of thrips within the last week. It might also suggest, with a lower certainty, that 10 percent damage could have occurred if the thrips feeding rate was higher than described in the model. In either case, the discrepancy indicates that the field in question ought to be sampled again. Information from the new sample would be used to update the model-based system so that the reasoning system's world would be once again consistent with the field situation.

Finally, once the discrepancy between field and model has been eliminated, the schedulers and planning systems are run again to determine a new plan for that field, given the new state of the world. Alternatively, the counterfactual reasoning system can be used again to determine if there is a way to return the crop to its original target trajectory. The target state of the field can be asserted as a counterfactual, and the counterfactual reasoning system then returns the set of modifications to the current world, i.e., the crop field, necessary to achieve the target path.

Intelligent safety systems and IPM. The classical application domain of model-based reasoning has been fault diagnosis. For example, the use of the counterfactual inference engine for model-based reasoning, as mentioned in Chap. 11, was developed to perform fault diagnosis on the large guns of naval ships. When a fault diagnosis system is linked with a real-time monitoring system as we have suggested above, the result is called a *safety system*; these systems are currently in use in nuclear facility management (Corcoran et al., 1981). When the fault diagnosis component uses model-based reasoning, the result has been dubbed, not surprisingly, an intelligent safety system (Klein and Finin, 1987).

Although the on-farm implementation of intelligent safety systems may be a decade away, we believe that they have the potential to revolutionize IPM and crop management. If we go back to the basic definition of IPM (Smith and Reynolds, 1966), as "a pest population management system that utilizes all suitable techniques in a compatible manner to reduce pest populations and maintain them at levels below those causing economic injury," we see that IPM practitioners must perform two basic tasks. First, they must be able to identify when injurious pest levels are likely to occur *before* they occur. Second, they must be able to modify the crop system to prevent economic damage. To perform these two functions, farmers will have to do the same things that industrial safety systems do: monitoring, comparing observed system behavior to what is expected and identifying dangerous incipient deviations, and making whatever corrections are required to prevent problems from causing unacceptable levels of damage.

Currently, farmers and pest control specialists act mainly in a responsive manner. Information is sampled relatively infrequently, and the expectation of crop growth and development is very vague. In the future, automated sensors and continually running computer systems will make the cost of gathering information trivial, and deep models of crop systems will be able to provide more specific predictions of crop growth. The use of intelligent safety systems will enable farmers to spend most of their valuable time sampling those fields that show potential problems. They will be able to anticipate problems and to react with progressive and preventive tactics that will in general be much less disruptive to the natural system.

12.2 Intelligent Databases on the Farm

In the futuristic scenario presented at the outset of this chapter, knowledge-based systems, simulation models, and remote computers would all have to share access to locally collected data, and remotely stored data would have to be just as accessible to local software modules. The same database used by the Soil Conservation Service would have to be queried by programs on the farmer's computer across a network. The information used in the planning and scheduling modules of the farm-level decision support system would have to be dynamically linked to the financial data used in farm accounting and management software to estimate production and machinery maintenance costs. A tremendous amount of information would be flowing into the farm computer every 15 minutes, updating the status of all the remote weather stations and soil and crop sensors on the farm, and sometimes that information would suggest that action must be taken. Furthermore, every time a decision is made or an action taken in a field, all the relevant databases would have to be updated to reflect the change.

How is all this to occur? The tasks involved in coordinating the interaction of multiple databases, in automatically processing and recording real-time data, and in automatically making inferences about information in a database come under the umbrella of intelligent database management. Intelligent databases are to traditional databases what model-based reasoning is to simulation modeling. In the same way that a simulation model written in a procedural code like Fortran can provide only its output to a knowledge-based system, a conventional database management system can only interact with a knowledge-based system through its query language and command protocol. The reasoning component of the system cannot directly access the data within. Ideally, the same knowledge representation scheme would be used by all the databases in an interdependent computer network as well as by the reasoning components in the system. In practice, solutions must be forged from an alloy of existing databases and newly developed intelligent systems.

There are two distinct issues that need to be addressed by intelligent databases. First, how can different computer programs, sometimes on different computer platforms, share access to data files efficiently? Second, how can reasoning capabilities be added to a database management system to enable that system to perform automated discovery, to detect errors in the data, and to respond to vague or incomplete queries?

12.2.1 Access to data files

The first of these issues goes back to the roots of database management systems. Consider the case of CIRMAN, an expert system that helps farmers decide whether or not to purchase federal crop insurance (Helms et al., 1990). This program requires extensive economic data as input, including production costs, machinery complement, debt structure, and other data that are also relevant to a tax program or an accounting package (Fig. 12.1). It is clearly inefficient for a farmer to have to duplicate all the data entry in order to use a tax advisor and CIRMAN. This example illustrates the need for a database management system (DBMS) (Fig. 12.2). These systems were designed to allow multiple uses of the same data. The DBMS takes care of data entry and maintenance and responds to a specific query language that can be used by multiple applications or users.

However, the situation shown in Fig. 12.2 has not become the status

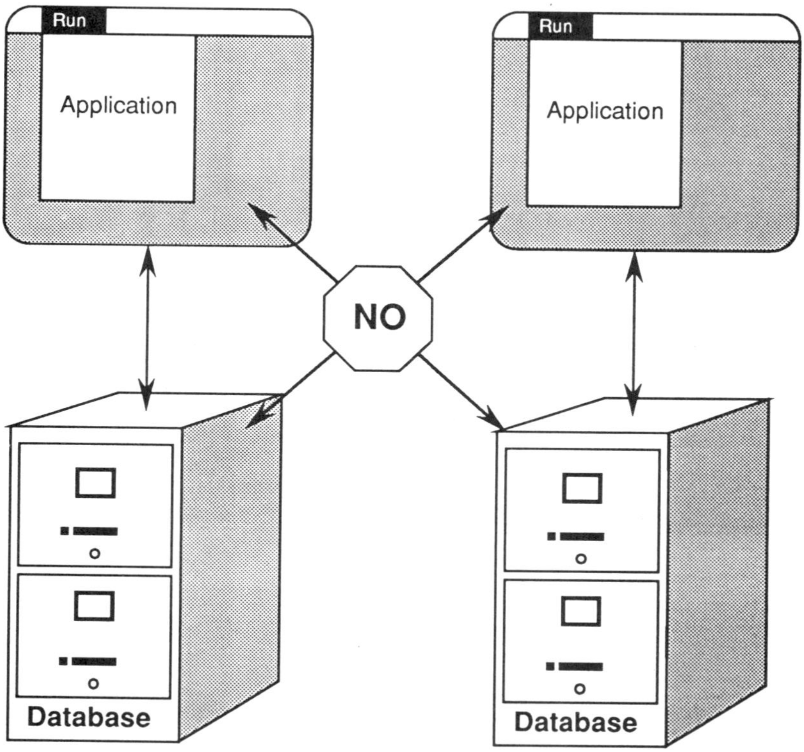

Figure 12.1 Access to data files in traditional, stand-alone application programs. Each application maintains its own database.

quo. Instead, there has been a proliferation of DBMS programs. A software developer who wants to use a DBMS to handle data files for an application has to choose which DBMS to support, with the knowledge that the user may not have access to any of the supported systems (Fig. 12.1). Many commercially available expert system shells, for example, support popular DBMS packages. However, there is no standard database protocol for agricultural information. Mainframe computers continue to use different systems than personal computers, and computers with different operating systems almost always store data in incompatible formats. It is probably still simpler for the designer of an expert system to build an application-specific DBMS into the application, e.g., the CARMS record management system for CIRMAN (Richardson et al., 1990), than to rely on the user to purchase and become facile with a third-party DBMS.

Nevertheless, if users and applications software are going to be able to access and share data with remote databases and programs, not only must the database scheme shown in Fig. 12.2 be accepted but another layer must also be added (Fig. 12.3). When a computer program

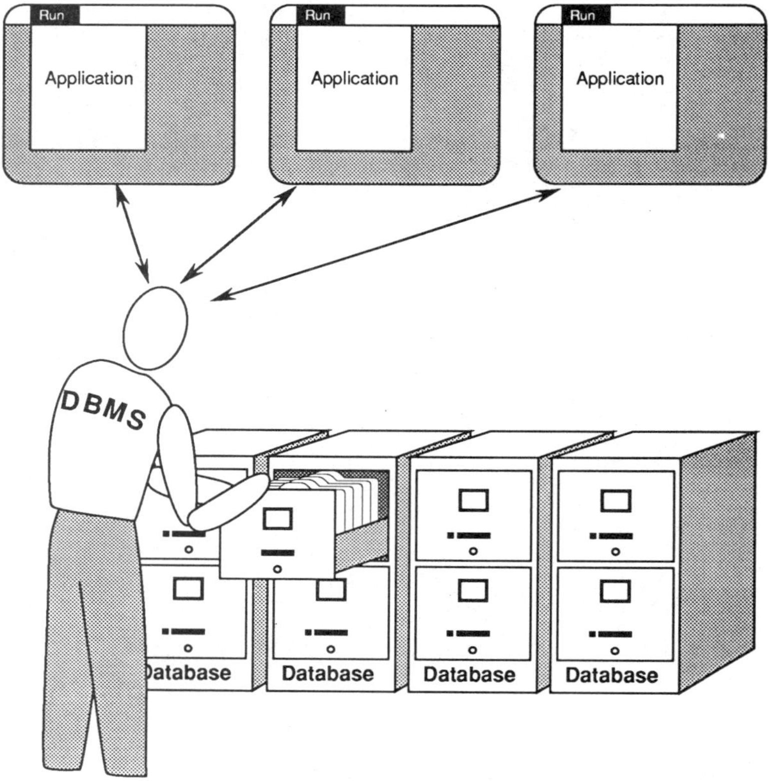

Figure 12.2 Idealized situation using a database management system (DBMS). The application programs all share a common database, each one accessing the data through the DBMS.

must access not just separate data files, but separate database systems that might reside on separate machines, the resulting system is called a *heterogeneous* or *distributed* database system (Gupta, 1989). These systems require interfaces for each individual database system in the distributed system, and they also include a global database manager.

All queries from users or from applications are sent to the global database manager using a single, global query language. The global database manager then determines which database in the distributed system can respond to the query and sends the query to that system. Thereupon, a local interface program translates the global query into the query language specific to the database with the information.

Data are returned over the same path. The data are first sent to the local interface, which translates the data into the global format. The data are then passed back to the global database manager, which routes the data back to the application or user that requested it. This

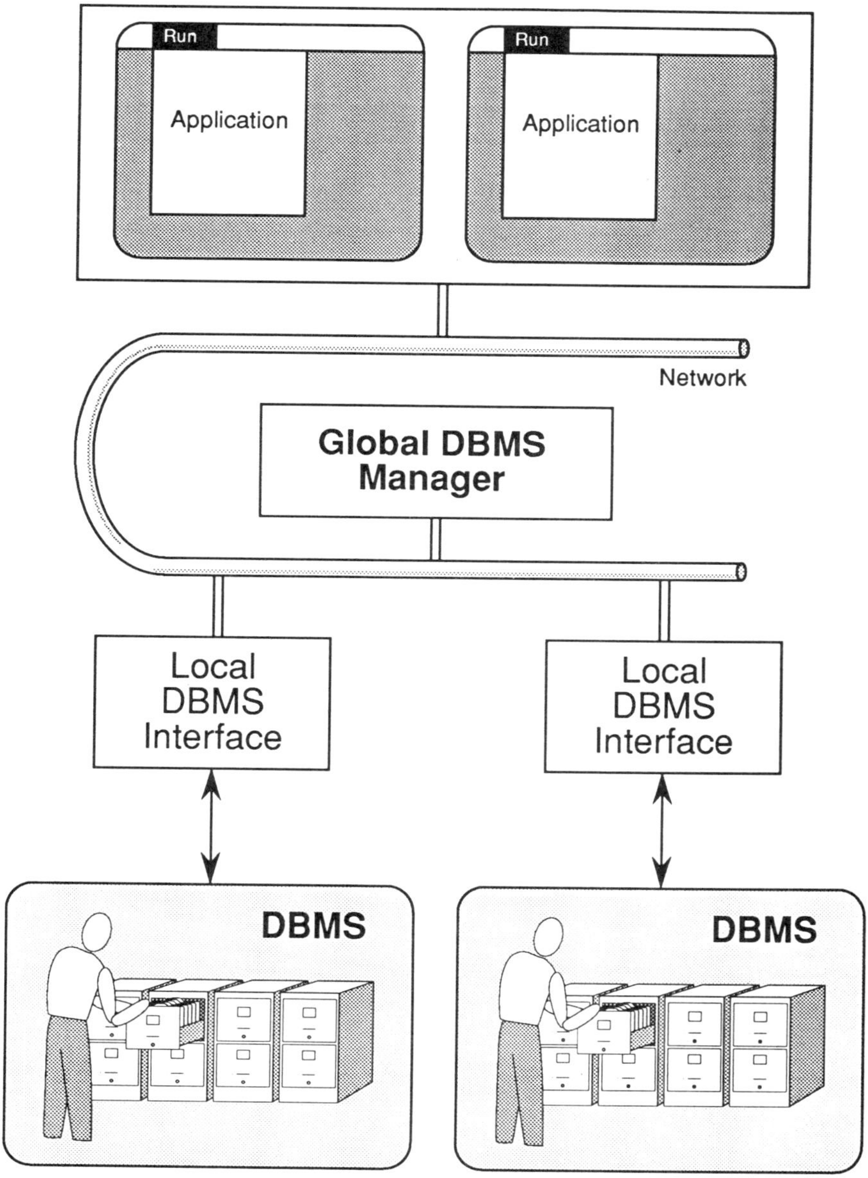

Figure 12.3 Schematic design of a heterogeneous database system. The application programs can be run at any point in the network, and all queries are sent through the global DBMS manager using a global query language.

plan requires that all application programs follow the global database syntax. At the top level, all information fits a single format and interface. At the database level, there can be multiple database formats, operating systems, and database management protocols.

Heterogeneous database systems solve some problems, but they are still ultimately limited by the functionality of the databases included in the individual system. For example, if all these databases are conventional databases, then the overall system will not have the power to reason about the data contained in them. To provide users with the full functionality of all the DBMS packages in the distributed system, the global database manager must be as powerful and flexible as the most sophisticated local DBMS. In general, such systems will not be able to take advantage of all the features of all the databases on the network.

12.2.2 Adding intelligence to databases

Having addressed the problem of accessing data files in a heterogeneous computer environment, we are still left with the problem of reasoning about the data in a database. In particular, we want our database system to be able to examine its data automatically, to report errors, and to discover patterns that may require attention. We also want to be able to ask for information from the database in terms that are more vague or general than is possible with conventional database systems. For instance, we would like to be able to ask the system if any fields are showing signs of unusually high mite infestations. The database must understand the concept *unusually high* and must recognize that the expectations for different fields will be different based on field history and neighboring crops.

Context dictionaries. To allow the system to reason, we need to add context to the information in the database. This is accomplished by two components of intelligent database systems, the *data dictionary* and the *concept dictionary* (Parsaye et al., 1989). The former contains information about the fields in the database, which we will call *slots* here for clarity.* The latter defines concepts in terms of the values stored in specific data fields. In the elementary case, the data dictionary merely tells the type of information, e.g., real number, integer, or character string, in each slot of the data record and the slot length. However, the data dictionary can also be a framelike knowledge rep-

---

*In an agricultural text there is unavoidable confusion between the two meanings of the word *field,* one being the piece of land on a farm in which a crop is grown and the other being one of the data-holding slots in a database record.

resentation description, providing a description of the relationships among the different slots in data records and other components of the problem domain. For example, assume that a row-crop database contains records of each field on a farm. One of the database slots holds the name of the cultivar, e.g., Coker-485, a soybean variety resistant to southern root-knot nematode (Anonymous, 1990). In a standard database system, a second slot would probably contain the generic crop name, soybeans, and a third would describe the cultivar's resistance to the nematode. In an intelligent database, the data dictionary might contain the information represented in the semantic net shown in Fig. 12.4. In that case, only the single data slot holding the name of the cultivar is needed for the database to be able to return a list of all soybean cultivars resistant to southern root-knot nematode.

The concept dictionary is used to define relationships that can be used to make inferences about the data entered into slots. Concepts can also be used to enable the system to respond to qualitative queries and to set up error conditions. For example, in our crop the concept *tall* could be defined in the concept dictionary to mean *height,* a slot, greater than one meter. We could then ask the database to find all fields in which corn plants are tall.

Complex concepts can be used to allow the system to make even more intelligent deductions. For example, suppose a farmer wants to

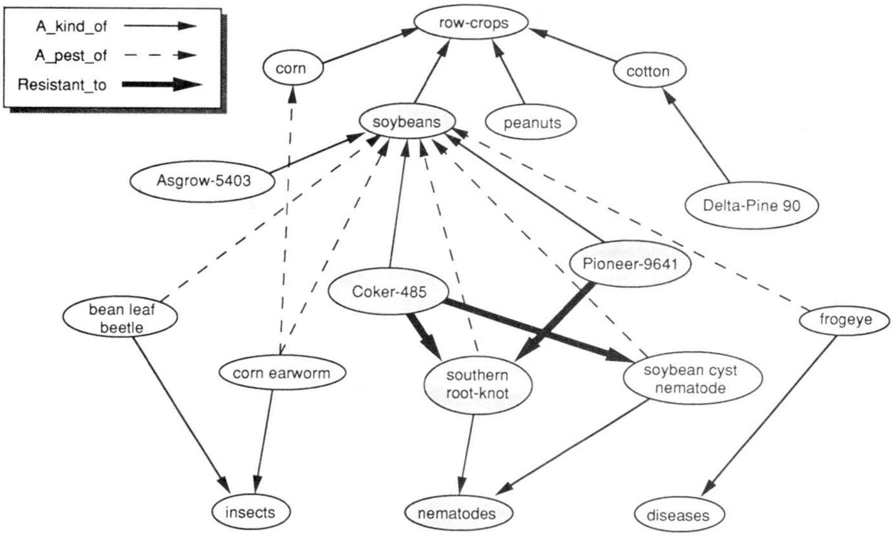

Figure 12.4 Semantic net showing information about soybean cultivars and their pests (information from Anonymous, 1990). Including this type of semantic information in an intelligent database's data dictionary provides context to the information in data records.

determine which soybean fields performed better than expected last year. Each soybean cultivar has an expected yield for the farm's general area and soil types, but expectations vary depending on the history of soybean cyst nematode in the field. With a conventional database, this would involve querying for the yields from all the soybean fields in the row-crop database as well as the cultivar and information detailing whether the cultivar was resistant or susceptible and whether or not the field itself has a history of soybean cyst nematode. Next, the farmer would have to look up in a table or another database the expected yield for each cultivar planted and then determine which fields showed significantly higher than expected yields. With an intelligent database, the concept dictionary could already contain a table showing the expected yields of each soybean cultivar given the presence or absence of soybean cyst nematode. It could also define the concept better-than-expected to mean greater than 105 percent of the expected yield. In this case, the farmer would have only to make a single query to the database to determine which fields performed well.

Automatic discovery. If a computer system is operating all day and night as described in Holt's scenario, there will be a considerable amount of time available to the computer, as it were, while the users are all busy or asleep. Holt conceived that this time would be appropriate for running simulation models, performing calculations and analyses, and for automated accessing of remote computers. It is also time that could be used by an intelligent database system to try to find some novel patterns in or to automatically generate rules from the data in data files.

The concept of a database automatically discovering relationships is very similar to the idea of rule induction by machine-learning algorithms discussed in Chap. 4. The database records function as would a large example table, and the discovery algorithm effectively looks for rules that hold across examples. The key difference in database systems is that there is no expectation that a complete rule base can be generated. In general, database discovery systems rely on a combination of rule induction and statistical analysis, so that the rules they produce are typically of the form

```
IF      field = West-12 AND
        crop = sorghum OR
        crop = cotton
THEN    yield = below expectation (CF 65)
```

The process involves forming hypotheses, for instance, that the values of two database slots are related, forming queries to gather evidence, and performing statistical analyses on the results. The hypotheses are then either confirmed if the statistical test indicates at least some threshold of likelihood, rejected, or modified, and the process is repeated.

Automated discovery has been used in agriculture to detect patterns in insecticide effectiveness (Parsaye et al., 1990) as well as in several other domains. Commercial discovery programs are available for personal computers that will operate on a variety of database formats, and automated discovery is an active area in the collaboration between AI researchers and statisticians.

Understanding inexact queries. We have already seen how the definition of qualitative concepts in the concept dictionary can be used to allow qualitative querying of a database. It would also be very helpful, at times, for the database system to search not only for the information requested directly but also for information that seems related. Human beings do this all the time when recalling experiences. For example, when a person is stranded in a remote rest stop off the highway because the car's motor will not start, the person might recall all the times in the past when the car's motor refused to start. The individual might also be reminded of other times being stranded near a highway. Similarly, if a farmer applied an insecticide on Tuesday to control a pest that is still abundant on Thursday, the farmer might wish to query a database system to see what circumstances in the past are related.

Traditional database queries use elementary pattern matching on specific slots of data file records (e.g., Date and White, 1988). What is needed to recall related information is a set of links, stored with each data file record, that describe the type of situation the data record describes. For example, data records that show a high population of a particular pest could be linked to a class of *high-pest-population* records. Such a class could then have subclasses for individual pests. Another link class could contain all field scouting records that preceded a chemical application. This concept of adding links to data records is exactly analogous to indexing in case-based reasoning systems (Chap. 8).

12.2.3 Database implementation in COTFLEX

There are only preliminary examples of truly intelligent databases in agricultural knowledge-based systems, though work is under way to develop these databases (Beck et al., 1988; Beck, 1989). Here we present the implementation of a database with dynamic linkages to an expert system as developed in COTFLEX (Stone and Toman, 1989).

While the system does not attempt to use DBMS to provide general access to its data (Fig. 12.1), it does provide a kind of data dictionary. That is, it includes a hierarchical representation of the objects described by the data so that the reasoning system can access and reason about the data.

Once again referring back to Donald Holt's scenario, expert systems for IPM are said to reside on county agent computers and are accessed by remote, portable computers in the field. While some aspects of IPM can be addressed by such a system, pest management in many cropping systems cannot be accomplished based only on rules of thumb and user observations in the field. Farm and field databases are required to allow the system to reason about trends and time-related phenomena. For example, proper evaluation of the severity of a current pest infestation depends to a large degree on what has happened in that field in the past. An obvious case involves the pest management of *Helicoverpa virescens* in cotton (see Chap. 7), which has exhibited widespread but inconsistent resistance to pyrethroid insecticides. Sampling data on the number and age structure of *H. virescens* larvae in a cotton field must be interpreted very differently depending on the previous history of the insect's presence and the previous use of insecticides. Likewise, when a farmer finds a population of pests that is very close to the recommended action threshold, the farmer should take into account the recent trends in infestation levels (as in Chap. 6). Thus, proper pest management often demands that the manager refer back to field history.

In the COTFLEX system for cotton management, field histories are recorded in a set of linked databases that can be accessed by both the reasoning component of the system and by the user independently (Stone and Toman, 1989). The database is built from a hierarchy of C-language structures that allows a framelike representation (Fig. 12.5). For each cotton field in the system, the field frame contains information on varietal type (cultivars for long or short growing seasons), field acreage, and soil type. The field frame also contains arrays of frames to record sampling information for each field.

The system achieves dynamic linkage between the database and the reasoning component by implementing separate, parallel structures in the reasoning system and by building an automatic interface between the two parallel data structures (Fig. 12.6). In the reasoning component, each knowledge base entry with a corresponding database object also has two facets or flags: an *up-to-date* flag that is set to *true* when the system is sure that there is not more up-to-date information in the database and an *is-needed* flag that is set to *true* when the value of the object is required for the inference process to continue. When the is-needed flag is set to *true,* the system queries the database auto-

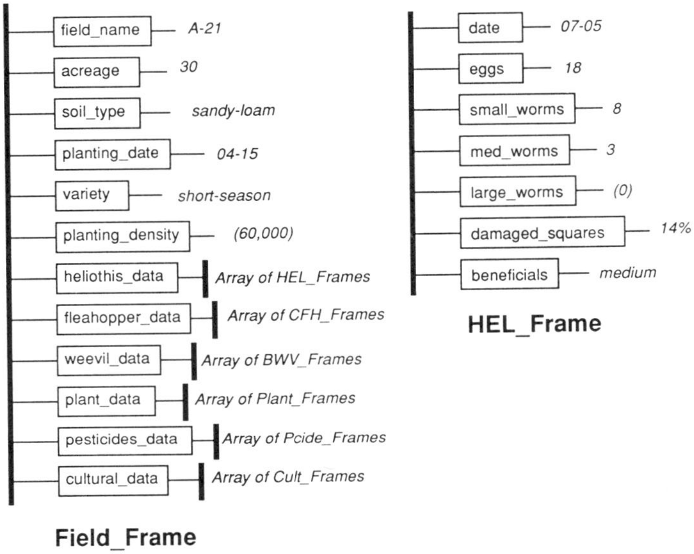

Figure 12.5 Two frames from the field database structure in the COTFLEX system. BWV refers to the boll weevil, HEL to the cotton bollworm and tobacco budworm (*Helicoverpa* spp.) complex, and CFH to the cotton fleahopper.

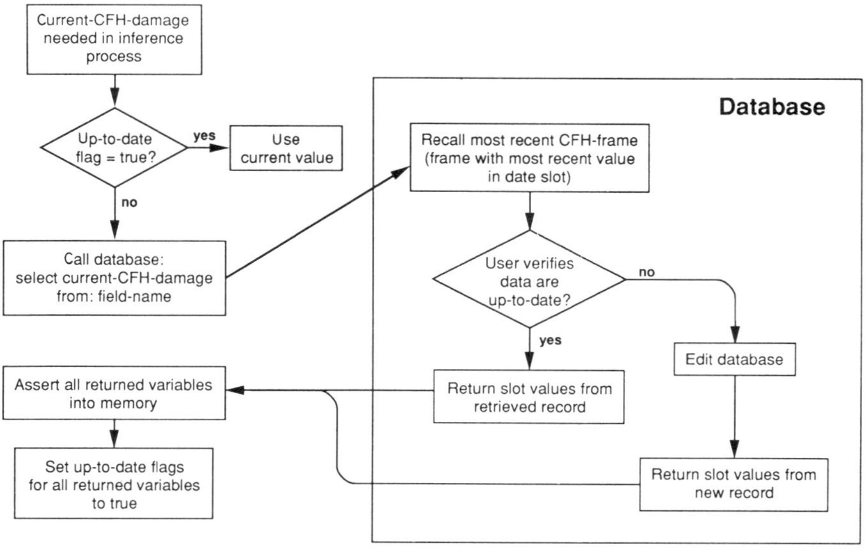

Figure 12.6 Linkage between database and knowledge base in the COTFLEX pest management advisor.

matically, for instance, for the most recent cotton fleahopper scouting record, which consists of the date and the percent infested cotton terminals. After the data are returned from the database, they are displayed to the user, who is asked whether the data retrieved are the most recent data available. If not, the system calls the database in edit mode, and the user is asked to bring the data file up to date. Finally, all the verified data in the most recent data record are returned to the reasoning system and assigned to the appropriate parallel data structure in the knowledge base. The up-to-date flags for all these are set to *true,* so that later, if another of the variables is needed in the same consultation, the system will know not to query the database.

Even though the linkage between the knowledge base and the database in COTFLEX is crude compared to a full-fledged intelligent database system, the advantages are similar. There is a uniform data interface presented to users, whether they are requesting advice about pest management in a field or updating the database directly from field scouting forms. Also, the inference engine can reason directly about objects whose values are stored in the database, since proper access to the data is taken care of automatically.

12.3 Planning in an Integrated System

Coordinated management of a farm's operations is impossible without shared access to up-to-date information. As we discussed, databases and intelligent databases can solve this problem. However, there remains the problem of coordinating the management of interdependent farm processes.

12.3.1 Integrated crop management

In developing a multiagent planning system it is necessary to take into account the ways in which different activities influence each other. Tate (1977) and Wilensky (1979) point out some of these ways. Some of the most important are *goal conflict,* in which two goals cannot be achieved simultaneously, *goal subsumption,* in which taking one action renders another irrelevant, and *necessary preconditions,* in which one goal must be accomplished prior to another. Since the goals in crop management are rather loosely defined, the interactions and influences are not so much between the goals as between the various actions that are scheduled. For example, a conflict occurs in the simultaneous scheduling of irrigation and pest scouting if the scout could not enter the field when it was wet due to irrigation, and subsumption occurs if, in scouting for one insect pest species, the farmer automatically also scouted for another. An example of necessary preconditions is not harvesting a cotton crop until it has been defoliated. In the construction of a schedule for crop management, other, more subtle forms of influence may occur as well. For example, there is evidence that some species of leaf-feeding insects may be more attracted to lush foliage. If this is the case, then nutrient application decisions may be influenced by the population densities of these insects.

The construction of knowledge bases for such a scheduler involves the cooperation of many different experts, all of whom are primarily specialists in one or a few fields with little knowledge of the details of other fields. Moreover, each knowledge base functions in a semi-autonomous manner, in the sense that a schedule for each action could be derived by itself; the actions only interact to the extent that they influence each other. We now consider how the activity frame concepts developed in Chaps. 8 and 9 can mediate this interaction. Suppose that the following three activities are to be integrated by the expert system: an insect pest, irrigation, and a plant pathogen. Suppose further that the three activities influence each other as shown in Fig. 12.7. If the field is wet because of an irrigation, then it cannot be scouted; if the plant pathogen is present, then irrigations must be less frequent, and if an insecticide application is to be made, then the field cannot be irrigated because of restrictions on the use of the insecticide.

The activity iteration process shown in Fig. 9.3 would function as follows. Suppose the knowledge bases are processed in the following

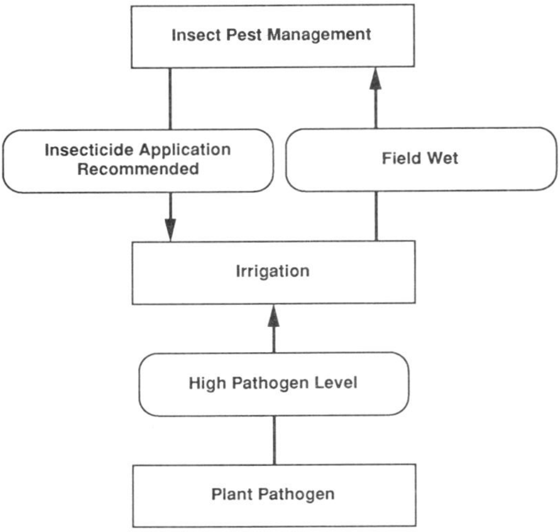

Figure 12.7 Interactions between three hypothetical activities in an agricultural management decision support system.

order: the insect pest, irrigation, and the plant pathogen. The order of processing cannot affect the outcome of the expert system consultation since this would violate domain independence. On the first loop each activity would automatically be processed, i.e., the critic automatically processes each activity once. Suppose that the insect pest activity determined that the level of infestation was below the economic threshold and scheduled a field scouting date, the irrigation activity generated an irrigation schedule, and the verticillium wilt module determined that there was a high level of verticillium wilt infestation. The pest module would not put anything on the blackboard; the irrigation module would put *Field wet* on it together with the dates on which this condition was true; and the plant pathogen module would put *High level of pathogen* on it. The activity critic would then begin the second loop. The insect pest module would read *Field wet* and check that the scouting date was not scheduled in conflict with a scheduled irrigation. The *Field wet* condition would then be marked as having been taken into account by the pest module. The critic would then move on to the irrigation activity and note the condition *High level of pathogen*. Irrigation would therefore be rescheduled. The critic would then move on to the plant pathogen, which has nothing on its influence list and would pass.

The rescheduling of irrigation would, if it actually changed the dates of one or more irrigations, wipe the pest activity's account mark from the *Field wet* condition, so on the next loop the pest activity would be processed again. Irrigation would pass since its influence list matched nothing new on the blackboard. The plant pathogen would then pass again. On the next loop, pest would pass, followed by the other two activities' passing, and the process would terminate.

A system such as this has the potential to create an infinite loop. Suppose that in the example of the preceding paragraph, the insect pest module recommended an insecticide application instead of a scouting. Suppose that the insecticide was to be applied by a ground rig, and that the irrigation system had scheduled an irrigation for the same day. Both actions cannot be done on the same day, and the two modules could bounce back and forth between themselves. In designing a system to deal with a situation such as this, it is wise to recall the original purpose of the management decision support system: to provide support for the decision maker, *not* to make the decisions! The best course of action in dealing with a conflict such as this is also the simplest: present all the available information to the user and let the user make the final choice.

The multiagent planner described in the previous paragraphs deals with the management of a single field. The next logical step is to integrate the management of several fields. For example, pest control advisors might want to arrange their schedules so that nearby fields are scouted for pests on the same day, or a farmer with several fields might want to arrange their irrigation in some particular order. Beyond these simple OR-style integrations, there are more complex biological interactions to consider. For example, the management of resistance in insect pests must often be handled on an area-wide basis. The development of expert systems that integrate the management of multiple fields require the direct linkage of the expert system with geographic information systems and large scale databases of the type described at the start of this chapter.

12.4 Conclusions

We recognize that this chapter presents a very optimistic view of the future and that this view depends on the successful implementation and coordination of a great deal of technology and the development of new knowledge in the form of robust deep models of cropping systems. Although the technology exists to do what has been described in this chapter, we have not mentioned either the cost of such a farm-level system or the speed with which the system could be expected to run today. Furthermore, most of the applications of object-oriented databases, model-based reasoning systems, and planning systems are in prototype stages in agriculture and other domains. There is considerable work to be done.

Nevertheless, the pace of computer technology cannot be ignored. Just a few years ago, there was no such thing as an IBM personal computer, desktop publishing, or hypertext. Visi-Calc was the king of spreadsheets, and floppy disks held 90 kilobytes of information. We have every reason to expect that the powerful microcomputers of today will seem archaic compared to the machines available in the future. In short, the types of computer programs that we have discussed in this chapter should not face any technical obstacles by the time they are ready to implement.

As we stated in Chap. 1, farming is an increasingly competitive and complex enterprise. For farms to remain viable, farmers will more and more have to make sound decisions in unfamiliar domains. Farmers, the custodians of more than 15 percent of our nation's land and the consumers of 85 percent of water in the United States (National Research Council, 1989), will have to become responsible environmentalists. They will have to adapt to new farming practices and restrictions on the chemical inputs so central to crop production since the 1940s.

Information and its dissemination in knowledge-based systems will play a key role in helping farmers deal with the challenges of the next decade and beyond. Many obstacles lie ahead including the most fundamental of all, the discovery of critical basic knowledge about cropping systems and how they function. However, the technology of knowledge-based systems can be used to give structure to that knowledge as it exists today and as it changes and is augmented in the future. In particular, knowledge-based systems provide the link between knowledge and decision making that is so critical to successful farm management.

References

Abbott, L. A., F. A. Bisby, and D. J. Rogers, *Taxonomic Analysis in Biology: Computers, Models, and Databases,* Columbia University Press, New York, 1985.

Aikins, J. S., "Prototypical knowledge for expert systems," *Artificial Intelligence* **20**: 163–210 (1983).

Allen, J. F., "Maintaining knowledge about temporal intervals," *Communications of the ACM* **26**:832–843 (1983).

Andronov, A. A., A. A. Vitt, and S. E. Khaikin, *Theory of Oscillators,* Addison-Wesley, Reading, Mass., 1966.

Anonymous, *1990 Soybean Production Guide,* Virginia Cooperative Extension Service, Blacksburg, Va., 1990.

Arnold, V. I., *Ordinary Differential Equations,* MIT Press, Cambridge, Mass., 1973.

Atkinson, W. D. and A. Gammerman, "An application of expert systems technology to biological identification," *Taxon* **36**:705–714 (1987).

Austin, S., "AI infusion: Getting your company involved," *AI Expert* **5** (**3**):48–53 (1990).

Avriel, M., *Nonlinear Programming,* Prentice-Hall, Englewood Cliff, N.J., 1976.

Backman, P. A., T. P. Mack, R. Rodriguez-Kabana, and D. A. Herbert, "A computerized integrated pest management model (AUSSIM) for soybeans grown in the southeastern United States," Proceedings of the World Soybean Research Conv. IV, Buenos Aires, 1989.

Bain, W. M., *Case-Based Reasoning: A Computer Model of Subjective Assessment,* Ph.D. thesis, Yale University, New Haven, Conn., 1986.

Baker, D. N., J. R. Lambert, and J. M. McKinion, *GOSSYM: A Simulator of Cotton Crop Growth and Yield,* South Carolina Agric. Exp. Sta., Clemson, S.C., 1983.

Barnette, J. A., "Computational methods for a mathematical theory of evidence," Proceedings of the Seventh International Joint Conference on Artificial Intelligence, Vancouver, B.C., 1981.

Bayes, T. R., "An essay towards solving a problem in the doctrine of chances," *Biometrika* **45**:293 (1958).

Beck, H. W., "Database management," in J. R. Barrett and D. D. Jones (eds.), *Knowledge Engineering in Agriculture,* American Society of Agricultural Engineers, St. Joseph, Mich., pp. 83–116, 1989.

Beck, H. W., P. Jones, and J. W. Jones, "SOYBUG: An expert system for soybean insect pest management," *Agricultural Systems* **30**:269–286 (1989).

Beck, H. W., P. Jones, and D. Watson, "An expert database for ornamental plant selection," *ASAE Technical Paper* #88–7530 (1988).

Bellman, R. E., *Dynamic Programming,* Princeton University Press, Princeton, N.J., 1957.

Bellman, R. E. and S. E. Dreyfus, *Applied Dynamic Programming,* Princeton University Press, Princeton, N.J., 1962.

Benedict, J. H., K. M. El-Zik, L. R. Oliver, P. A. Roberts, and L. T. Wilson, "Economic injury levels and thresholds for pests of cotton," in R. E. Frisbie, K. M. El-Zik, and L. T. Wilson (eds.), *Integrated Pest Management Systems and Cotton Production,* Wiley-Interscience, pp. 121–153, 1989.

Berryman, A. A., "Rising tide of technology," *Bull. Entomol. Soc. Am.* **33**(**1**):2 (1987).
Bertsekas, D. P., *Dynamic Programming and Stochastic Control,* Academic Press, New York, 1976.
Blancard, D., A. Bonnet, and A. Coléno, "TOM: Un système expert en maladies des tomates," *P. H. M. Revue Horticole* **261**:7–13 (1986).
Blonningen, P., "Weed control in sugar weed cropping," Proceedings of the International DLG-Congress for Computer Technology, Frankfurt, June 19–22, 1988.
Bobrow, D. G. and M. J. Stefik, "Perspectives on artificial intelligence programming," *Science* **231**:951–957 (1986).
Boose, J. H., "Personal construct theory and the transfer of human expertise," in T. O'Shea (ed.), *Advances in Artificial Intelligence,* pp. 51–60, 1985.
Boose, J. H. and J. Bradshaw, "Expertise transfer and complex problems: Using AQUINAS as a knowledge-acquisition workbench for knowledge-based systems," *International Journal of Man-Machine Studies* **26**:3–28 (1987).
Box, G. E. P. and G. M. Jenkins, *Time Series Analysis: Forecasting and Control,* Holden-Day, San Francisco, 1976.
Bratko, I., I. Mozetic, and N. Navrac, *KARDIO: A Study in Deep and Qualitative Knowledge for Expert Systems,* MIT Press, Cambridge, Mass., 1989.
Brown, L. J., R. W. McClendon, and J. W. Jones, "Cotton and insect management simulation model," in R. L. Ridgeway, E. P. Lloyd, and W. H. Cross (eds.), *Cotton Insect Management with Special Reference to the Boll Weevil,* U.S. Department of Agriculture, Washington, D.C., pp. 437–479, 1983.
Bryson, A. E. and Y.-C. Ho, *Applied Optimal Control,* Blaisdell, Waltham, Mass., 1969.
Buchanan, B. G., D. Barstow, R. Bechtel, W. C. Bennett, C. Kulikowski, T. Mitchell, and D. A. Waterman, "Constructing an expert system," in F. Hayes-Roth, D. A. Waterman, and D. B. Lenat (eds.), *Building Expert Systems,* Addison-Wesley, Reading, Mass., pp. 127–168, 1983.
Buchanan, B. G. and E. H. Shortliffe, *Rule-Based Expert Systems,* Addison-Wesley, Reading, Mass., 1984.
Carpenter, T. E., "Epidemiologic modeling using a microcomputer spreadsheet package," *American Journal of Epidemiology* **120**:943–951 (1984).
Cervo, R., G. Maracchi, and F. Miglietta, "Diagnosis of pests and diseases in olives," Proceedings of the International DLG-Congress for Computer Technology, Frankfurt, June 19–22, 1988.
Cheeseman, P., "Probabilistic vs. fuzzy reasoning," in L. N. Kanal and J. F. Lemmer (eds.), *Uncertainty in Artificial Intelligence,* North-Holland, New York, pp. 85–102, 1986.
Chen, C. T., *Introduction to Linear System Theory,* Holt, Rinehart and Winston, New York, 1970.
Clancey, W. J., "The epistemology of a rule-based expert system: A framework for explanation," *Artificial Intelligence* **20**:215–251 (1983).
Clancey, W. J., "Viewing knowledge bases as qualitative models," *IEEE Expert* **4**(**2**):9–23 (1989).
Cohen, P. R., *Heuristic Reasoning about Uncertainty: An Artificial Intelligence Approach,* Pitman, Boston, 1985.
Colomb, R. M., B. Hearn, K. Brook, B. Jansen, N. Ashburner, and M. Clarke, *SIRATAC: Expert Decision Support for Cotton Pest Management,* CSIRO, North Ryde, NSW 2113, Australia, 1988.
Conway, G. R., *Pest and Pathogen Control: Strategic, Tactical, and Policy Models,* John Wiley and Sons, New York, 1984.
Corcoran, W., D. Finnicum, R. Hubbard III, C. Musick, and P. Waltzer, "Nuclear power plant safety functions," *Nuclear Safety* **22**:20–35 (1981).
Dale, N. and D. Orshalick, *Introduction to PASCAL and Structured Design,* D. C. Heath, Lexington, Mass., 1983.
Dallwitz, M. J. and T. A. Paine, *User's Guide to the DELTA System,* CSIRO Division of Entomology, Canberra, Australia, 1986.
Date, C. J. and C. J. White, *A Guide to DB2,* 2d ed., Addison-Wesley, Reading, Mass., 1988.
Davis, R. and D. B. Lenat, *Knowledge-Based Systems in Artificial Intelligence,* McGraw-Hill, New York, 1982.

Davis, R. and R. Smith, "Negotiation as a metaphor for distributed problem solving," *Artificial Intelligence* **20**:63–109 (1983).
Dawkins, R., "Hierarchical organization: A candidate principle for ethology." P. P. G. Bateson and R. A. Hinde (eds.), *Growing Points in Ethology,* Cambridge University Press, Cambridge, pp. 7–54, 1976.
de Kleer, J. and J. S. Brown, "A qualitative physics based on confluences," *Artificial Intelligence* **24**:7–83 (1984).
de Kleer, J. and J. S. Brown, "Theories of causal ordering," *Artificial Intelligence* **29**:33–63 (1986).
Dempster, A. P., "A generalization of bayesian inference," *Journal of the Royal Statistical Society Series B* **30**:205–247 (1968).
Dhar, V. and H. E. Pople, "Rule-based versus structure-based models for explaining and generating expert behavior," *Communications of the ACM* **30**:542–555 (1987).
Draper, N. and H. Smith, *Applied Regression Analysis,* John Wiley, New York, 1981.
Dreyfus, H. L. and S. E. Dreyfus, *Mind Over Machine: The Power of Human Intuition and Expertise in the Area of the Computer,* Free Press, New York, 1986.
Dreyfus, S. E. and A. M. Law, *The Art and Theory of Dynamic Programming,* Academic Press, New York, 1977.
Dubois, D. and H. Prade, *Fuzzy Sets and Systems: Theory and Applications,* Academic Press, New York, 1980.
Duda, R. O. and P. E. Hart, *Pattern Classification and Scene Analysis,* Wiley-Interscience, New York, 1973.
Duda, R. O. and J. G. Gaschnig, "Knowledge-based expert systems come of age," *Byte* **6(9)**:238–280 (1981).
Duda, R. O., P. E. Hart, and N. J. Nilsson, "Subjective bayesian methods for rule-based inference systems," *Proceedings of the National Computer Conference (AFIPS)* **45**: 1075–1082 (1976).
Dudley, N. J., D. T. Howell, and W. F. Masgrove, "Optimal intraseasonal water allocation," *Water Resources Research* **7**:770–788 (1971).
Emlen, J. M., *Population Biology,* Macmillan, New York, 1984.
Engel, B. A., R. H. Thieme, and A. D. Whittaker, "Representation and reasoning," in J. R. Barrett and D. D. Jones (eds.), *Knowledge Engineering in Agriculture,* American Society of Agricultural Engineers, St. Joseph, Mich., pp. 47–81, 1989.
Erman, L. D., F. Hayes-Roth, V. R. Lesser, and D. R. Reddy, "The Hearsay-II speech-understanding system: Integrating knowledge to resolve uncertainty," *Computing Surveys* **12**:213–253 (1980).
Farrow, R. A. and J. C. Daly, "Long-range movements as an adaptive strategy in the genus *Heliothis* (Lepidoptera: Noctuidae): A review of its occurrence and detection in four pest species." *Australian Journal of Zoology* **35**:1–24 (1987).
Faught, W. S., "Applications of AI in engineering," *Computer* **19(7)**:17–27 (1986).
Feder, G., "Pesticides, information, and pest management under uncertainty," *American Journal of Agricultural Economics* **61**:97–103 (1979).
Feinstein, A. R., "The haze of Bayes, the aerial palaces of decision analysis, and the computerized Ouija board," *Clinical Pharmacology and Therapeutics* **21**:482–496 (1977).
Fereres, E., D. W. Henderson, W. O. Pruitt, W. F. Richardson, and R. S. Ayers, *Basic Irrigation Scheduling,* Division of Agriculture and Natural Resources, University of California, Oakland, Calif., 1981.
Fermanian, T. W., R. S. Michalski, and B. Katz, "AgAssistant: A tool for developing expert systems," Proceedings of the International DLG-Congress for Computer Technology, Frankfurt, June 19–22, 1988.
Fikes, R. and T. Kehler, "The role of frame-based representation in reasoning," *Comm. ACM* **28(9)**:904–920 (1985).
Fikes, R. E. and N. J. Nilsson, "STRIPS: A new approach to the application of theorem proving to problem solving," *Artificial Intelligence* **2**:189–208 (1971).
Fisher, R. A., *The Design of Experiments,* Oliver and Boyd, London, 1935.
Flynn, R. P., W. L. Roller, and H. M. Keener, "A decision model for nutrient management in controlled environment agriculture," *Agricultural Systems* **31**:35–54 (1989).
Forbus, K. D., "Qualitative process theory," *Artificial Intelligence* **24**:85–168 (1984).

Forget, P. M., J. Lebbe, H. Puig, R. Vignes, and M. Hideux, "Microcomputer-aided identification: An application to trees from French Guiana," *Botanical Journal of the Linnean Society* **93**:205–223 (1986).

France, J. and J. H. M. Thornely, *Mathematical Models in Agriculture,* Butterworths, London, 1984.

Fransella, F. and D. Bannister, *A Manual for Repertory Grid Technique,* Academic Press, New York, 1977.

Free, C. A., J. R. Beddington, and J. H. Lawton, "On the inadequacy of simple models of mutual interference for parasitism and predation," *Journal of Animal Ecology* **46**: 543–554 (1977).

French, S., *Decision Theory: An Introduction to the Mathematics of Rationality,* Ellis Horwood Ltd., New York, 1986.

Frey, P. W., "A bit-mapped classifier," *Byte* **11**(**12**):161–172 (1986).

Frisbie, R. E. and P. L. Adkisson, *Integrated Pest Management on Major Agricultural Systems,* Texas Agricultural Experiment Station, College Station, Tex., 1986.

Gale, W. A., "REX review," in W. A. Gale (ed.), *Artificial Intelligence and Statistics,* Addison-Wesley, Reading, Mass., pp. 173–228, 1986.

Gelb, A., *Applied Optimal Estimation,* MIT Press, Cambridge, Mass., 1974.

Getz, W. M., "Feedback control for a class of pest management problems modeled by a noninteractively distributed system," *Journal of Optimization Theory and Applications* **38**:513–523 (1982).

Getz, W. M., "Interfacing biology and systems analysis in pest management," in M. Mangel, J. R. Carey, and R. E. Plant (eds.), *Pest Control: Operations and Systems Analysis in Fruit Fly Management,* Springer-Verlag, New York, pp. 301–314, 1986.

Getz, W. M. and A. P. Gutierrez, "A perspective on systems analysis in crop production and insect pest management," *Annual Reviews in Entomology* **27**:447–466 (1982).

Ginsberg, A., S. M. Weiss, and P. Politakis, "Automatic knowledge base refinement for classification systems," *Artificial Intelligence* **35**:197–226 (1988).

Ginsberg, M., "Counterfactuals," *Artificial Intelligence* **30**:35–79 (1986).

Glasstone, S., *Textbook of Physical Chemistry,* Van Nostrand, Princeton, N.J., 1946.

Goh, B. S., G. Leitmann, and T. L. Vincent, "Optimal control of a prey-predator system," *Mathematical Biosciences* **19**:263–286 (1974).

Goldberg, A. and D. Robson, *Smalltalk-80: The Language and Its Implementation,* Addison-Wesley Publishing Company, Reading, Mass., 1983.

Goodell, P. B., R. E. Plant, T. A. Kerby, J. F. Strand, L. T. Wilson, L. Zelinski, J. A. Young, A. Horrocks, R. D. Corbett, and R. Vargas, "CALEX/Cotton: An integrated expert system for crop production and management in California cotton," *California Agriculture* **44**(**5**):18–21 (1990).

Gordon, J. and E. H. Shortliffe, "The Dempster-Shafer theory of evidence," in B. G. Buchanan and E. H. Shortliffe, (eds.), *Rule-Based Expert Systems,* Addison-Wesley, Reading, Mass., pp. 272—292, 1984.

Grimes, D. W., H. Yamada, and S. W. Hughes, "Climate-normalized cotton leaf water potentials for irrigation scheduling," *Agricultural Water Management* **12**:293–304 (1987).

Grimes, D. W. and K. M. El-Zik, *Water Management for Cotton,* Division of Agriculture and Natural Resources, University of California, Oakland, Calif., 1982.

Gupta, A., *Integration of Information Systems: Bridging Heterogeneous Databases,* IEEE Press, New York, 1989.

Gutierrez, A. P., J. U. Baumgaertner, and C. G. Summers, "Multitrophic models of predator-prey energetics," *Canadian Entomologist* **116**:923–963 (1984).

Gutierrez, A. P., R. Daxl, G. L. Quant, and L. A. Falcon, "Estimating economic thresholds for bollworm, *Heliothis zea* Boddie, and boll weevil, *Anthonomus grandis* Boheman, damage in Nicaraguan cotton, *Gossypium hirsutum* L." *Environmental Entomology* **10**:872–879 (1981).

Gutierrez, A. P., M. A. Pizzamiglio, W. J. Santos, R. Tennyson, and A. M. Villacorta, "A general distributed delay time-varying life table plant population model: Cotton (*Gossypium hirsutum* L.) growth and development as an example," *Environmental Entomology* **26**:231–249 (1984).

Gutierrez, A. P., Y. Wang, and R. Daxl, "The interaction of cotton and boll weevil

(Coleoptera: Curculionidae)—a study of coadaptation," *Canadian Entomologist* **111**: 357–366 (1979a).
Gutierrez, A. P., Y. Wang, and U. Regev, "An optimization model for *Lygus hesperus* (Heteroptera: Miridae) damage in cotton: The economic threshold revisited," *Canadian Entomologist* **111**:41–54 (1979b).
Gyllenberg, H., "A general method for deriving determination schemes for random collections of microbial isolates," *Annales Academiae Scientiarum Fennicae Ser. A; IV. Biologica* **69**:5–23 (1963).
Hall, W. A. and W. S. Butcher, "Optimal timing of irrigation," *Journal of the Irrigation and Drainage Division, Proceedings of the ASCE* **94**:267–275 (1968).
Harper, J. L., *The Population Biology of Plants,* Academic Press, New York, 1977.
Hart, H., *Knowledge Acquisition for Expert Systems,* McGraw-Hill, New York, 1986.
Hartstack, A. W., J. D. Lopez, R. A. Muller, W. L. Sterling, E. G. King, J. A. Witz, and A. C. Eversull, "Evidence of long range migration of *Heliothis zea* (Boddie) into Texas and Arkansas," *Southwestern Entomologist* **7**:188–201 (1982).
Hartstack, A. W., J. A. Witz, J. P. Hollingsworth, R. L. Ridgeway, and J. D. Lopez, "MOTHZV-2: A computer simulation of *Heliothis zea* and *Heliothis virescens* population dynamics," *USDA ARS-S-127* (1976).
Hasling, D. W., W. J. Clancey, and G. Rennels, "Strategic explanations for a diagnostic consultation system," *International Journal of Man-Machine Studies* **20**:3–19 (1984).
Hass, N. and G. G. Hendrix, "Learning by being told: Acquiring knowledge for information management," in R. S. Michalski, J. G. Carbonell, and T. M. Mitchell (eds.), *Machine Learning: An Artificial Intelligence Approach,* Tioga Publishing Co., Palo Alto, Calif., pp. 405–428, 1983.
Hassell, M. P., *The Dynamics of Arthropod Predator-Prey Systems,* Princeton University Press, Princeton, N.J., 1978.
Hassell, M. P. and G. C. Varley, "New inductive population model for insect parasites and its bearing on biological control," *Nature* **223**:1133–1136 (1969).
Hawkins, D. M., *Identification of Outliers,* Chapman and Hall, London, 1980.
Head, R. B., "Report of the cotton insect loss committee of the thirty-fifth annual conference on cotton insect research and control," in *Proceedings of the Beltwide Cotton Production and Research Conferences,* National Cotton Council of America, Memphis, Tenn., p. 182, 1985.
Headley, J. C., "Defining the economic threshold," in *Pest Control Strategies for the Future,* National Academy of Science, Washington, D.C., pp. 100–108, 1972.
Helms, G., J. W. Richardson, M. J. Cochran, and M. E. Rister, "A farm-level expert simulation system to aid farmers in selecting among crop insurance strategies," *Computers and Electronics in Agriculture* **4(3)**:169–190 (1990).
Helms, G., J. W. Richardson, T. O. Knight, M. E. Rister, and N. D. Stone, "COTFLEX: An expert system to provide cropmix and farm policy advice to producers," in *Proceedings of the Beltwide Cotton Production and Research Conferences,* National Cotton Council of America, Memphis, Tenn., pp. 197–201, 1987.
Hendler, J., A. Tate, and M. Drummond, "AI planning: Systems and techniques," *AI Magazine* **11(2)**:61–77 (1990).
Himmelblau, D. M., *Process Analysis by Statistical Methods,* Sterling Swift, Manchaca, Tex., 1970.
Hink, R. F., "How humans process uncertain knowledge: An introduction," *AI Magazine* **8(3)**:41–54 (1987).
Hogarth, R. M., *Judgement and Choice,* Wiley-Interscience, New York, 1980.
Holland, J. H., "Escaping brittleness: The possibilities of general purpose learning algorithms applied to parallel rule-based systems," in R. S. Michalski, J. G. Carbonell, and T. M. Mitchell (eds.), *Machine Learning: An Artificial Intelligence Approach,* vol. 2, Tioga Publishing Co., Palo Alto, Calif., pp. 593–624, 1986.
Holling, C. S., "Some characteristics of simple types of predation and parasitism," *Canadian Entomologist* **91**:385–398 (1959).
Holt, D. A., "Computers in production agriculture," *Science* **228**:422–427 (1985).
Hopper, K. R. and S. B. Stark, "A simulation model for making decisions about *Heliothis* control," in *Proceedings of the Beltwide Cotton Production and Research Conferences,* National Cotton Council of America, Memphis, Tenn., pp. 286–290, 1987.

Hoshi, T. and T. Kozai, "Disease and pest diagnosis for tomatoes," Proceedings International DLG-Congress for Computer Technology, Frankfurt, June 19–22, 1988, pp. 457–472, 1988.

Huffaker, C. B., *New Technology of Pest Control,* Wiley, New York, 1980.

Ives, P. M., L. T. Wilson, P. O. Cull, W. A. Palmer, C. Haywood, N. J. Thomson, A. B. Hearn, and A. G. L. Wilson, "Field use of SIRATAC: An Australian computer-based pest management system for cotton," *Protection Ecology* **6**:1–21 (1984).

Iwasaki, Y. and H. A. Simon, "Causality in device behavior," *Artificial Intelligence* **29**: 3–32 (1986a).

Iwasaki, Y. and H. A Simon, "Theories of causal ordering: Reply to de Kleer and Brown," *Artificial Intelligence* **29**:63–72 (1986b).

Jones, J. W., P. Jones, and P. A. Everett, "Combining expert systems and agricultural models: A case study," *Trans. of the ASAE* **30**:1308–1314 (1987).

Jones, P., J. W. Jones, P. A. Everett, and H. Beck, *Knowledge Acquisition: A Case History of an Insect Control Expert System,* American Society of Agricultural Engineers, St. Joseph, Mich., 1986.

Kahneman, D., P. Slovic, and A. Tversky, *Judgement under Uncertainty: Heuristics and Biases,* Cambridge University Press, Cambridge, 1982.

Karnopp, D. and R. Rosenberg, *System Dynamics: A Unified Approach,* Wiley, New York, 1975.

Katchalsky, A. and P. F. Curran, *Nonequilibrium Thermodynamics in Biophysics,* Harvard University Press, Cambridge, Mass., 1965.

Kelly, G. A., *The Psychology of Personal Constructs,* Norton, New York, 1955.

Klein, D. and T. Finin, "What's in a deep model? A characterization of knowledge depth in intelligent safety systems," Proceedings of the Tenth International Joint Conference on Artificial Intelligence, Menlo Park, Calif., pp. 559–562, 1987.

Konolige, K. and N. J. Nilsson, "Multi-agent planning systems," Proceedings of the First National Conference on Artificial Intelligence, American Association for Artificial Intelligence, Menlo Park, Calif., pp. 138–142, 1980.

Kovac, M. P. and W. J. Davis, "Neural mechanisms underlying behavioral choice in *Pleurobranchia,*" *Journal of Neurophysiology* **43**:469–487 (1980).

Kuipers, B., "Commonsense reasoning about causality: Deriving behavior from structure," *Artificial Intelligence* **24**:169–203 (1984).

Kuipers, B., "Qualitative simulation," *Artificial Intelligence* **29**:289–338 (1986).

Kunz, J., R. Fallat, D. McClung, J. Osborn, B. Votteri, H. Nii, J. Aikins, L. Fagan, and E. Feigenbaum, *A Physiological Rule-Based System for Interpreting Pulmonary Function Test Results,* Department of Computer Science, Stanford University, Stanford, Calif., 1978.

Langlotz, C. and E. H. Shortliffe, "Logical and decision-theoretic methods for planning under uncertainty," *Artificial Intelligence* **10(1)**:39–48 (1989).

Larsen, R. J. and M. L. Marx, *An Introduction to Mathematical Statistics and Its Applications,* Prentice-Hall, Englewood Cliffs, N.J., 1981.

Law, A. M. and W. D. Kelton, *Simulation Modeling and Analysis,* McGraw-Hill, New York, 1982.

Le Renard, J., "SEVP—Diagnosis in plant protection for 17 types of crop," Proceedings International DLG-Congress for Computer Technology, Frankfurt, June 19–22, 1988.

Legaspi, B. A. C., W. L. Sterling, A. W. Hartstack, and D. A. Dean, "Testing the interactions of pest-predator-plant components of the TEXCIM model," *Environmental Entomology* **18**:157–163 (1989).

Lemmon, H. E., "COMAX: An expert system for cotton crop management." *Science* **233**: 29–33 (1986).

Lewis, D., *Counterfactuals,* Harvard University Press, Cambridge, Mass., 1973.

Lindley, D. V., *Making Decisions,* Wiley-Interscience, New York, 1971.

Lindley, D. V., *The Probability Approach to the Treatment of Uncertainty in Artificial Intelligence and Expert Systems,* Institute for Reliability and Risk Analysis, George Washington University, Washington, D.C., 1985.

Liu, H. and T. W. Fermanian, "An expert system for planning turfgrass establishment," Illinois Turfgrass Research Report, Hort Series **51** (1988).

Loh, D. K., K. J. English, Y. Choo, Y. Chu, and M. Sun, "Integrated natural resource automation—the knowledge shop," in *Proceedings: Resource Technology 88,* American Society for Photogrammetry and Remote Sensing, Falls Church, Va., pp. 158–167, 1988.

Lotka, A. J., *Elements of Physical Biology,* Williams and Wilkins, Baltimore, 1925.

Lucas, H. C., Jr., *The Analysis, Design, and Implementation of Information Systems,* 2d Ed, McGraw-Hill, New York, 1981.

Makela, M. E., N. D. Stone, and S. B. Vinson, "Host-parasitoid population dynamics in a heterogeneous environment," in T. Henson (ed.), *Proceedings SCS Multiconference on AI and Simulation: The Diversity of Applications,* Society for Computer Simulation, San Diego, pp. 228–233, 1988.

Mann, S. H., "Mathematical models for the control of pest populations," *Biometrics* **27**: 357–368 (1971).

Manna, Z. and A. Waldinger, *The Logical Basis for Computer Programming: Deductive Reasoning,* vol. 1, Addison-Wesley, Reading, Mass., 1985.

Manola, F., "Object-oriented databases," *AI Expert* **5**(**3**):26–37 (1990).

Maran, L. and D. R. Pike, "WEAS—Plant protection advisory system," Proceedings International DLG-Congress for Computer Technology, Frankfurt, June 19–22, 1988.

Marcot, B., "Testing your knowledge base," *AI Expert* **2**(**8**):42–47 (1987).

May, R. M. and G. F. Oster, "Bifurcations and Dynamic Complexity in Simple Ecological Models," *The American Naturalist* **110**:573–599 (1976).

McCarl, B. A., "Economics of integrated pest management: An interpretive review of the literature," *Oregon Agricultural Experiment Station Special Report* **636** (1981).

McClendon, R. W., W. D. Batchelor, and J. W. Jones, *Insect Pest Management with an Expert System Coupled Crop Model,* American Society of Agricultural Engineers, St. Joseph, Mich., 1987.

McGraw, K. L. and K. Harbison-Briggs, *Knowledge Acquisition: Principles and Guidelines,* Prentice-Hall, Englewood Cliffs, N.J., 1989.

McKinion, J. M., D. N. Baker, F. D. Whisler, J. R. Lambert, J. A. Landivar, and G. P. Mullendore, "Application of the GOSSYM/COMAX system to cotton crop management," *Agricultural Systems* **31**:55–65 (1989).

Meidner, H. and D. W. Sheriff, *Water and Plants,* Halsted, New York, 1976.

Merkhofer, M. W., *Decision Science and Social Risk Management,* D. Reidel Publishing Co., Boston, Mass., 1987.

Metcalf, Z. P., "The construction of keys," *Systematic Zoology* **3**:38–45 (1954).

Meyer, C. R., "Integrating expert systems with modern soil erosion prediction simulation (abstract), *AI Applic. in Nat. Res. Mgmt.* **4**(**2**):126 (1990).

Michalski, R. S., "Pattern recognition as rule-guided inductive inference," *Transactions on Pattern Analysis and Machine Intelligence* PAMI-2, pp. 349–361 (1980).

Michalski, R. S., J. G. Carbonell, and T. M. Mitchell, *Machine Learning: An Artificial Intelligence Approach,* Tioga Publishing Co., Palo Alto, Calif., 1983.

Michalski, R. S., J. G. Carbonell, and T. M. Mitchell, *Machine Learning: An Artificial Intelligence Approach,* vol. 2, Tioga Publishing Co., Palo Alto, Calif., 1986.

Michalski, R. S. and R. L. Chilausky, "Learning by being told and learning from examples: An experimental comparison of the two methods of knowledge acquisition in the context of developing an expert system for soybean disease diagnosis," *International Journal of Policy Analysis and Information Systems* **4**:125–161 (1980).

Michalski, R. S., J. H. Davis, V. S. Bisht, and J. B. Sinclair, "A computer-based advisory system for diagnosing soybean diseases in Illinois," *Plant Disease* **4**:459–463 (1983).

Michalski, R. S., J. H. Davis, V. S. Bisht, and J. B. Sinclair, "PLANT/ds: An expert consulting system for the diagnosis of soybean diseases," in L. Steels and J. A. Campbell (eds.), *Progress in Artificial Intelligence,* Halstead Press, New York, pp. 257–270, 1985.

Minsky, M., "A framework for representing knowledge," in P. H. Winston (ed.) *The Psychology of Computer Vision,* McGraw-Hill, New York, pp. 211–271, 1975.

Morse, L. E., "Recent advances in the theory and practice of biological specimen identification," in R. J. Pankhurst (ed.), *Biological Identification with Computers,* 1975.

Mumford, J. D. and G. A. Norton, "Economics of decision making in pest management," *Annual Review of Entomology* **29**:157–174 (1984).

Myers, R. H., *Classical and Modern Regression with Applications,* Duxbury Press, Boston, Mass., 1986.

National Research Council, *Alternative Agriculture,* National Academy Press, Washington, D.C., 1989.
Newell, I. M., "Construction and use of tabular keys," *Pacific Insects* **12**:25–37 (1970).
Nguyen, T. J., W. A. Perkins, T. J. Laffey, and D. Pecora, "Knowledge base verification," *AI Magazine* **8**(**3**):69–75 (1987).
Nicholson, A. J. and V. A. Bailey, "The balance of animal populations, Part I." *Proceedings of the Zoological Society of London,* pp. 551–598 (1935).
Nicolosi, E. and M. S. Leaning, "Qualitative simulation of compartmental systems," in C. Cobelli and L. Mariani (eds.), *Modelling and Control in Biomedical Systems,* Pergamon, New York, pp. 577–582, 1989.
Nilsson, N. J., *Problem-Solving Methods in Artificial Intelligence,* McGraw-Hill, New York, 1971.
Nisbett, A., *Konrad Lorenz,* J. M. Dent and Sons, London, 1976.
Noble, B. and J. W. Daniel, *Applied Linear Algebra,* Prentice-Hall, Englewood Cliffs, N.J., 1973.
Norgaard, R. B., "The economics of improving pesticide use," *Annual Review of Entomology* **21**:45–60 (1976).
O'Keefe, R. M., O. Balci, and E. P. Smith, "Validating expert system performance," *IEEE Expert* **2**(**4**):81–90 (1987).
Osborne, D. V., "A numerical representation for taxonomic keys," *New Phytologist* **62**: 35–43 (1963).
Ostergard, M., "CALED: An intelligent editor for CALEX knowledge bases," *AI Applications in Resource Management* (1990).
Packard, J. M., L. J. Folse, N. D. Stone, M. E. Makela, and R. N. Coulson, "Applications of artificial intelligence to animal behavior," in M. Bekoff and D. Jamieson (eds.), *Interpretation and Explanation in the Study of Animal Behavior,* vol. II, Westview Press, Boulder, Colo., pp. 147–191, 1990.
Pankhurst, R. J., "A computer program for generating diagnostic keys," *The Computer Journal* **13**:145–151 (1970).
Pankhurst, R. J., *Biological Identification,* Edward Arnold, Ltd., London, 1978.
Pankhurst, R. J., "Medical diagnosis in developing countries," *Computers in Biology and Medicine* **10**:69–82 (1980).
Pankhurst, R. J., *Online Identification Program, Version 4,* British Museum of Natural History, London, 1984.
Parsaye, K., M. Chignell, S. Khoshafian, and H. Wong, *Intelligent Databases: Object-Oriented, Deductive Hypermedia Technologies,* John Wiley and Sons, New York, 1989.
Parsaye, K., M. Chignell, S. Khoshafian, and H. Wong, "Intelligent Databases," *AI Expert* **5**(**3**):38–47 (1990).
Pearl, J., "How to do with probabilities what people say you can't," Second Conference on Artificial Intelligence Applications, 1985.
Pearl, J., "Reverend Bayes on inference engines: a distributed hierarchical approach," Proceedings of the National Conference on Artificial Intelligence, 1982.
Pearl, R., *The Rate of Living,* Knopf, New York, 1928.
Pedigo, L. P., S. H. Hutchins, and L. G. Higley, "Economic injury levels in theory and practice," *Annual Review of Entomology* **31**:341–368 (1986).
Penning de Vries, F. W. T. and H. H. Van Laar, *Simulation of Plant Growth and Crop Production,* Centre for Agricultural Pub. and Doc., Wageningen, Netherlands, 1982.
Plant, R. E., "An integrated expert decision support system for agricultural management," *Agricultural Systems* **29**:49–66 (1989).
Plant, R. E., "Uncertainty and the Economic Threshold," *Journal of Economic Entomology* **79**:1–6 (1986).
Plant, R. E., T. A. Kerby, L. T. Wilson, L. J. Zelinski, and P. B. Goodell, "Using knowledge-based regression for forecasting in CALEX," *AI Applications in Resource Management* **4**(**2**):66–72 (1990).
Plant, R. E., M. Mangel, and L. E. Flynn, "Multiseasonal management of an agricultural pest II: The economic optimization problem," *Journal of Environmental Economics and Management* **12**:45–61 (1985).
Plant, R. E., L. J. Zelinski, P. B. Goodell, T. A. Kerby, L. T. Wilson, and F. G. Zalom,

"CALEX: An integrated expert decision support system for farm management," in F. S. Zazueta and A. B. Bottcher (eds.), Proceedings of the International Conference on Computers in Extension, Orlando, Fla., 1988.

Podolsky, A. S., *New Phenology: Elements of Mathematical Forecasting in Ecology,* John Wiley and Sons, New York, 1984.

Politakis, P. and S. M. Weiss, "Using empirical analysis to refine expert system knowledge bases," *Artificial Intelligence* **22**:23–48 (1984).

Pople, H. E., "Heuristic methods for imposing structure on ill-structured problems: The structure of medical diagnosis," in P. Szolvits et al. (eds.), *Artificial Intelligence in Medicine,* Westview Press, Boulder, Colo., pp. 119–190, 1982.

Post, E., "Formal reductions of the general combinatorial problem," *American Journal of Mathematics* **65**:197–268 (1943).

Raiffa, H. and R. Schlaifer, *Applied Statistical Decision Theory,* Harvard Business School, Cambridge, Mass., 1961.

Ray, P. M., *The Living Plant,* Holt, Rinehart and Winston, New York, 1972.

Rescher, N., *Hypothetical Reasoning,* North-Holland, Amsterdam, 1964.

Rhenals, A. E. and R. Bras, "The irrigation scheduling problem and evapotranspiration uncertainty," *Water Resources Research* **17**:1328–1338 (1981).

Rich, E., *Artificial Intelligence*, McGraw-Hill, New York, 1983.

Richardson, C. W. and D. A. Wright, *WGEN: A Model for Generating Daily Weather Records,* USDA, 1984.

Richardson, J. W. and C. J. Nixon, *The Farm Level Income and Policy Simulation Model: FLIPSIM,* Texas Agricultural Experiment Station, College Station, Tex., 1985.

Richardson, J. W., P. T. Zimmel, D. C. Gerloff, M. J. Cochran, and S. C. Goh, "CARMS: Computer-assisted records management system for expert simulation system." *AI Applications in Natural Resource Management,* **4**(**2**):85–93 (1990).

Riesbeck, C. K. and R. C. Schank, *Inside Case-Based Reasoning,* Lawrence Erlbaum Associates, Hillsdale, N.J., 1989.

Roach, J. W., F. Eichelman, and J. D. Whitehead, "A coherence logic for counterfactual reasoning," 1990 (unpublished).

Robinson, J. V. and J. W. Stewart, *Management of Cotton Insects in the Southern, Eastern, and Blackland Areas of Texas,* Texas A & M University, College Station, Tex., 1987.

Saarenmaa, H., "Model-based reasoning in ecology and natural resource management," in *Proceedings: Resource Technology 88,* American Society for Photogrammetry and Remote Sensing, Falls Church, Va., pp. 141–157, 1988.

Sacerdoti, E. D., *A Structure for Plans and Behavior,* Elsevier, New York, 1975.

Sacks, E., "A dynamic systems perspective on qualitative simulation," *Artificial Intelligence* **42**:349–362 (1990).

Saunders, M. C., C. W. Haeseler, J. W. Travis, B. J. Miller, R. N. Coulson, K. D. Loh, and N. D. Stone, "GRAPES: An expert system for viticulture in Pennsylvania," *AI Applic. in Nat. Res. Mgmt.* **1**(**2**):13–20 (1987).

Savage, L. J., *The Foundation of Statistics,* Wiley, New York, 1954.

Schank, R. C., *Dynamic Memory: A Theory of Learning in Computers and People,* Cambridge University Press, Cambridge, 1982.

Schank, R. C. and R. Abelson, *Scripts, Plans, Goals, and Understanding,* Lawrence Erlbaum Associates, Hillsdale, N.J., 1977.

Schaub, L. P. and N. D. Stone, "Embedding an expert system into a stochastic simulation model for the analysis of control strategies for *Heliothis* in Texas," in *Proceedings of the Beltwide Cotton Production and Research Conferences,* National Cotton Council of America, Memphis, Tenn., pp. 220–222, 1989.

Schneider, J. C., R. T. Roush, W. F. Kitten and M. L. Laster, "Movement of *Heliothis virescens* (Lepidoptera: Noctuidae) in Mississippi in the spring: Implications for areawide management," *Environmental Entomology* **18**:438–446 (1989).

Schroderbeck, P., *Management Systems,* 2d. ed, John Wiley & Sons, New York, 1971.

Scott, A. C., W. J. Clancy, R. Davis, and E. H. Shortliffe, "Methods for generating explanations," in B. G. Buchanan and E. H. Shortliffe (eds.), *Rule-Based Expert Systems,* Addison-Wesley, Reading, Mass., pp. 338–362, 1984.

Sequeira, R. A., N. D. Stone, M. Cochran, and K. M. El-Zik, "Inclusion of plant structure and quality into a distributed delay cotton fruiting model," in *Proceedings of the Beltwide Cotton Production and Research Conferences,* National Cotton Council of America, Memphis, Tenn., pp. 156–159, 1989.

Shafer, G., *A Mathematical Theory of Evidence,* Princeton University Press, Princeton, N.J., 1976.

Shafer, G. and A. Tversky, "Languages and designs for probability judgement," *Cognitive Science* **9**:309–339 (1985).

Shaw, M. L. G., *On Becoming a Personal Scientist,* Academic Press, New York, 1980.

Shoemaker, C. A., "Application of dynamic programming and other optimization methods in pest management," *IEEE Transactions on Automatic Control* **26**:1125–1132 (1981).

Shoemaker, C. A. and D. W. Onstad, "Optimization analysis of the integration of biological, cultural, and chemical control of alfalfa weevil (Coleoptera: Curculionidae), *Environmental Entomology* **12**:286–295 (1983).

Shortliffe, E. H., *Computer-Based Medical Consultations: MYCIN,* American Elsevier, New York, 1976.

Shroger, J., S. Young, and S. Cox, *Wheat Whiz User's Guide,* Cooperative Extension Service, Kansas State University, Manhattan, Kan., 1987.

Simon, H. A. and N. Rescher, "Cause and counterfactual," *Philosophy of Science* **33**:323–340 (1966).

Smith, C., M. E. Rister, A. D. Kosterboer, E. F. Eastin, R. J. Mayer, M. E. Sterle, R. N. Coulson, N. D. Stone, J. A. Jackman, J. W. Stansel, J. Wheeler, M. J. Lin, J. Toland, J. Jordan, R. Rupani, and T. W. Toman, *Weed Control Advisor Beta 1.0 Version: Mini-User's Manual (Revised),* Texas A&M University, College Station, Tex., 1989.

Smith, R. F. and H. T. Reynolds, "Principles, definitions and scope of integrated control measures in Baden-Wurttemberg, West Germany," *Z. Angew. Entomol.* **77**:398–401 (1966).

Sokal, R. R. and R. J. Rohlf, *Biometry: The Principles and Practice of Statistics in Biological Research,* W. H. Freeman and Company, San Francisco, 1969.

Sokal, R. R. and P. H. A. Sneath, *Principles of Numerical Taxonomy,* W.H. Freeman, San Francisco, 1963.

Spiegelhalter, D. J. and R. P. Knill-Jones, "Statistical and knowledge-based approaches to clinical decision-support systems, with an application in gastroenterology," *Journal of the Royal Statistical Society, Series A* **147**:35–77 (1984).

Staddon, J. E. R., *Adaptive Behavior and Learning,* Cambridge University Press, Cambridge, 1983.

Starfield, A. M., B. P. Farm, and R. H. Taylor, "A rule-based ecological model for the management of an estuarine lake," *Ecological Modelling* **46**:107–119 (1989).

Stefik, J. M. and D. G. Bobrow, "Object-oriented programming: Themes and variations," *AI Magazine* **6**:40–62 (1986).

Sterling, W. L. and A. W. Hartstack, "TEXCIM—a fleahopper-*Heliothis*-plant model for on-farm use," in *Proceedings of the Beltwide Cotton Production and Research Conferences,* National Cotton Council of America, Memphis, Tenn., pp. 258–259, 1987.

Sterling, W. L., L. T. Wilson, A. P. Gutierrez, D. R. Rummel, J. R. Phillips, N. D. Stone, and J. H. Benedict, "Strategies and tactics for managing insects and mites," in R. E. Frisbie, K. M. El-Zik, and L. T. Wilson (eds.), *Integrated Pest Management Systems and Cotton Production,* Wiley, New York, pp. 267–325, 1989.

Stern, V. M., R. F. Smith, R. van den Brosch, and K. S. Hagen, "The integrated control concept," *Hilgardia* **29**:81–101 (1959).

Stewart, S. D., W. L. Sterling, and A. W. Hartstack, *Age, Loctation, Compensation, and the Economic Value of Cotton Fruit,* Texas Agricultural Experiment Station, College Station, Tex., 1989.

Stinner, R. E., R. L. Rabb, and J. R. Bradley, "Population dynamics of *Heliothis zea* (Boddie) and *H. virescens* (F.) in North Carolina: A simulation model," *Environmental Entomology* **3**:163–168 (1974).

Stone, N. D., "Knowledge-based systems as a unifying paradigm for IPM," in *Proceedings of the National Integrated Pest Management Symposium/Workshop,* New York State Agricultural Experiment Station, Geneva, N.Y., 1989.

Stone, N. D., "Chaos in an individual-level predator-prey model," *Natural Resource Modeling,* vol. 4, no. 4 (1990).

Stone, N. D., R. N. Coulson, R. E. Frisbie, and D. K. Loh, "Expert systems in entomology: Three approaches to problem solving," *Bulletin of the Entomological Society of America* **32**:161–166 (1986a).

Stone, N. D., R. E. Frisbie, J. W. Richardson, and C. Sansone, "COTFLEX, a modular expert system that synthesizes biological and economic analyses: The pest management advisor as an example," in *Proceedings of the Beltwide Cotton Production and Research Conferences,* National Cotton Council of America, Memphis, Tenn., 1987.

Stone, N. D. and A. P. Gutierrez, "Pink bollworm control in southwestern desert cotton. I. A field oriented simulation model," *Hilgardia* **54(9)**:1–24 (1986).

Stone, N. D., A. P. Gutierrez, W. M. Getz, and R. Norgaard, "Pink bollworm control in southwestern desert cotton. III. Strategies for control: An economic simulation study," *Hilgardia* **54(9)**:42–56 (1986b).

Stone, N. D., D. R. Rummel, S. Carroll, M. E. Makela, and R. E. Frisbie, "Simulation of boll weevil (Coleoptera: Curculionidae) spring emergence and overwintering survival in the Texas Rolling Plains," *Environmental Entomology* **19(1)**:91–98 (1990).

Stone, N. D. and L. P. Schaub, "A hybrid expert system/simulation model for the analysis of pest management strategies," *AI Applications in Natural Resource Management* **4(2)**:17–26 (1990).

Stone, N. D. and T. W. Toman, "A dynamically linked expert-database system for decision support in Texas cotton production," *Computers and Electronics in Agriculture* **4(2)**:139–148 (1989).

Suwa, M., A. C. Scott, and E. H. Shortliffe, "An approach to verifying completeness and consistency in a rule-based expert system," *AI Magazine* **3(4)**:16–21 (1982).

Swartout, W. R., "XPLAIN: A system for creating and explaining expert consulting programs," *Artificial Intelligence* **21**:285–325 (1983).

Tanimoto, S. L., *The Elements of Artificial Intelligence,* Computer Science Press, Rockville, Md., 1987.

Tate, A., "Generating project networks," Proceedings of the Fifth International Joint Conference on Artificial Intelligence, Cambridge, Mass., 1977.

Thompson, W. R., "La théorie mathématique de l'action des parasites entomophages et le facteur du hasard," *Annals Fac. Sci. Marseille* **2**:69–89 (1924).

Toates, F. M., *Motivational Behavior,* Cambridge University Press, Cambridge, 1986.

van Melle, W., E. H. Shortliffe, and B. G. Buchanan, "EMYCIN: A knowledge engineer's tool for constructing rule-based expert systems," in B. G. Buchanan and E. H. Shortliffe (eds.), *Rule-Based Expert Systems,* Addison-Wesley, Reading, Mass., pp. 302–313, 1984.

Vere, S. A., "Planning in time: Windows and durations for activities and goals," *IEEE Transactions on Pattern Analysis and Machine Intelligence* PAMI-5, pp. 246–261 (1983).

Walters, C., *Adaptive Management of Renewable Resources,* McGraw-Hill, New York, 1986.

Wang, Y., A. P. Gutierrez, G. Oster, and R. Daxl, "A population model for plant growth and development: Coupling cotton-herbivore interaction," *Canadian Entomologist* **109**:1359–1374 (1977).

Waterman, D. A., *A Guide to Expert Systems,* Addison-Wesley, Reading, Mass., 1986.

Watt, K. E. F., "Mathematical models for use in insect pest control," *Canadian Entomologist* **93**(Suppl.):19 (1961).

Weiner, J. L., "BLAH, a system which explains its reasoning," *Artificial Intelligence* **15**: 19–48 (1980).

Weiss, S., M. Kulikowski, and S. Amarel, "A model-based method for computer-aided medical decision making," *Artificial Intelligence* **11**:145–172 (1978).

Weld, D. S. and J. de Kleer, *Readings in Qualitative Reasoning about Physical Systems,* Morgan Kaufmann, Los Altos, Calif., 1989.

Whitehead, J. D. and J. W. Roach, "HOIST: A second generation expert system based on qualitative physics," *AI Magazine* **11(3)**:108–119 (1990).

Wilensky, R., "Understanding complex situations," Proceedings of the Sixth International Joint Conference on Artificial Intelligence, Tokyo, 1979.

Wilson, J. B. and T. R. Partridge, "Interactive plant identification," *Taxon* **35**:1–12 (1986).

Wilson, L. T., "Changing perspectives on the use of simulation models in IPM programs," Proceedings of the National IPM Symposium/Workshop, 1989.

Wilson, L. T. and W. W. Barnett, "Degree-days: An aid in crop and pest management," *California Agriculture* **37**(**1**):4–6 (1983).

Wilson, L. T., D. Gonzalez, and R. E. Plant, "Predicting sampling frequency and economic status of spider mites on cotton," Proceedings of the Beltwide Cotton Production Research Conferences, Las Vegas, Nev., 1986.

Wilson, L. T. and A. P. Gutierrez, "Fruit predation submodel: Heliothis larvae feeding upon cotton fruiting structures," *Hilgardia* **48**(**2**):24–36 (1980).

Winston, P. H., *Artificial Intelligence,* Addison-Wesley, Reading, Mass., 1984.

Winston, P. H. and B. K. P. Horn, *LISP,* Addison-Wesley, Reading, Mass., 1981.

Woolley, J. B. and N. D. Stone, "Application of artificial intelligence to systematics: SYSTEX—a prototype expert system for species identification," *Systematic Zoology* **36**(**3**):248–267 (1987).

Yakowitz, S., "Dynamic programming applications in water resources," *Water Resources Research* **18**:673–696 (1982).

Zadeh, L. A., "Fuzzy Sets," *Information and Control* **8**:338–353 (1965).

Zadeh, L. A., "Fuzzy sets as a basis for a theory of possibility," *Fuzzy Sets and Systems* **1**:3–28 (1978).

Zadeh, L. A., *On the Validity of Dempster's Rule of Combination of Evidence,* University of California, Berkeley, 1979.

Zadeh, L. A., "The role of fuzzy logic in the management of uncertainty in expert systems," *Fuzzy Sets and Systems* **11**:199–227 (1983).

Zadeh, L. A. and C. A. Desoer, *Linear System Theory: The State Space Approach,* Van Nostrand Reinhold, New York, 1970.

Author Index

Subject Index

ABOUT THE AUTHORS

RICHARD E. PLANT, formerly a Professor of Mathematics, is a Professor in the Department of Agronomy and Range Science at the University of California, Davis. He is a member of the American Society of Agricultural Engineers, the American Association of Artificial Intelligence, and the American Society of Agronomy. He has published numerous papers on agricultural applications of mathematics, simulation modeling, and knowledge-based systems.

NICHOLAS D. STONE is Associate Professor of Entomology at Virginia Polytechnic Institute and State University in Blacksburg. He was a founding member of the Knowledge Engineering Laboratory at Texas A & M University, a center devoted to the application of knowledge-based systems to agriculture, and he has been actively involved in establishing the Center for Computer-Aided Decision Making in the College of Agriculture at VPI & SU. A member of the Entomological Society of America, the American Association of Artificial Intelligence, and the International Society of Ecological Modellers, he has published numerous papers on knowledge-based systems in agriculture. He has presented invited lectures on the subject in Barcelona, Helsinki, and Montreal, and in 1990 served as an expert panelist for the Office of Technology Assessment in the area of knowledge-based systems for crops.